T0207175

Lecture Notes in Computer Science 13899

Founding Editors

Gerhard Goos
Juris Hartmanis

Editorial Board Members

Elisa Bertino, *Purdue University, West Lafayette, IN, USA*
Wen Gao, *Peking University, Beijing, China*
Bernhard Steffen ⓘ, *TU Dortmund University, Dortmund, Germany*
Moti Yung ⓘ, *Columbia University, New York, NY, USA*

The series Lecture Notes in Computer Science (LNCS), including its subseries Lecture Notes in Artificial Intelligence (LNAI) and Lecture Notes in Bioinformatics (LNBI), has established itself as a medium for the publication of new developments in computer science and information technology research, teaching, and education.

LNCS enjoys close cooperation with the computer science R & D community, the series counts many renowned academics among its volume editors and paper authors, and collaborates with prestigious societies. Its mission is to serve this international community by providing an invaluable service, mainly focused on the publication of conference and workshop proceedings and postproceedings. LNCS commenced publication in 1973.

Anna Frid · Robert Mercaş
Editors

Combinatorics on Words

14th International Conference, WORDS 2023
Umeå, Sweden, June 12–16, 2023
Proceedings

 Springer

Editors
Anna Frid
Aix-Marseille University
Marseille, France

Robert Mercaş 🆔
Loughborough University
Loughborough, UK

ISSN 0302-9743   ISSN 1611-3349 (electronic)
Lecture Notes in Computer Science
ISBN 978-3-031-33179-4   ISBN 978-3-031-33180-0 (eBook)
https://doi.org/10.1007/978-3-031-33180-0

© The Editor(s) (if applicable) and The Author(s), under exclusive license
to Springer Nature Switzerland AG 2023
This work is subject to copyright. All rights are reserved by the Publisher, whether the whole or part of
the material is concerned, specifically the rights of translation, reprinting, reuse of illustrations, recitation,
broadcasting, reproduction on microfilms or in any other physical way, and transmission or information
storage and retrieval, electronic adaptation, computer software, or by similar or dissimilar methodology now
known or hereafter developed.
The use of general descriptive names, registered names, trademarks, service marks, etc. in this publication
does not imply, even in the absence of a specific statement, that such names are exempt from the relevant
protective laws and regulations and therefore free for general use.
The publisher, the authors, and the editors are safe to assume that the advice and information in this book
are believed to be true and accurate at the date of publication. Neither the publisher nor the authors or the
editors give a warranty, expressed or implied, with respect to the material contained herein or for any errors
or omissions that may have been made. The publisher remains neutral with regard to jurisdictional claims in
published maps and institutional affiliations.

This Springer imprint is published by the registered company Springer Nature Switzerland AG
The registered company address is: Gewerbestrasse 11, 6330 Cham, Switzerland

Preface

This volume of Lecture Notes in Computer Science contains the proceedings of the 14th International Conference WORDS, which was organized by the Department of Computing Science at Umeå University and took place, one week before Midsummer, during June 12–16, 2023 in Umeå, Sweden. For this edition, the conference was held in conjunction with the International Conference on Developments in Language Theory (DLT 2023).

WORDS is the main conference series devoted to the mathematical theory of words. In particular, the combinatorial, algebraic and algorithmic aspects of words are emphasized. Motivations may also come from other domains such as theoretical computer science, bioinformatics, digital geometry, symbolic dynamics, numeration systems, text processing, number theory, etc.

The conference WORDS takes place every two years. The first conference of the series was held in Rouen, France in 1997. Since then, the locations of WORDS conferences have been: Rouen, France (1999, 2021), Palermo, Italy (2001), Turku, Finland (2003 and 2013), Montréal, Canada (2005 and 2017), Marseille, France (2007), Salerno, Italy (2009), Prague, Czech Republic (2011), Kiel, Germany (2015), and Loughborough, UK (2019).

Like all in-person international events, WORDS conferences have suffered a painful interruption because of the Covid-19 pandemic, as the 2021 meeting was held mostly online. Now two years later we have a first attempt to return things to a relatively normal state, even though international travel is still more difficult than it used to be, because of both the remnants of covid restrictions and the terrible Russian aggression against Ukraine. We just hope that soon enough organisation issues will get back to being the most important trouble for researchers. Till then, we are fully compassionate toward those in hardships.

For the current edition, there were 28 submissions, from 12 countries, and each of them was single-blind reviewed by two or three referees. The selection process was undertaken by the Program Committee with the help of the generous reviewers. From these submissions, 19 papers were selected to be published and presented at WORDS.

In addition to the contributed presentations, WORDS featured two specific invited talks, as well as another three invited talks that it shared with DLT 2023:

- Mélodie Andrieu (University of the Littoral Opal Coast, France): *"Minimal Complexities for Infinite Words Written with d Letters"*,
- Shuo Li (University of Quebec at Montreal, Canada): *"On the number of squares in a finite word"*,
- Émilie Charlier (University of Liège, Belgium): *"Alternate base numeration systems"*,
- Jussi Karlgren (Spotify, Sweden): *"When the Map is More Exact than the Terrain"*,
- Markus Schmid (Humboldt-Universität in Berlin, Germany): *"On Structural Tractability Parameters for Hard String Problems"*.

The present volume also includes the papers of three of these invited talks.

We take this opportunity to warmly thank all the invited speakers and all the authors for their contributions. We are also grateful to all Program Committee members and the additional reviewers for their hard work that led to the selection of papers published in this volume. We would like to also express our gratitude to Florin Manea for all of his hard work.

The reviewing process was facilitated by the EasyChair conference system, kindly sponsored by the local organizers. Special thanks are due to the Lecture Notes in Computer Science team at Springer for having granted us the opportunity to publish this special issue devoted to WORDS 2023 and for their help during the process. We are also grateful for the kind support received from the journal Algorithms from MDPI that led to the award of various prizes during the conference.

Finally, we are much obliged to the local organizers, who greatly contributed to the success of the conference: Martin Berglund, Johanna Björklund, Frank Drewes and Lena Strobl. Our warmest thanks for their assistance in the organization of the event!

WORDS 2023 was financially supported by

- Department of Computing Science, Umeå University
- Umeå municipality, Region Västerbotten and Umeå University
- MDPI (Multidisciplinary Digital Publishing Institute)

April 2023   Anna Frid
                                                            Robert Mercaş

Organization

Program Committee Chairs

Anna Frid   Aix-Marseille University, France
Robert Mercaş   Loughborough University, UK

Steering Committee

Valérie Berthé   IRIF, Paris Cité University, France
Srečko Brlek   University of Quebec at Montreal, Canada
Julien Cassaigne   Institut de Mathématiques de Marseille, France
Maxime Crochemore Gustave Eiffel University, France
Anna Frid   Aix-Marseille University, France
Juhani Karhumäki   University of Turku, Finland
Jean Néraud   University of Rouen, France
Dirk Nowotka   Kiel University, Germany
Edita Pelantová   FNSPE Czech Technical University in Prague,
            Czech Republic
Dominique Perrin   Gustave Eiffel University, France
Daniel Reidenbach   University of Keele, UK
Antonio Restivo   University of Palermo, Italy
Christophe Reutenauer University of Quebec at Montreal, Canada
Jeffrey Shallit   University of Waterloo, Canada
Mikhail Volkov   Ural Federal University, Russia

Program Committee

Giulia Bernardini   University of Trieste, Italy
Joel Day   Loughborough University, UK
L'ubomíra Dvořáková FNSPE Czech Technical University in Prague,
            Czech Republic
Gabriele Fici   University of Palermo, Italy
Pamela Fleischmann Kiel University, Germany
Bryna Kra   Northwestern University, USA
Narad Rampersad   University of Winnipeg, Canada
Marinella Sciortino   University of Palermo, Italy

Arseny Shur   Bar Ilan University, Israel
Manon Stipulanti   University of Liège, Belgium
Markus Whiteland   University of Liège, Belgium

Additional Reviewers

Duncan Adamson   Johan Kopra
Christopher Cabezas   Tore Koß
James Currie   Florin Manea
Alessandro De Luca   Yuto Nakashima
Francesco Dolce   Edita Pelantová
Szilárd Zsolt Fazekas   Jarkko Peltomäki
Daniel Gabric   Gloria Pietropolli
France Gheeraert   Antoine Renard
Lukas Haschke   Gwenaël Richomme
Štěpán Holub   Ville Salo
Jonas Höfer   Stefan Siemer
Annika Huch   Pierre Stas
Shunsuke Inenaga   Reem Yassawi
Artur Jez   Max Wiedenhöft

Contents

Invited Papers

Minimal Complexities for Infinite Words Written with d Letters

Mélodie Andrieu[1]([✉]) and Léo Vivion[2]

[1] Université Littoral Côte d'Opale, UR 2597, LMPA, Laboratoire de Mathématiques Pures et Appliquées, Joseph Liouville, 62100 Calais, France
`melodie.andrieu@univ-littoral.fr`
[2] Aix-Marseille Université, CNRS, Centrale Marseille, Institut de mathématiques de Marseille, I2M - UMR 7373, 13453 Marseille, France

Abstract. In this extended abstract, we discuss the minimal subword complexity and the minimal abelian complexity functions for infinite d-ary words. This leads us to answer a question of Rauzy from 1983: cubic billiard words are a good generalization of Sturmian words for the abelian complexity.

1 General Motivation

Sturmian words (1940) form a class of infinite words over the binary alphabet, which sheds light on the remarkable interactions between combinatorics, dynamical systems, and number theory. These interactions are reflected in the various ways to define them (see the historic paper [MH40], or refer to the book [Lot97] for a modern introduction). For instance, Sturmian words are equivalently

- words with subword complexity $n + 1$, *i.e.*, admitting exactly $n + 1$ factors of length n for all n (a *factor* of w of length n is a subword of w consisting of n consecutive letters);
- binary aperiodic words with imbalance equal to 1: all factors of a given length contain, up to one, the same number of occurrences of 1s (and thus, up to one as well, the same number of 2s);
- the symbolic trajectories of a ball in a square billiard, launched with a momentum with rationally independent components.

They give rise to several generalizations over the d-letter alphabet for $d \geq 3$, depending on the considered definition: Arnoux-Rauzy ([AR91]) and episturmian words ([JP02, GJ09]), other words associated with d-dimensional continued fraction algorithms (see [Sch00, Ber11, BD14]), interval exchange transformations ([Via06, Yoc07]), polygonal ([Tab05, CHT02] and references therein) or cubic billiard words ([AMST94a]). A large program, initiated by Rauzy in the 80s ([Rau83, Rau84, Rau85]), is to determine which properties are still equivalent in higher dimensions, and which are not.

The present manuscript is part of this program. It mostly focuses on one of the dynamical representations of Sturmian words: as words generated by a

© The Author(s), under exclusive license to Springer Nature Switzerland AG 2023
A. Frid and R. Mercaş (Eds.): WORDS 2023, LNCS 13899, pp. 3–13, 2023.
https://doi.org/10.1007/978-3-031-33180-0_1

billiard on a square table, which generalizes itself to a billiard in the cube, and in the cube of dimension d; and on two combinatorial quantities which characterize Sturmian words: the subword complexity and the abelian complexity. The manuscript is directed by the question: are the subword and abelian complexities of cubic billiard words in dimension d minimal among the complexities of d-ary words, as is the case when $d = 2$?

The manuscript is organized as follows. Hypercubic billiard words are defined in Sect. 2. Section 3 is devoted to the subword complexity. Section 4 is devoted to the abelian complexity.

2   Hypercubic Billiard Words

We start by defining cubic billiard words in dimension d, which we generically call *hypercubic billiard words*.

Let $d \geq 1$. The set $\mathbb{R}^d$ is equipped with the usual inner product and canonical basis $(e_1, ..., e_d)$. We are interested in the sequence of the faces successively hit by a billiard ball, initially located in $x \in [0, 1]^d$, which is given an initial momentum $\theta = (\theta_1, ..., \theta_d) \in (\mathbb{R} \backslash \{0\})^d$. We will use the letter k to code a hit against one of the two faces orthogonal to the vector e_k. The sequence that we thus obtain, denoted by $w(x, \theta)$, is an infinite word over the alphabet $\{1, ..., d\}$, that we call a *cubic billiard word in dimension d* (Fig. 1).

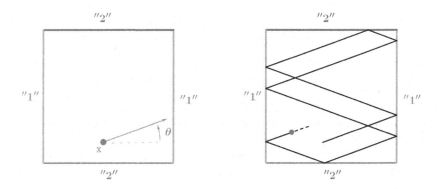

Fig. 1. The billiard word $w(x, \theta)$ starts with 11211121...

Remark. This definition omits to specify what happens when the trajectory of the ball intersects several faces *simultaneously* (for instance, for $d = 2$, when the ball hits a corner of the square). It has no importance: first, once the momentum θ is set up, the set of all starting positions $x \in [0, 1]^d$ for which such an event occurs has Lebesgue measure 0; secondly, for these rare "bad" starting positions, there still exists a reasonable coding strategy (not detailed here) that endows the words obtained with the same properties as regular hypercubic billiard words. In particular, all the results stated in this manuscript (noticeably Theorems 2, 3, 8, 9, and Corollary 1) are *true for all hypercubic billiard words*, including those generated by "bad" starting positions.

In the sequel, we focus on the billiard words generated by momenta θ with **rationally independent components**. The reason for this choice is that square billiard words generated by momenta with rationally independent components *are exactly* Sturmian words. Note that the rational independence of the momentum's components is equivalent to the *minimality* of one/all trajectories: balls launched with such momenta pass arbitrarily close to every point in the hypercube.

3 Subword Complexity of Hypercubic Billiard Words

Most of the literature defines *Sturmian words* as right-infinite words with subword complexity $p : n \mapsto n+1$. We recall that the *subword complexity* of w (often simply called *complexity*) is the function that counts the number of factors of length n in w. This function is intricately linked with the notion of **entropy in dynamics**, which measures how well a trajectory can be predicted from the knowledge of a small portion of it. A fundamental fact is that $n \mapsto n + 1$ is the **smallest complexity** function for a "non-trivial" word.

Theorem 1 (Morse and Hedlund, 1938 [MH38]). *Let w be an infinite word. If there exists an integer m such that $p(m) \leq m$, then the word w is ultimately periodic, and its subword complexity p is ultimately constant.*

The particular equality $p(1) = 2$ tells us that Sturmian words are written with exactly 2 letters. A natural question, when it comes to generalizing Sturmian words, is the following one.

Question 1. Let $d \geq 3$. What is the subword complexity of cubic billiard words in dimension d? Is this function the smallest subword complexity for non-ultimately periodic words written with exactly d letters?

3.1 Subword Complexity of Hypercubic Billiard Words

The first part of the question has been fully answered by Arnoux, Mauduit, Shiokawa, Tamura, Baryshnikov, and Bédaride between 1994 and 2009. First, when $d = 3$, we have the following expression, which was originally conjectured by Rauzy in 1983 ([Rau83]).

Theorem 2 (Arnoux, Mauduit, Shiokawa, and Tamura, 1994 [AMST94a] [AMST94b]; corrected by Bédaride [Béd03]).
Let $x \in [0,1]^3$, and $\theta \in \mathbb{R}^3$ with rationally independent components. If the inverses $\theta_1^{-1}, \theta_2^{-1}$ and θ_3^{-1} are also rationally independent, then the number of distinct factors of length n in the cubic billiard word $w(x,\theta)$ is

$$n^2 + n + 1.$$

This formula extends itself to arbitrary dimensions in the following way.

Theorem 3 (Bédaride [Béd09]).
Let $d \geq 1$, $x \in [0,1]^d$, and $\theta \in \mathbb{R}^d$ with rationally independent components. If $d \geq 3$, assume, moreover, that for any three indices i, j, k, the numbers θ_i^{-1}, θ_j^{-1} and θ_k^{-1} are rationally independent (♣). Then the number of distinct factors of length n in the billiard word $w(x, \theta)$ is

$$\sum_{k=0}^{\min(n,d-1)} k! \binom{n}{k} \binom{d-1}{k}.$$

Borel [Bor06] and Bédaride [Béd07] showed that the condition (♣) is necessary.

This general formula was first conjectured by Arnoux, Mauduit, Shiokawa, and Tamura in [AMST94a], and proved one year later, in 1995, under a strictly stronger – and thereby non optimal – condition by Baryshnikov [Bar95].

3.2   Minimal Subword Complexity for *d*-ary Words

Regarding the second part of Question 1 ("is this function the minimal subword complexity for non-ultimately periodic words written with exactly d letters?"), one quickly convinces oneself that the question is poorly stated. Indeed, the *uninteresting ternary* word $w = 3 \cdot w'$, where w' is any Sturmian word on the alphabet $\{1, 2\}$, is non-ultimately periodic, and has subword complexity $n \mapsto n + 2$. It is not difficult to prove that this complexity is the lowest for a non-ultimately periodic ternary word. Note that there exist other families of words with subword complexity $n \mapsto n + 2$. They are classified in [Ale95] (see also the reference [KT07], which is easier to find.)

In our opinion, an essential property of Sturmian words is that their letter frequencies are **rationally independent** (this is a manifestation of the action of the continued fraction). We recall that the *frequency* of a letter a in a right-infinite word w is the limit, if it exists, of the proportion of as in growing prefixes of w. By contrast to Sturmian words, the frequencies of letters of ultimately periodic words are rational, and thereby, as soon as there are at least two letters, rationally dependent. Thus, Sturmian words are exactly words with minimal complexity function among binary words whose letter frequencies – which exist by a classical argument of [Bos84] – are rationally independent. From this perspective, our question can be better formulated.

Question 2. Let $d \geq 3$. What is the smallest subword complexity for a d-ary word with rationally independent frequencies of letters?

This question was answered by Tijdeman in a mostly forgotten article from 1999.

Theorem 4 (Tijdeman, 1999, [Tij99]). *Let $d \geq 1$.*
(1) Let w be a d-ary word that admits frequencies of letters. If there exists an integer m such that $p(m) \leq (d-1)m$, where p denotes the subword complexity of w, then the frequencies of letters in w are rationally dependent.
(2) There exist d-ary words, with rationally independent letter frequencies, whose subword complexity is $p : n \mapsto (d-1)n + 1$.

In other words, the smallest subword complexity of d-ary words, with rationally independent letter frequencies, is $p : n \mapsto (d-1)n + 1$.

To our knowledge, there is no classification of the words, with rationally independent letter frequencies, which realize this minimal complexity. However, some families of such words are known, as illustrated below.

On the first hand, one finds the words associated with the interval exchange transformations defined by a partition of $[0, 1)$ into d sub-intervals, with rationally independent lengths, and by the cyclic permutation $(2, 3, ..., d, 1)$ (this is the example exhibited in [Tij99]). To learn more about interval exchange transformations, we refer to the lectures [Via06, Yoc07].

On the other hand, there are the words associated with certain multidimensional continued fraction algorithms. Noticeably, for $d = 3$, one finds the primitive C-adic words ([CLL17]), and the Arnoux-Rauzy words, for which the rational independence of the letter frequencies was only proven two years ago (see [And21b] or [DHS22]). More generally, for $d \geq 2$, one finds the strict episturmian words over the d-letter alphabet.

Theorem 5 (Andrieu, *in preparation*, but a proof can be found in [And21a]). *Let $d \geq 2$. The frequencies of letters of any strict episturmian word over the d-letter alphabet are rationally independent.*

Note that the fact that strict episturmian words have subword complexity $n \mapsto (d-1)n + 1$ stems from their definition. To learn more about episturmian words (starting with their definition), we refer to the survey [GJ09].

Coming back to billiard words and to Question 1, one observes that:
- the frequencies of letters of a hypercubic billiard word are rationally independent if, and only if, the components of its momentum θ are rationally independent. Indeed, these two vectors are equal up to a dilatation, and a change in the signs of some components.
- BUT *its subword complexity function is far from being minimal* when $d \geq 3$. Indeed, one checks that the complexity of a cubic billiard word in dimension d, under the assumption (♣), is polynomial with degree $d - 1$.

3.3   A Long-Standing Conjecture

We conclude this section with a long-standing conjecture, which aims at generalizing Theorem 1 to bi-dimensional words (instead of d-ary words, as we did).

Conjecture 1 (Nivat, 1997, [Niv97]). Let $\mathcal{A}$ be an alphabet, and w an infinite bi-dimensional word over $\mathcal{A}$, *i.e.*, an element of $\mathcal{A}^{\mathbb{Z}^2}$ (Fig. 2). If there exist $n, m \in \mathbb{N}$ such that the number of distinct $n \times m$ patterns is at most nm, then the word w is periodic in some direction.

Here, "periodic in some direction" means that there exists $t \in \mathbb{Z}^2$ such that for every $z \in \mathbb{Z}^2$, the letters at positions z and $z+t$ coincide. A lot of efforts have been, and continue to be, given towards Conjecture 1, which remains *unsolved* at the time of writing ([ST02, EKM03, QZ04, CK15, CK16, KS20, KM19]).

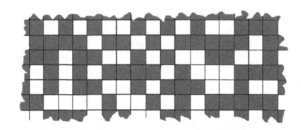

Fig. 2. Example of a bi-dimensional word over the alphabet $\{\square, \blacksquare\}$.

4   Abelian Complexity of Hypercubic Billiard Words

In this section, we consider a variation in the definition of complexity.

Two finite words u and v are *abelian equivalent* if u is the anagram of v (e.g., *twelveplusone* and *elevenplustwo*). The *abelian complexity* of an infinite word w is the function that counts the number of non-abelian equivalent factors of length n, for all n. For instance, for $n \geq 1$, the ultimately constant word $w = 21222222...$ admits exactly two non abelian-equivalent factors of length n: the one which contains 1, and the one written with the letter 2 only. For an introduction to abelian complexity, or, more generally, to abelian combinatorics on words, we refer to the recent survey [FP23].

It turns out that Sturmian words are also characterized by a remarkable abelian complexity.

Theorem 6 (derived from Coven and Hedlund, 1973, [CH73]). *A word is Sturmian if and only if its abelian complexity is constant, equal to 2, for all lengths $n \geq 1$, and if its frequencies of letters (which exist) are rationally independent.*

Interestingly, this is the smallest abelian complexity for non-periodic words. Indeed, it is not difficult to see that if, in an infinite word w, all factors of length n are abelian equivalent, then w is periodic with period n.

Thus, a natural question, first asked by Rauzy in 1983, is the following: does there exist a generalization of Sturmian words (among which are Arnoux-Rauzy and episturmian words, words associated with other multidimensional continued fractions, hypercubic billiard words, etc.), which somewhat preserves such a nice abelian complexity.

Question 3 (Rauzy, 1983, [Rau83] (Section 6), free translation from French). Does there exist a class of infinite words over the alphabet $\{1, 2, 3\}$ whose abelian complexity is constant, equal to 3? It seems that no, except perhaps for very particular frequencies of letters. What is, then, the "good" generalization of the Coven and Hedlund's theorem?

4.1 Proving the Conjecture of Rauzy

Question 3 splits into a conjecture, and a question. We start by examining the conjecture.

Conjecture 2. Under a condition $\mathfrak{A}$ (to be determined), there exists no ternary word w with abelian complexity equal to 3 for all $n \geq 1$.

With no condition $\mathfrak{A}$, Conjecture 2 is false. Indeed, one readily checks that the ternary words:

$$w_1 = 12333333333... \qquad \text{and} \qquad w_2 = 1 \cdot w_2',$$

where w_2' is any Sturmian word over the alphabet $\{2, 3\}$, admit exactly 3 non-abelian equivalent factors of length n, for all $n \geq 1$. One may ask the word w to be uniformly recurrent (a word w is *uniformly recurrent* if each factor appears infinitely often, with bounded gaps). The motivation for choosing this condition is that Sturmian words are uniformly recurrent. Furthermore, with this additional condition, the trivial counterexamples w_1 and w_2 are eliminated.

However, even with the condition $\mathfrak{A}$: "w must be uniformly recurrent", Conjecture 2 remains false [RSZ11], and the counterexamples exhibited by Richomme, Saari, and Zamboni are no good generalizations of Sturmian words.

For the reason stated in Sect. 3, on the strength of Tijdeman's theorem (Theorem 4), and following our reformulation of Coven and Hedlund's theorem (Theorem 6), we think that the condition $\mathfrak{A}$ should be: "the frequencies of letters are rationally independent". Under this condition $\mathfrak{A}$, Conjecture 2 *turns out to be true*.

Theorem 7 (Andrieu, Vivion, *in preparation*). *Let $w \in \{1, 2, 3\}^{\mathbb{N}}$ be a word with rationally independent letter frequencies. Then there exists an integer $n \geq 1$ such that w admits at least 4 non-abelian equivalent factors of length n.*

It remains to answer the second part of Question 3: what is, then, the good generalization of the Morse and Hedlund's theorem? We could, for instance, look for a class of ternary words with rationally independent letter frequencies and abelian complexity constant, equal to 4. Of course, this would only be possible for $n \geq 2$.

4.2   Abelian Complexity of Hypercubic Billiard Words

Abelian complexity has been extensively studied since 2007 ([KT07]), in particular after the progress of Richomme, Saari, and Zamboni towards Question 3 ([RSZ11]), see for instance [Saa09, RSZ10, CR11, BBT11, MR13, Tur13, BSFR14, BSCRF14], [Tur15, BSSW16, LCWW17, CW19, KR19, Whi19, Sha21], etc. Experience showed that, even for friendly words (for instance, the Tribonacci word), **the abelian complexity can be a surprisingly tricky function**, and one generally obtains bounds instead of exact values.

Question 4. What can be said of the abelian complexity of hypercubic billiard words? Can we explicitly compute this function for some easy instances?

The answer is positive, *beyond hope*: not only is it possible to compute the abelian complexity of all hypercubic billiard words generated by momenta with rationally independent components, but this function, remarkably simple, turns out to be the same for all of them (which is, remember the condition (♣), noticeably not the case for the subword complexity, whenever $d \geq 3$).

Theorem 8 (Andrieu, Vivion, *in preparation*). *Let $d \geq 1$, $x \in [0,1]^d$, and $\theta \in \mathbb{R}^d$ with rationally independent components. Let $n \in \mathbb{N}$. The number of non-abelian equivalent factors of length n in the billiard word $w(x,\theta)$ is*

$$\sum_{k=0}^{\min(n,d-1)} \binom{d-1}{k}.$$

Corollary 1. *Let $d \geq 1$, $x \in [0,1]^d$, and $\theta \in \mathbb{R}^d$ with rationally independent components. Let $n \geq d-1$. There are exactly 2^{d-1} non-abelian equivalent factors of length n in $w(x,\theta)$.*

In other words, the abelian complexity of cubic billiard words in dimension d is ultimately constant, equal to 2^{d-1}. For $d = 3$, cubic billiard words have an abelian complexity constant, equal to 4, for all $n \geq 2$. **This result answers Question** 3: cubic billiard words generated by momenta with rationally independent components (or, equivalently, and very interestingly, those with rationally independent letter frequencies) are a "good" generalization of Sturmian words, from the point of view of abelian complexity.

We conclude this subsection with a stronger result: not only do we know the number of non-abelian equivalent factors of length n of a hypercubic billiard word, but we are in fact able to compute them easily.

Theorem 9 (Andrieu, Vivion, *in preparation*). *Let $d \geq 1$, $x \in [0,1]^d$, and $\theta \in \mathbb{R}^d$ with rationally independent components. Let $n \in \mathbb{N}$. The set of equivalence classes of length n factors of $w(x,\theta)$, for the relation "being abelian equivalent", can be computed from the sole knowledge of the first $n-1$ letters of the billiard word $w(\mathbf{0},\theta)$.*

This contrasts sharply with the difficulty of listing the factors of a given length in a hypercubic billiard word. For instance, in dimension 3, and for the parameters $x = (0, 0, 0)$ and $\theta = (\sqrt{2}, \sqrt{3}, \sqrt{5})$, we know that the billiard word $w(x, \theta)$ admits 43 factors of length 6 (it is a quick computation from Theorem 3); nonetheless, they do not all appear in the prefix of length 10000000 of $w(x, \theta)$ ([AL22]).

4.3   Another Old Question

We conclude this extended abstract with a long-standing question, first asked by Arnoux, Mauduit, Shiokawa, and Tamura in 1994: *why* is the formula of the subword complexity (given in Theorem 3) symmetric in n (length of factors) and $d - 1$ (cardinal of the alphabet minus one)?

Question 5 (Arnoux, Mauduit, Shiokawa, Tamura, 1994, [AMST94a]). Can we give a meaningful bijective map between the factors of length n of a hypercubic billiard word over the alphabet $\{1, \ldots, d\}$, and the factors of length $d - 1$ of a particular/any hypercubic billiard word over the alphabet $\{1, \ldots, n + 1\}$?

To our knowledge, this question is *still open* at the time of writing.

Acknowledgements. We would like to express our gratitude to Boris Solomyak and to Julien Cassaigne for numerous exciting discussions on this topic; and to Shalom Eliahou for his kind help in improving the quality of the manuscript.

References

[AL22]   Andrieu, M., Labbé, S.: Sagemath function "billiard_nd" (2022). https://gitlab.com/seblabbe/slabbe/-/blob/develop/slabbe/billiard_nD.py. Accessed: 01 Dec 2022

[Ale95]   Alessandri, P.: Classification et représentation des mots de complexité $n + 2$. Université Aix-Marseille II, Rapport technique (1995)

[AMST94a]   Arnoux, P., Mauduit, C., Shiokawa, I., Tamura, J.-I.: Complexity of sequences defined by billiard in the cube. Bulletin de la Société Mathématique de France **122**(1), 1–12 (1994)

[AMST94b]   Arnoux, P., Mauduit, C., Shiokawa, I., Tamura, J.-I.: Rauzy's conjecture on billiards in the cube. Tokyo J. Math. **17**(1), 211–218 (1994)

[And21a]   Andrieu, M.: Exceptional trajectories in the symbolic dynamics of multi-dimensional continued fraction algorithms (chapter 4). PhD dissertation, Aix-Marseille Université (2021)

[And21b]   Andrieu, M.: A Rauzy fractal unbounded in all directions of the plane. Comptes Rendus. Mathématique **359**, 399–407 (2021)

[AR91]   Arnoux, P., Rauzy, G.: Représentation géométrique de suites de complexité 2n+1. Bulletin de la Société Mathématique de France **119**, 199–215 (1991)

[Bar95]   Baryshnikov, Y.: Complexity of trajectories in rectangular billiards. Commun. Math. Phys. **174**(1), 43–56 (1995)

[BBT11]   Balková, L., Břinda, K., Turek, O.: Abelian complexity of infinite words associated with quadratic Parry numbers. Theor. Comput. Sci. **412**(45), 6252–6260 (2011)

[BD14]   Berthé, V., Delecroix, V.: Beyond substitutive dynamical systems: S-adic expansions. In: Lecture Note 'Kokyuroku Bessatu', pp. 81–123 (2014)

[Béd03] Bédaride, N.: Billiard complexity in rational polyhedra. Regul. Chaotic Dyn. **8**(1), 97–104 (2003)

[Béd07] Bédaride, N.: Classification of rotations on the torus $\mathbb{T}^2$. Theor. Comput. Sci. **385**(1), 214–225 (2007)

[Béd09] Bédaride, N.: Directional complexity of the hypercubic billiard. Disc. Math. **309**(8), 2053–2066 (2009)

[Ber11] Berthé, V.: Multidimensional Euclidean algorithms, numeration and substitutions. Integers [electronic only] (2011)

[Bor06] Borel, J.P.: Complexity of degenerate three dimensional billiard words. In: Developments in Language Theory, pp. 386–396 (2006)

[Bos84] Boshernitzan, M.: A unique ergodicity of minimal symbolic flows with linear block growth. J. d'Analyse Mathématique **44**, 77–96 (1984)

[BSCRF14] Blanchet-Sadri, F., Currie, J., Rampersad, N., Fox, N.: Abelian complexity of fixed point of morphism $0 \mapsto 012, 1 \mapsto 02, 2 \mapsto 1$. Integers **14** (2014)

[BSFR14] Blanchet-Sadri, F., Fox, N., Rampersad, N.: On the asymptotic abelian complexity of morphic words. Adv. Appl. Math. **61**, 46–84 (2014)

[BSSW16] Blanchet-Sadri, F., Seita, D., Wise, D.: Computing abelian complexity of binary uniform morphic words. Theor. Comput. Sci. **640**, 41–51 (2016)

[CH73]   Coven, E.M., Hedlund, G.A.: Sequences with minimal block growth. Math. Syst. Theory **7**, 138–153 (1973)

[CHT02] Cassaigne, J., Hubert, P., Troubetzkoy, S.: Complexity and growth for polygonal billiards. Annales de l'Institut Fourier **52**(3), 835–847 (2002)

[CK15]   Cyr, V., Kra, B.: Nonexpansive $\mathbb{Z}^2$-subdynamics and Nivat's conjecture. Trans. Am. Math. Soc. **367**(9), 6487–6537 (2015)

[CK16]   Cassaigne, J., Kaboré, I.: Abelian complexity and frequencies of letters in infinite words. Int. J. Found. Comput. Sci. **27**(05), 631–649 (2016)

[CLL17] Cassaigne, J., Labbé, S., Leroy, J.: A set of sequences of complexity $2n+1$. In: Brlek, S., Dolce, F., Reutenauer, C., Vandomme, É. (eds.) WORDS 2017. LNCS, vol. 10432, pp. 144–156. Springer, Cham (2017). https://doi.org/10.1007/978-3-319-66396-8_14

[CR11]   Currie, J., Rampersad, N.: Recurrent words with constant abelian complexity. Adv. Appl. Math. **47**(1), 116–124 (2011)

[CW19]   Chen, J., Wen, Z.-X.: On the abelian complexity of generalized Thue-Morse sequences. Theor. Comput. Sci. **780**, 66–73 (2019)

[DHS22] Dynnikov, I., Hubert, P., Skripchenko, A.: Dynamical systems around the Rauzy gasket and their ergodic properties. Int. Math. Res. Not. **2023**(8), 6461–6503 (2022)

[EKM03] Epifanio, C., Koskas, M., Mignosi, F.: On a conjecture on bidimensional words. Theor. Comput. Sci. **299**, 123–150 (2003)

[FP23]   Fici, G., Puzynina, S.: Abelian combinatorics on words: a survey. Comput. Sci. Rev. **47**, 100532 (2023)

[GJ09]   Glen, A., Justin, J.: Episturmian words: a survey. RAIRO - Theor. Inf. Appl. **43**(3), 403–442 (2009)

[JP02]   Justin, J., Pirillo, G.: Episturmian words and episturmian morphisms. Theor. Comput. Sci. **276**(1), 281–313 (2002)

[KM19]   Kari, J., Moutot, E.: Nivat's conjecture and pattern complexity in algebraic subshifts. Theor. Comput. Sci. **777**, 379–386 (2019)

[KR19]   Kaye, I., Rampersad, N.: The abelian complexity of infinite words and the Frobenius problem. arxiv1907.08247 (2019)

[KS20]   Kari, J., Szabados, M.: An algebraic geometric approach to Nivat's conjecture. Inf. Comput. **271**, 104481 (2020)

[KT07]   Kaboré, I., Tapsoba, T.: Combinatoire de mots récurrents de complexité $n + 2$. RAIRO - Theor. Inf. Appl. **41**(4), 425–446 (2007)

[LCWW17] Lü, X., Chen, J., Wen, Z.-X., Wen, W.: On the abelian complexity of the Rudin-Shapiro sequence. J. Math. Anal. Appl. **451**(2), 822–838 (2017)

[Lot97] Lothaire, M.: Combinatorics on Words, 2nd edn. Cambridge Mathematical Library, Cambridge (1997)

[MH38]   Morse, M., Hedlund, G.A.: Symbolic dynamics. Am. J. Math. **60**(4), 815–866 (1938)

[MH40]   Morse, M., Hedlund, G.A.: Symbolic dynamics II. Sturmian trajectories. Am. J. Math. **62**(1), 1–42 (1940)

[MR13]   Madill, B., Rampersad, N.: The abelian complexity of the paperfolding word. Disc. Math. **313**(7), 831–838 (2013)

[Niv97] Nivat, M.: Invited talk at ICALP, Bologna (1997)

[QZ04]   Quas, A., Zamboni, L.: Periodicity and local complexity. Theor. Comput. Sci. **319**, 229–240 (2004)

[Rau84] Rauzy, G.: Des mots en arithmétique. Publications du Département de mathématiques (Lyon) **6B**, 103–113 (1984)

[Rau85] Rauzy, G.: Mots infinis en arithmetique. In: Nivat, M., Perrin, D. (eds.) LITP 1984. LNCS, vol. 192, pp. 164–171. Springer, Heidelberg (1985). https://doi.org/10.1007/3-540-15641-0_32

[Rau83] Rauzy, G.: Suites à termes dans un alphabet fini. Séminaire de Théorie des Nombres de Bordeaux, pp. 1–16 (1982–1983)

[RSZ10] Richomme, G., Saari, K., Zamboni, L.Q.: Balance and abelian complexity of the Tribonacci word. Adv. Appl. Math. **45**(2), 212–231 (2010)

[RSZ11] Richomme, G., Saari, K., Zamboni, L.Q.: Abelian complexity of minimal subshifts. J. Lond. Math. Soc. **83**(1), 79–95 (2011)

[Saa09] Saarela, A.: Ultimately constant abelian complexity of infinite words. J. Automata Lang. Comb. **14**(3), 255–258 (2009)

[Sch00] Schweiger, F.: Multidimensional Continued Fractions. Oxford University Press, Cambridge (2000)

[Sha21] Shallit, J.: Abelian complexity and synchronization. Integers **21** (2021)

[ST02]   Sander, J.W., Tijdeman, R.: The rectangle complexity of functions on two-dimensional lattices. Theor. Comput. Sci. **270**, 857–863 (2002)

[Tab05] Tabachnikov, S.: Geometry and Billiards. Mathematics Advanced Study Semesters. American Mathematical Society, Providence, R.I. (2005)

[Tij99] Tijdeman, R.: On the minimal complexity of infinite words. Indagationes Mathematicae **10**, 123–129 (1999)

[Tur13] Turek, O.: Abelian complexity and abelian co-decomposition. Theor. Comput. Sci. **469**, 77–91 (2013)

[Tur15] Turek, O.: Abelian complexity function of the Tribonacci word. J. Integer Seq. **18**, 15.3.4 (2015)

[Via06] Viana, M.: Ergodic theory of interval exchange maps. Revista Matemática Complutense **19** (2006)

[Whi19] Whiteland, M.A.: Asymptotic abelian complexities of certain morphic binary words. J. Automata Lang. Comb. **24**(1), 89–114 (2019)

[Yoc07] Yoccoz, J.C.: Interval exchange maps and translation surfaces. Lecture (Pisa) (2007)

Alternate Base Numeration Systems

Émilie Charlier[(✉)] [ID]

Department of Mathematics, University of Liège, Liège, Belgium
echarlier@uliege.be

Abstract. Alternate base numeration systems generalize real base numeration systems as defined by Renyi. Real numbers are represented using a finite number of bases periodically. Such systems naturally appear when considering linear numeration systems without a dominant root. As it happens, many classical results generalize to these numeration systems with multiple bases but some don't. This is a survey of the work done so far concerning combinatorial, algebraic and dynamical aspects. This study has been led in collaboration with several co-authors : Célia Cisternino, Karma Dajani, Savinien Kreczman, Zuzana Masáková and Edita Pelantová.

Keywords: Cantor real bases · Alternate bases · Subshifts · Parry numbers · Pisot numbers · Automata · Normalization · Dynamical systems · Ergodic theory

1 Introduction

Expansions of nonnegative real numbers x with respect to a real base $\beta > 1$ are sequences of integer digits $(a_n)_{n \geq 0}$ such that $x = \sum_{n=0}^{\infty} \frac{a_n}{\beta^{n+1}}$. A distinguished expansion of a given $x \in [0,1]$, denoted $d_\beta(x) = (d_n)_{n \geq 0}$ and called the β-expansion of x, is computed by the greedy algorithm: set $r_0 = x$ and for all $n \geq 0$, let $d_n = \lfloor \beta r_n \rfloor$ and $r_{n+1} = \beta r_n - d_n$. These numeration systems, introduced by Rényi [30], are extensively studied under various points of view and we can only cite a few of the many possible references [3,15,27,29,33].

In parallel, other numeration systems are also widely studied, this time to represent nonnegative integers. We choose an increasing integer sequence $U = (U_n)_{n \geq 0}$ such that $U_0 = 1$ and the quotients between consecutive terms $\frac{U_{n+1}}{U_n}$ are bounded. A nonnegative integer x is then represented by a finite sequence of integer digits $a_0 \cdots a_{\ell-1}$ such that $x = \sum_{n=0}^{\ell-1} a_n U_{\ell-1-n}$. Again, we distinguish the expansion that is obtained thanks to the greedy algorithm: let $\ell \geq 0$ be maximal such that $x < U_\ell$ and set $r_0 = x$; then for $n \in \{0, \ldots, \ell-1\}$, set $d_n = \lfloor \frac{r_n}{U_{\ell-1-n}} \rfloor$ and $r_{n+1} = r_n - d_n U_{\ell-1-n}$. The so-obtained expansion $d_0 \cdots d_{\ell-1}$ is called the U-expansion of x. Similarly, the literature about U-expansions of nonnegative integers is vast, see [4,5,12,13,18,21,26] for the most topic-related ones.

FNRS grant J.0034.22.

© The Author(s), under exclusive license to Springer Nature Switzerland AG 2023
A. Frid and R. Mercaş (Eds.): WORDS 2023, LNCS 13899, pp. 14–34, 2023.
https://doi.org/10.1007/978-3-031-33180-0_2

There exists an intimate link between β-expansions and U-expansions. Its study goes back to the work [4] of Bertrand-Mathis. The case where the base sequence U has a dominant root $\beta > 1$ is quite well understood [21]. More precisely, we say that U has the dominant root $\beta > 1$ whenever $\lim_{n \to \infty} \frac{U_{n+1}}{U_n} = \beta$. In particular, for sufficiently large n, the U-expansions of $U_n - 1$ share long common prefixes with specific expansions of 1 with respect to the real base β. In the case where the base sequence U has no dominant root, a similar phenomenon occurs with respect to expansions of 1 in a numeration system given by an *alternate base* $(\beta_0, \ldots, \beta_{p-1})$ associated with U for some well defined $p \geq 1$: for each $i \in \{0, \ldots, p-1\}$, we have $\beta_i = \lim_{n \to \infty} \frac{U_{np-i}}{U_{np-i-1}}$. This discovery was the original motivation for the study of alternate base expansions of real numbers. It turns out that a lot of classical results concerning β-expansions of real numbers extend to this new framework, and sometimes, to the even more general framework of *Cantor real bases*. The purpose of this survey is to give an overview of the results obtained so far in these generalized numeration systems. The study of linear numeration systems without a dominant root will be treated separately, in a subsequent paper.

2 Cantor Real Bases and Alternate Bases

Cantor expansions of real numbers were originally introduced by Cantor in 1869 [7]. A real number $x \in [0, 1)$ is represented via a base sequence $(b_n)_{n \geq 0}$ of integers greater than or equal to 2 as follows:

$$x = \sum_{n=0}^{\infty} \frac{a_n}{\prod_{k=0}^{n} b_k}$$

where for each $n \geq 0$, the digit a_n belongs to $\{0, \ldots, b_n - 1\}$. Many studies are devoted to Cantor series; see [16,20,22,31] to cite just a few.

In [8], we introduced series expansions of real numbers that are based on a sequence $\beta = (\beta_n)_{n \geq 0}$ of real numbers greater than 1. We call such a base sequence β a Cantor real base, and we talk about β-expansions. In doing so, we generalize both Cantor series and real base expansions. The same framework was introduced simultaneously in [6]. Moreover, other kinds of expansions with multiple real bases were also recently studied, see [23,25,28].

2.1 Cantor Real Bases

Let $\beta = (\beta_n)_{n \geq 0}$ be a sequence of real numbers greater than 1 such that $\prod_{n=0}^{\infty} \beta_n = \infty$. We call such a sequence β a *Cantor real base*. We define the β-value (partial) map $\mathrm{val}_\beta \colon (\mathbb{R}_{\geq 0})^{\mathbb{N}} \to \mathbb{R}_{\geq 0}$ by

$$\mathrm{val}_\beta(a) = \sum_{n=0}^{\infty} \frac{a_n}{\prod_{k=0}^{n} \beta_k} \tag{1}$$

for any sequence $a = (a_n)_{n \geq 0}$ over $\mathbb{R}_{\geq 0}$, provided that the series converges. If $\mathrm{val}_{\beta}(a) = x$ then we say that a is an *expansion* of x in base β. By taking $\beta_n = \beta$ for all $n \geq 0$, we recover Rényi expansions [30].

We will need to represent real numbers not only in a fixed Cantor real base β but also in all Cantor real bases obtained by shifting β. We define

$$\beta^{(n)} = (\beta_n, \beta_{n+1}, \ldots) \quad \text{for all } n \geq 0.$$

In particular $\beta^{(0)} = \beta$. We will also need to consider shifted sequences. The *shift operator* is given by

$$\sigma \colon A^{\mathbb{N}} \to A^{\mathbb{N}}, \ (a_n)_{n \geq 0} \mapsto (a_{n+1})_{n \geq 0}$$

(where A is any given set).

As a first result, we mention a characterization of those sequences $a \in (\mathbb{R}_{\geq 0})^{\mathbb{N}}$ for which there exists a Cantor real base β such that $\mathrm{val}_{\beta}(a) = 1$. Note that, unlike what happens in the real base case [27], given a sequence a, the equation $\mathrm{val}_{\beta}(a) = 1$ admits more than one Cantor real base β as a solution in general.

Theorem 1 ([8]). *Let $a = (a_n)_{n \geq 0}$ be a sequence over $\mathbb{R}_{\geq 0}$. There exists a Cantor real base β such that $\mathrm{val}_{\beta}(a) = 1$ if and only if $\sum_{n=0}^{\infty} a_n > 1$.*

2.2 The Greedy Algorithm

A distinguished expansion of a given $x \in [0, 1]$ is obtained thanks to the *greedy algorithm*. We first set $r_0 = x$. Then for all $n \geq 0$, we compute $d_n = \lfloor \beta_n r_n \rfloor$ and $r_{n+1} = \beta_n r_n - d_n$. The obtained expansion is denoted by $d_{\beta}(x) = (d_n)_{n \geq 0}$ and is called the β-*expansion* of x. Thus for all $\ell \geq 0$, one has

$$x = \sum_{n=0}^{\ell} \frac{d_n}{\prod_{k=0}^{n} \beta_k} + \frac{r_\ell}{\prod_{k=0}^{\ell} \beta_k}$$

where $r_\ell \in [0, 1)$. Note that since a Cantor real base satisfies $\prod_{n=0}^{\infty} \beta_n = \infty$, the latter equality implies the convergence of the greedy algorithm. We let A_{β} denote the (possibly infinite) alphabet $\{0, \ldots, \sup_{n \geq 0} \lceil \beta_n \rceil - 1\}$ and D_{β} denote the subset of $A_{\beta}^{\mathbb{N}}$ of the β-expansions of the real numbers in the interval $[0, 1)$, that is, $D_{\beta} = \{d_{\beta}(x) \colon x \in [0, 1)\}$.

We can also express the greedy digits d_n and remainders r_n thanks to the β_n-transformations. For $\beta > 1$, the β-transformation is the map

$$T_{\beta} \colon [0, 1) \to [0, 1), \ x \mapsto \beta x - \lfloor \beta x \rfloor.$$

Then for all $x \in [0, 1)$ and $n \geq 0$, we have

$$d_n = \lfloor \beta_n \left(T_{\beta_{n-1}} \circ \cdots \circ T_{\beta_0}(x) \right) \rfloor \quad \text{and} \quad r_n = T_{\beta_{n-1}} \circ \cdots \circ T_{\beta_0}(x).$$

We call an *alternate base* a periodic Cantor real base: there exists $p \geq 1$ such that for all $n \geq 0$, we have $\beta_n = \beta_{n+p}$. In this case we simply write $\beta = (\beta_0, \ldots, \beta_{p-1})$ and the integer p is called the *length* of the alternate base β.

Example 1. – Any sequence $\boldsymbol{\beta} = (\beta_n)_{n\geq 0}$ of real numbers greater than 1 that takes only finitely many values is a Cantor real base since in this case, the condition $\prod_{n=0}^{\infty} \beta_n = \infty$ is trivially satisfied.

– For $n \geq 0$, let $\alpha_n = 1 + \frac{1}{2^{n+1}}$ and $\beta_n = 2 + \frac{1}{2^{n+1}}$. The sequence $\boldsymbol{\alpha} = (\alpha_n)_{n\geq 0}$ is not a Cantor real base since $\prod_{n=0}^{\infty} \alpha_n < \infty$. If we perform the greedy algorithm on $x = 1$ for the sequence $\boldsymbol{\alpha}$, we obtain the sequence of digits 10^ω (where the ω notation means an infinite repetition), which is clearly not an expansion of 1 with respect to $\boldsymbol{\alpha}$. However, the sequence $\boldsymbol{\beta} = (\beta_n)_{n\geq 0}$ is indeed a Cantor real base since $\prod_{n=0}^{\infty} \beta_n = \infty$.

– If there exists $n \geq 0$ such that β_n is an integer (without any restriction on the other β_m), then $d_{\boldsymbol{\beta}^{(n)}}(1) = \beta_n 0^\omega$.

– Let $\alpha = \frac{1+\sqrt{13}}{2}$ and $\beta = \frac{5+\sqrt{13}}{6}$.

  • For the alternate base $\boldsymbol{\beta} = (\alpha, \beta)$, we get that $d_{\boldsymbol{\beta}^{(0)}}(1) = 2010^\omega$ and $d_{\boldsymbol{\beta}^{(1)}}(1) = 110^\omega$.

  • Consider $\boldsymbol{\beta} = (\beta_n)_{n\geq 0}$ the Cantor real base defined by

$$\beta_n = \begin{cases} \alpha, & \text{if } \mathrm{rep}_2(n) \text{ has an even number of 1's;} \\ \beta, & \text{otherwise} \end{cases}$$

where $\mathrm{rep}_2(n)$ is the binary expansion of n. We get the Thue-Morse sequence $\boldsymbol{\beta} = (\alpha, \beta, \beta, \alpha, \beta, \alpha, \alpha, \beta, \ldots)$ over the alphabet $\{\alpha, \beta\}$. We compute $d_{\boldsymbol{\beta}^{(0)}}(1) = 20010110^\omega$, $d_{\boldsymbol{\beta}^{(1)}}(1) = 1010110^\omega$ and $d_{\boldsymbol{\beta}^{(2)}}(1) = 110^\omega$. Note that since the base is aperiodic, these computations give us no information on $d_{\boldsymbol{\beta}^{(n)}}(1)$ for $n \geq 3$.

– Let $\varphi = \frac{1+\sqrt{5}}{2}$ be the Golden Ratio and let $\boldsymbol{\beta} = (3, \varphi, \varphi)$. We have $d_{\boldsymbol{\beta}^{(0)}}(1) = 30^\omega$, $d_{\boldsymbol{\beta}^{(1)}}(1) = 110^\omega$ and $d_{\boldsymbol{\beta}^{(2)}}(1) = 1(110)^\omega$.

– For the alternate base $\boldsymbol{\beta} = (\sqrt{6}, 3, \frac{2+\sqrt{6}}{3})$, we have that $d_{\boldsymbol{\beta}^{(0)}}(1) = 2(10)^\omega$, $d_{\boldsymbol{\beta}^{(1)}}(1) = 30^\omega$ and $d_{\boldsymbol{\beta}^{(2)}}(1) = 110020^\omega$. This shows that the $\boldsymbol{\beta}$-expansion of 1 can have a period less than the length of the base.

The classical properties of the β-expansion theory are still valid for Cantor real bases. Until the end of this section, unless otherwise stated, we consider a fixed Cantor real base $\boldsymbol{\beta} = (\beta_n)_{n\geq 0}$.

Proposition 1. *For all $x \in [0,1)$ and all integers $n \geq 0$, we have*

$$\sigma^n \circ d_{\boldsymbol{\beta}}(x) = d_{\boldsymbol{\beta}^{(n)}} \circ T_{\beta_{n-1}} \circ \cdots \circ T_{\beta_0}(x).$$

Proposition 2. *For all sequences a over $\mathbb{N}$ and all $x \in [0,1]$, we have $a = d_{\boldsymbol{\beta}}(x)$ if and only if $\mathrm{val}_{\boldsymbol{\beta}}(a) = x$ and*

$$\sum_{n=\ell+1}^{\infty} \frac{a_n}{\prod_{k=0}^n \beta_k} < \frac{1}{\prod_{k=0}^{\ell} \beta_k} \quad \text{for all } \ell \geq 0.$$

Proposition 3. *Let a be any expansion of a real number x in $[0,1]$ in base $\boldsymbol{\beta}$. Then the following four assertions are equivalent.*

1. *The sequence a is the β-expansion of x.*
2. *For all $n \geq 1$, we have $\mathrm{val}_{\beta^{(n)}}(\sigma^n(a)) < 1$.*
3. *The sequence $\sigma(a)$ belongs to $D_{\beta^{(1)}}$.*
4. *For all $n \geq 1$, the sequence $\sigma^n(a)$ belongs to $D_{\beta^{(n)}}$.*

Proposition 4. *A sequence a over $\mathbb{N}$ belongs to the set D_β if and only if $\mathrm{val}_{\beta^{(n)}}(\sigma^n(a)) < 1$ for all $n \geq 0$.*

Proposition 5. *The β-expansion of a real number $x \in [0,1]$ is lexicographically maximal among all expansions of x in base β.*

Proposition 6. *The function $d_\beta \colon [0,1] \to \mathbb{N}^{\mathbb{N}}$ is increasing: for all $x, y \in [0,1]$, we have $x < y \iff d_\beta(x) <_{\mathrm{lex}} d_\beta(y)$.*

Corollary 1. *If a is a sequence over $\mathbb{N}$ such that $\mathrm{val}_\beta(a) \leq 1$, then $a \leq_{\mathrm{lex}} d_\beta(1)$. In particular, the sequence $d_\beta(1)$ is lexicographically maximal among all expansions of all real numbers in $[0,1]$ in base β.*

Rényi expansions satisfies the property that considering two real bases α and β, one has $\alpha < \beta$ if and only if $d_\alpha(1) < d_\beta(1)$ [27]. The following proposition is a generalization of a weaker version of this property.

Proposition 7. *Let $\alpha = (\alpha_n)_{n \geq 0}$ and $\beta = (\beta_n)_{n \geq 0}$ be Cantor real bases such that $\prod_{k=0}^n \alpha_k \leq \prod_{k=0}^n \beta_k$ for all $n \geq 0$. Then $d_\alpha(x) \leq_{\mathrm{lex}} d_\beta(x)$ for all $x \in [0,1]$.*

However, it is not true that $d_\alpha(1) <_{\mathrm{lex}} d_\beta(1)$ implies that $\prod_{i=0}^n \alpha_i \leq \prod_{i=0}^n \beta_i$ for all $n \geq 0$, as the following example shows. The same example shows that the lexicographic order on the Cantor real bases is not sufficient either. Here, the term lexicographic order refers to the following order: $\alpha < \beta$ whenever there exists $\ell \geq 0$ such that $\alpha_n = \beta_n$ for $n < \ell$ and $\alpha_\ell < \beta_\ell$.

Example 2. Let $\alpha = (2 + \sqrt{3}, 2)$ and $\beta = (2 + \sqrt{2}, 5)$. Then $d_\alpha(1) = 31^\omega$ and $d_\beta(1)$ starts with the prefix 32, hence $d_\alpha(1) <_{\mathrm{lex}} d_\beta(1)$.

2.3 Quasi-Greedy Expansions and Admissible Sequences

An expansion is said to be *finite* if is ultimately zero, and *infinite* otherwise. The *length* of a finite expansion is the length of the longest prefix ending in a non-zero digit. In the finite case, we usually omit to write the tail of zeros.

When the β-expansion of 1 is finite, we modify it in order to obtain an infinite expansion of 1 that is lexicographically maximal among all infinite expansions of 1. The obtained expansion is denoted by $d_\beta^*(1)$ and is called the *quasi-greedy β-expansion* of 1. It is defined recursively as follows:

$$d_\beta^*(1) = \begin{cases} d_\beta(1), & \text{if } d_\beta(1) \text{ is infinite;} \\ d_0 \cdots d_{n-2}(d_{n-1} - 1)d_{\beta^{(n)}}^*(1), & \text{if } d_\beta(1) = d_0 \cdots d_{n-1} \text{ with } d_{n-1} > 0. \end{cases}$$

Example 3. – When $\beta = (\beta, \beta, \ldots)$, we recover the usual definition of the quasi-greedy β-expansion.

- Let $\beta = (\beta_0, \ldots, \beta_{p-1})$ be an alternate base such that each β_i is an integer. Then for all $i \in \{0, \ldots, p-1\}$, we have $d_{\beta^{(i)}}(1) = \beta_i 0^\omega$ and

$$d_{\beta^{(i)}}^*(1) = ((\beta_i - 1) \cdots (\beta_{p-1} - 1)(\beta_0 - 1) \ldots (\beta_{i-1} - 1))^\omega.$$

- Let $\beta = (\frac{1+\sqrt{13}}{2}, \frac{5+\sqrt{13}}{6})$. Since $d_{\beta^{(0)}}(1) = 201$ and $d_{\beta^{(1)}}(1) = 11$, we obtain $d_{\beta^{(1)}}^*(1) = (10)^\omega$ and $d_{\beta^{(0)}}^*(1) = 200 d_{\beta^{(1)}}^*(1) = 200(10)^\omega = 20(01)^\omega$.
- For $\beta = (3, \varphi, \varphi)$, we directly have that $d_{\beta^{(2)}}^*(1) = d_{\beta^{(2)}}(1) = 1(110)^\omega$. In order to compute $d_{\beta^{(0)}}^*(1)$ and $d_{\beta^{(1)}}^*(1)$, we need to go through the definition several times since $d_{\beta^{(0)}}(1) = 3$ and $d_{\beta^{(1)}}(1) = 11$ are finite. We compute $d_{\beta^{(0)}}^*(1) = 2d_{\beta^{(1)}}^*(1) = 210d_{\beta^{(0)}}^*(1) = (210)^\omega$ and $d_{\beta^{(1)}}^*(1) = 10d_{\beta^{(0)}}^*(1) = 10(210)^\omega = (102)^\omega$.
- Consider $\beta = (3, \beta, \beta, \beta, \beta, \ldots)$ where $\beta = \sqrt{6(2 + \sqrt{6})}$. We get $d_{\beta^{(0)}}(1) = 3$ and $d_{\beta^{(1)}}(1) = d_\beta(1)$ is infinite not ultimately periodic [2]. Therefore, the quasi-greedy expansion $d_{\beta^{(0)}}^*(1) = 2d_{\beta^{(1)}}^*(1)$ is not ultimately periodic.

The following propositions list the main properties of the quasi-greedy β-expansion of 1.

Proposition 8. *The quasi-greedy β-expansion of 1 is an expansion of 1 in base β, i.e., we have* $\mathrm{val}_\beta(d_\beta^*(1)) = 1$.

Proposition 9. *If a is a sequence over $\mathbb{N}$ such that $\mathrm{val}_\beta(a) < 1$, then $a <_{\text{lex}} d_\beta^*(1)$. Furthermore, $d_\beta^*(1)$ is lexicographically maximal among all infinite expansions of all real numbers in $[0, 1]$ in base β.*

Proposition 10. *The quasi-greedy β-expansion of 1 can also be obtained as the following limit:* $d_\beta^*(1) = \lim_{x \to 1^-} d_\beta(x)$.

In [29], Parry characterized those sequences over $\mathbb{N}$ that belong to D_β. Such sequences are sometimes called β-*admissible sequences*. Analogously, sequences in $D_{\boldsymbol{\beta}}$ are said to be the $\boldsymbol{\beta}$-*admissible sequences*. The following theorem generalizes Parry's theorem to Cantor real bases.

Theorem 2 ([8]). *A sequence a over $\mathbb{N}$ belongs to $D_{\boldsymbol{\beta}}$ if and only if $\sigma^n(a) <_{\text{lex}} d_{\beta^{(n)}}^*(1)$ for all $n \geq 0$.*

Example 4. Let $\boldsymbol{\beta} = (3, \varphi, \varphi)$. We obtain from Theorem 2 that $a = 210(110)^\omega$ is the $\boldsymbol{\beta}$-expansion of some $x \in [0, 1)$ since $d_{\beta^{(0)}}^*(1) = (210)^\omega$, $d_{\beta^{(1)}}^*(1) = (102)^\omega$ and $d_{\beta^{(2)}}^*(1) = 1(110)^\omega$. This x is given by $\mathrm{val}_\beta(a) = \frac{19+9\sqrt{5}}{3(7+3\sqrt{5})}$.

We then obtain a characterization of the $\boldsymbol{\beta}$-expansions of a real number x in the interval $[0, 1]$ among all its expansions in base $\boldsymbol{\beta}$.

Theorem 3. ([8]). *An expansion a of some real number $x \in [0, 1]$ in base $\boldsymbol{\beta}$ is its $\boldsymbol{\beta}$-expansion if and only if $\sigma^n(a) <_{\text{lex}} d_{\beta^{(n)}}^*(1)$ for all $n \geq 1$.*

Example 5. Consider $\beta = (\frac{16+5\sqrt{10}}{9}, 9)$. Then $d_{\beta^{(0)}}(1) = d^*_{\beta^{(0)}}(1) = 34(27)^\omega$, $d_{\beta^{(1)}}(1) = 9$ and $d^*_{\beta^{(1)}}(1) = 834(27)^\omega$. For all $n \geq 1$, we have $\sigma^{2n}(34(27)^\omega) <_{\text{lex}} d^*_{\beta^{(0)}}(1)$ and $\sigma^{2n-1}(34(27)^\omega) <_{\text{lex}} d^*_{\beta^{(1)}}(1)$ as prescribed by Theorem 3.

In comparison with the real base expansion theory, considering a Cantor real base β and a sequence a over $\mathbb{N}$, Theorem 3 does not provide us with a purely combinatorial condition to check whether a is the β-expansion of 1. More details will be given in Sect. 3.2, where we will see that even though an improvement of this result in the context of alternate bases can be proved, a purely combinatorial condition cannot exist.

3   Combinatorial Properties of Alternate Base Expansions

Recall that an alternate base is a periodic Cantor real base. The aim of this section is to discuss some results that are specific to these particular Cantor real bases.

In Theorem 1, we gave a characterization of those sequences $a \in (\mathbb{R}_{\geq 0})^\mathbb{N}$ for which there exists a Cantor real base β such that $\text{val}_\beta(a) = 1$. Here, we are interested in the stronger condition of the existence of an alternate base β satisfying $\text{val}_\beta(a) = 1$.

Theorem 4 ([8]). *Let a be a sequence over $\mathbb{R}_{\geq 0}$ such that $a_n \in O(n^d)$ for some $d \in \mathbb{N}$ and let $p \in \mathbb{N}_{\geq 1}$. There exists an alternate base β of length p such that $\text{val}_\beta(a) = 1$ if and only if $\sum_{n=0}^\infty a_n > 1$. If moreover $p \geq 2$, then there exist uncountably many such alternate bases.*

From now on, we let $\beta = (\beta_0, \ldots, \beta_{p-1})$ be a fixed alternate base.

3.1   Greedy Alternate Expansions

The greedy and the quasi-greedy β-expansions of 1 enjoy specific properties whenever β is an alternate base.

Proposition 11. *The β-expansion of 1 is not purely periodic.*

In the framework of β-expansions, a real base β is called a *Parry number* whenever the quasi-greedy β-expansion of 1 is ultimately periodic. In the context of alternate bases, in order to have an ultimately periodic quasi-greedy β-expansion of 1, one might think at first that the product $\delta = \prod_{i=0}^{p-1} \beta_i$ should be a Parry number since by grouping terms p by p in the sum

$$\frac{a_0}{\beta_0} + \frac{a_1}{\beta_0\beta_1} + \frac{a_2}{\beta_0\beta_1\beta_2} + \cdots$$

we get an expansion of the kind

$$\frac{c_0}{\delta} + \frac{c_1}{\delta^2} + \frac{c_2}{\delta^3} + \cdots .$$

But in the previous expression, the numerators are no longer integers. The following example shows that this intuition is not correct, even whenever all quasi-greedy $\beta^{(i)}$-expansions of 1 are ultimately periodic for $i \in \{0, \ldots, p-1\}$.

Example 6. Consider again the Parry alternate base $\beta = (3, \varphi, \varphi)$. As previously seen, all the corresponding quasi-greedy expansions of 1 are ultimately periodic. However, let us show that the product $\delta = 3\varphi^2$ is not a Parry number, and moreover, none of its powers $\delta^n = (3\varphi^2)^n$ is. It is a well-known property of the Golden Ratio that $\varphi^n = f_n\varphi + f_{n-1}$ for all $n \geq 1$, where $(f_n)_{n\geq 0} = (0, 1, 1, 2, 3, 5, 8, \ldots)$ is the Fibonacci sequence starting with the initial conditions $0, 1$. Therefore, denoting $\overline{\varphi} = \frac{1-\sqrt{5}}{2}$, for all $n \geq 1$, the minimal polynomial of $(3\varphi^2)^n$ can be computed as

$$(X - 3^n(f_{2n}\varphi + f_{2n-1}))(X - 3^n(f_{2n}\overline{\varphi} + f_{2n-1}))$$
$$= X^2 - 3^n(f_{2n} + 2f_{2n-1})X + 3^{2n}(-f_{2n}^2 + f_{2n}f_{2n-1} + f_{2n-1}^2)$$
$$= X^2 - 3^n(f_{2n+1} + f_{2n-1})X + 3^{2n},$$

since it can be easily verified by induction that we have $-f_n^2 + f_n f_{n-1} + f_{n-1}^2 = (-1)^n$ for all $n \geq 1$. We can also check that $f_{n+1} + f_{n-1} \leq 3^{\frac{n}{2}}$ for all $n \geq 1$. But the quadratic Parry numbers are known to be roots of polynomials of the form $X^2 - aX - b$ with $a \geq b \geq 1$ or of the form $X^2 - aX + b$ with $a - 2 \geq b \geq 1$ [2]. Therefore, we get that $(3\varphi^2)^n$ is not a Parry number for all $n \geq 1$.

Proposition 12. *The quasi-greedy expansion $d_\beta^*(1)$ is ultimately periodic if and only if either an ultimately periodic expansion is reached or only finite expansions are involved within the first p recursive calls to the definition of $d_\beta^*(1)$.*

Ultimately periodic β-expansions will be investigated further in Sects. 3.3, 4.1, 4.3 and 4.4.

3.2 Admissible Sequences in Alternate Bases

The condition given in Theorem 3 does not allow us to check whether a given expansion of 1 is the β-expansion of 1 without effectively computing the quasi-greedy β-expansion of 1, and hence the β-expansion of 1 itself. The following result provides us with such a condition in the case of alternate bases, provided that we are given the quasi-greedy $\beta^{(i)}$-expansions of 1 for $i \in \{1, \ldots, p-1\}$. Note that the shifted sequences starting in positions that are multiple of p are compared with the sequence a itself and not with $d_\beta^*(1)$ as in Theorem 3.

Proposition 13. *An expansion a of 1 in the alternate base β is the β-expansion of 1 if and only if $\sigma^{pm}(a) <_{\text{lex}} a$ for all $m \geq 1$ and $\sigma^{pm+i}(a) <_{\text{lex}} d_{\beta^{(i)}}^*(1)$ for all $m \geq 0$ and $i \in \{1, \ldots, p-1\}$.*

We have seen in Theorem 4 that considering a sequence a over $\mathbb{N}$, there may exist more than one alternate base β of a given length such that $\text{val}_\beta(a) = 1$. Moreover, among all of these alternate bases, it may be that some are such that a is greedy and others are such that a is not. Thus, a purely combinatorial condition for checking whether an expansion is greedy cannot exist.

Example 7. Consider $a = 2(10)^\omega$. Then $\mathrm{val}_\alpha(a) = \mathrm{val}_\beta(a) = 1$ for both $\boldsymbol{\alpha} = (1 + \varphi, 2)$ and $\boldsymbol{\beta} = (\frac{31}{10}, \frac{420}{341})$. It can be checked that $d_\alpha(1) = a$ and $d_\beta(1) \neq a$.

Furthermore, a sequence a can be greedy for more than one alternate base.

Example 8. The sequence 110^ω is the expansion of 1 with respect to the three alternate bases φ, $(\frac{5+\sqrt{13}}{6}, \frac{1+\sqrt{13}}{2})$ and $(1.7, \frac{1}{0.7})$.

3.3   The Alternate β-Shift

Let us recall some definitions from symbolic dynamics. For a finite alphabet A, a subset of $A^\mathbb{N}$ is called a *subshift* of $A^\mathbb{N}$ if it is shift-invariant and closed with respect to the product topology. For a subset S of $A^\mathbb{N}$, we let $\mathrm{Fac}(S)$ denote the set of all finite factors of all elements in S. A subshift S of $A^\mathbb{N}$ is said to be *sofic* if the language $\mathrm{Fac}(S) \subset A^*$ is accepted by a finite automaton.

In this section, we generalize the notion of β-shift to the context of alternate bases, and study its properties. First, we let S_β denote the topological closure of D_β with respect to the product topology.

Proposition 14. *A sequence a over $\mathbb{N}$ belongs to S_β if and only if $\sigma^n(a) \leq_{\mathrm{lex}} d^*_{\beta^{(n)}}(1)$ for all $n \geq 0$.*

Proposition 15. *Let $a, b \in S_\beta$.*

1. *If $a <_{\mathrm{lex}} b$ then $\mathrm{val}_\beta(a) \leq \mathrm{val}_\beta(b)$.*
2. *If $\mathrm{val}_\beta(a) < \mathrm{val}_\beta(b)$ then $a <_{\mathrm{lex}} b$.*

Proposition 16. *For all $n \geq 0$, if $w \in S_{\beta^{(n)}}$ then $\sigma(w) \in S_{\beta^{(n+1)}}$.*

Since the set S_β is not shift-invariant, we rather consider the set

$$\Sigma_\beta = \bigcup_{i=0}^{p-1} S_{\beta^{(i)}}.$$

Proposition 17. *The sets Σ_β is closed and shift-invariant.*

The subset Σ_β is thus a subshift of $A_\beta^\mathbb{N}$, which we call the *β-shift*.

Proposition 18. *We have $\mathrm{Fac}(D_\beta) = \mathrm{Fac}(S_\beta) = \mathrm{Fac}(\Sigma_\beta)$.*

We define *Parry alternate bases* as the alternate bases $\boldsymbol{\beta}$ such that all $d^*_{\beta^{(i)}}(1)$ are ultimately periodic for $i \in \{0, \ldots, p-1\}$. We will see that, analogously to what happens for real base expansions, Parry alternate bases are exactly those alternate bases giving rise to a sofic β-shift, which justifies the terminology.

For a Parry alternate base $\boldsymbol{\beta}$, we define a deterministic finite automaton $\mathcal{A}_\beta = (Q, I, F, A_\beta, \delta)$. Without loss of generality, we can consider that the involved periods are all multiples of the length p of the base. Thus, let us write

$$d^*_{\beta^{(i)}}(1) = t_{i,0} \cdots t_{i,m_i-1}(t_{i,m_i} \cdots t_{i,m_i+n_i p-1})^\omega.$$

Then the set of states is $Q = \{0, \ldots, p-1\} \times \{0, \ldots, m_i + n_i p - 1\}$. The set I of initial states and the set F of final states are defined as $I = \{0, \ldots, p-1\} \times \{0\}$ and $F = Q$. The (partial) transition function $\delta \colon Q \times A_\beta \to Q$ of the automaton $\mathcal{A}_\beta$ is defined as follows. For each state $(i, k) \in Q$, we have

$$\delta((i, k), t_{i,k}) = \begin{cases} (i, k+1), & \text{if } k < m_i + n_i p - 1; \\ (i, m_i), & \text{otherwise} \end{cases}$$

and $\delta((i, k), s) = ((i + k + 1) \bmod p, 0)$ for all $s \in \{0, \ldots, t_{i,k} - 1\}$. By using Theorem 2 and Proposition 18, we get the following result.

Proposition 19. *The automaton $\mathcal{A}_\beta$ accepts the language* $\mathrm{Fac}(\Sigma_\beta)$.

This implies that the β-shift associated with a Parry alternate base is sofic. As it turns out, the converse is also true, so that we obtain the following result extending a result of Bertrand-Mathis for real bases [3]. Proving this result turned out to be much more difficult than the original result for $p = 1$.

Theorem 5 ([8]). *The alternate β-shift is sofic if and only if β is a Parry alternate base.*

Example 9. The finite automaton of Fig. 1 accepts the set of factors of elements in the β-shift for $\beta = (\frac{1+\sqrt{13}}{2}, \frac{5+\sqrt{13}}{6})$; also see Examples 1 and 3.

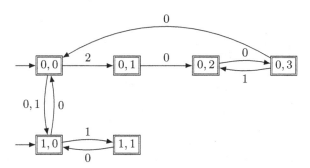

Fig. 1. A deterministic automaton accepting $\mathrm{Fac}(\Sigma_\beta)$ for $\beta = (\frac{1+\sqrt{13}}{2}, \frac{5+\sqrt{13}}{6})$.

Interestingly, some new phenomena occur in the extended framework of alternate bases when looking at subshifts of finite type. Recall that a subshift is said to be of *finite type* if its minimal set of forbidden factors is finite. For $p = 1$, it is well known that the β-shift is of finite type if and only if $d_\beta(1)$ is finite [3]. However, this result does not generalize to $p \geq 2$ since for the alternate base $\beta = (\frac{1+\sqrt{13}}{2}, \frac{5+\sqrt{13}}{6})$, we get $d^*_{\beta^{(0)}}(1) = 20(01)^\omega$ and $d^*_{\beta^{(1)}}(1) = (10)^\omega$, thus we see that all words in $2(00)^*2$ are factors avoided by Σ_β. Therefore, even though the $\beta^{(i)}$-expansions of 1 are finite for $i \in \{1, 2\}$ as we have seen in Example 1, the associated β-shift is not of finite type.

4   Algebraic Properties of Alternate Base Expansions

In this section, we report on the two works [10,11] were we studied the algebraic properties of alternate bases $\boldsymbol{\beta} = (\beta_0, \ldots, \beta_{p-1})$.

The property of being a Parry alternate base was defined in the previous section from a combinatorial point of view. Here we provide some algebraic necessary/sufficient conditions to have a Parry alternate base. Then, in the (stricly) stronger situation where the product $\delta = \prod_{i=0}^{p-1} \beta_i$ is a Pisot number and all the bases β_i belong to the extended field $\mathbb{Q}(\delta)$, we will be able to say much more. On the one hand, we obtain generalizations of some results of Schmidt [33] giving rise to an elementary proof of the original result. On the other hand, we obtain an analogue of Frougny's result [18] concerning normalization of real bases expansions: under these assumptions, the normalization function is computable by a finite Büchi automaton, and furthermore, we effectively construct such an automaton. Results on normalization will be presented in Sect. 5.

4.1   A Necessary Condition for Being a Parry Alternate Base

The following result gives a necessary condition on $\boldsymbol{\beta} = (\beta_0, \ldots, \beta_{p-1})$ to be a Parry alternate base, that is, to have eventually periodic $\boldsymbol{\beta}^{(i)}$-expansions of 1 for all $i \in \{0, \ldots, p-1\}$.

Theorem 6 ([11]). *If $\boldsymbol{\beta}$ is a Parry alternate base, then δ is an algebraic integer and $\beta_i \in \mathbb{Q}(\delta)$ for all $i \in \{0, \ldots, p-1\}$.*

The condition that the bases must be expressible as rational functions of the product δ is a phenomenon that does not show for Rényi numeration systems. Indeed, this condition is trivially satisfied whenever $p = 1$. Therefore, we see once again that new ideas and techniques are necessary in order to understand the properties of alternate bases.

4.2   Alternate Spectrum

An important tool in our study is the spectrum of numeration systems associated with alternate bases. The spectrum of a real number $\delta > 1$ and an alphabet $A \subset \mathbb{Z}$ was introduced by Erdős et al. [16] and further studied in [1,17]. For our purposes, we use a generalized concept with $\delta \in \mathbb{C}$ and $A \subset \mathbb{C}$ and study its topological properties.

From now on, we consider a p-tuple $\boldsymbol{D} = (D_0, \ldots, D_{p-1})$ where each D_i is an alphabet of integers containing 0. We use the convention that $D_n = D_{n \bmod p}$ and $\boldsymbol{D}^{(n)} = (D_n, \ldots, D_{n+p-1})$ for all $n \geq 0$. Grouping terms p by p, the left-hand side of (1) can be written as

$$\sum_{m=0}^{+\infty} \frac{\sum_{i=0}^{p-1} a_{mp+i}\beta_{i+1} \cdots \beta_{p-1}}{\delta^{m+1}}$$

where $\delta = \prod_{i=0}^{p-1} \beta_i$. If we add the constraint that each letter a_n belongs to D_n, then we obtain an expansion in base δ over the alphabet

$$\mathcal{D} = \left\{ \sum_{i=0}^{p-1} a_i \beta_{i+1} \cdots \beta_{p-1} : \forall i \in \{0, \ldots, p-1\},\ a_i \in D_i \right\}.$$

We define the *alternate spectrum* to be the set

$$X^{\mathcal{D}}(\delta) = \left\{ \sum_{n=0}^{\ell-1} c_n \delta^{\ell-1-n} : \ell \geq 0,\ c_0, c_1, \ldots, c_{\ell-1} \in \mathcal{D} \right\}.$$

For the sake of simplicity, for each $i \in \{0, \ldots, p-1\}$, we let X_i denote the spectrum built from the shifted base $\boldsymbol{\beta}^{(i)}$ and the shifted p-tuple of alphabets $\boldsymbol{D}^{(i)}$. In particular, we have $X_0 = X^{\mathcal{D}}(\delta)$.

Lemma 1. *For each $i \in \{0, \ldots, p-1\}$, we have $X_i \cdot \beta_i + D_i = X_{i+1}$ where it is understood that $X_p = X_0$.*

4.3   A Sufficient Condition for Being a Parry Alternate Base

In this section, we present a sufficient condition for $\boldsymbol{\beta}$ to be a Parry alternate base. We proceed in two steps, by studying the properties of the spectrum.

Proposition 20. *If $D_i \supseteq \{-\lfloor \beta_i \rfloor, \ldots, \lfloor \beta_i \rfloor\}$ for all $i \in \{0, \ldots, p-1\}$ and the spectrum $X^{\mathcal{D}}(\delta)$ has no accumulation point in $\mathbb{R}$, then $\boldsymbol{\beta}$ is a Parry alternate base.*

Proposition 21. *If δ is a Pisot number and $\beta_i \in \mathbb{Q}(\delta)$ for all $i \in \{0, \ldots, p-1\}$ then the spectrum $X^{\mathcal{D}}(\delta)$ has no accumulation point in $\mathbb{R}$.*

As a consequence, we get the following theorem, which for the case $p = 1$ is a well-known result of Schmidt [33]. In Sect. 4.4, we will present an alternative method for proving this result.

Theorem 7 ([11]). *If δ is a Pisot number and $\beta_i \in \mathbb{Q}(\delta)$ for all $i \in \{0, \ldots, p-1\}$ then $\boldsymbol{\beta}$ is a Parry alternate base.*

Let us make several remarks concerning this theorem. First, the condition of δ being a Pisot number is neither sufficient nor necessary for $\boldsymbol{\beta}$ to be a Parry alternate base. Indeed, it is not necessary even for $p = 1$ since there exist Parry numbers which are not Pisot. To see that it is not sufficient for $p \geq 2$, consider the alternate base $\boldsymbol{\beta} = (\sqrt{\beta}, \sqrt{\beta})$ where β is the smallest Pisot number. The product δ is the Pisot number β. However, the β-expansion of 1 is equal to $d_{\sqrt{\beta}}(1)$, which is known to be aperiodic.

Furthermore, the bases $\beta_0, \ldots, \beta_{p-1}$ need not be algebraic integers in order to have a Parry alternate base since, for instance, for the Parry alternate base $\boldsymbol{\beta} = (\frac{1+\sqrt{13}}{2}, \frac{5+\sqrt{13}}{6})$, the second base $\frac{5+\sqrt{13}}{6}$ is not an algebraic integer.

4.4   Ultimately Periodic Alternate Base Expansions

Here we present two results from [10] generalizing results on ultimately periodic Renyi expansions [33]. Recall that a *Salem number* is an algebraic integer greater than 1 whose Galois conjugates lie inside the unit disk with at least one of them on the unit circle. Thus, the set of algebraic integers greater than 1 whose Galois conjugates lie inside the unit disk is partioned into the Pisot numbers and the Salem numbers. We study the set $\mathrm{Per}(\beta) = \{x \in [0,1) : d_\beta(x) \text{ is ultimately periodic}\}$.

Theorem 8 ([10]).

1. *If $\mathbb{Q} \cap [0,1) \subseteq \cap_{i=0}^{p-1} \mathrm{Per}(\beta^{(i)})$ then $\beta_0, \ldots, \beta_{p-1} \in \mathbb{Q}(\delta)$ and δ is either a Pisot number or a Salem number.*
2. *If δ is a Pisot number and $\beta_0, \ldots, \beta_{p-1} \in \mathbb{Q}(\delta)$ then $\mathrm{Per}(\beta) = \mathbb{Q}(\delta) \cap [0,1)$.*

It is interesting to note that when adding the hypothesis $p = 1$ in our proof from [10], we obtain a much shorter proof than Schmidt's original one from [33].

Another particularly nice point is that we recover Theorem 7 as a corollary of Theorem 8, whereas the proof technique developed in [10] is independent from the first proof of Theorem 7 since it does not make use of the properties of the spectrum.

Another consequence of Theorem 8 is the following well-known property of Pisot numbers.

Corollary 2. *If β is a Pisot number then $\beta \in \mathbb{Q}(\beta^p)$ for all integer $p \geq 1$.*

The common proof of this result makes use of algebraic tools such as matrix diagonalization or the Kronecker theorem stating that if the roots of a monic polynomial with integers coefficients all lie inside the unit disc then they must be either zero or roots of unity. No such argument is used in [10].

The second generalization of Schmidt's results we obtain is the following.

Theorem 9 ([10]). *If δ is an algebraic integer that is neither a Pisot number nor a Salem number then $\mathrm{Per}(\beta) \cap \mathbb{Q}$ is nowhere dense in $[0,1)$.*

5   Normalization of Alternate Base Expansions

The *normalization function* $\nu_{\beta,D} : (\cup_{i=0}^{p-1} D_i)^{\mathbb{N}} \to (\cup_{i=0}^{p-1} \{0, \ldots, \lceil \beta_i \rceil - 1\})^{\mathbb{N}}$ is the partial function mapping any expansion $a \in \prod_{n=0}^{\infty} D_n$ of a real number $x \in [0,1)$ in base β to the β-expansion of x. We say that $\nu_{\beta,D}$ is *computable by a finite automaton* if there exists a finite Büchi automaton accepting the set

$$\left\{ (u,v) \in \prod_{n=0}^{\infty} (D_n \times A_{\beta_n}) : \mathrm{val}_\beta(u) = \mathrm{val}_\beta(v) \text{ and } \exists x \in [0,1), \ v = d_\beta(x) \right\}.$$

Büchi automata are defined as classical automata except for the acceptance criterion which has to be adapted in order to deal with infinite words: an infinite word is accepted if it labels an initial path going infinitely many times through a final state. A major difference between Büchi and classical automata (i.e., accepting finite words) is that a set of infinite words accepted by a finite Büchi automaton is not necessarily accepted by a deterministic one.

5.1   Zero Automaton

Let us now generalize the notion of zero automaton introduced by Frougny in [18] to the context of alternate bases. The aim is to define a deterministic Büchi automaton accepting the set

$$Z(\boldsymbol{\beta}, \boldsymbol{D}) = \left\{ (a_n)_{n \geq 0} \in \prod_{n=0}^{+\infty} D_n : \sum_{n=0}^{+\infty} \frac{a_n}{\prod_{k=0}^{n} \beta_k} = 0 \right\}$$

of all expansions of zero the n-th digit of which belongs to the alphabet D_n. This will be one of the key ingredient in order to compute the normalization function by using a finite (two-tape) Büchi automaton.

We define

$$M = \sum_{n=0}^{+\infty} \frac{\max(D_n)}{\prod_{k=0}^{n} \beta_k} \qquad \text{and} \qquad m = \sum_{n=0}^{+\infty} \frac{\min(D_n)}{\prod_{k=0}^{n} \beta_k}$$

where $\max(D_n)$ and $\min(D_n)$ respectively denote the maximal and minimal digit in the alphabet D_n. Then for each $i \in \{0, \dots, p-1\}$, we let M_i and m_i denote the numbers M and m corresponding to the shifted base $\beta^{(i)}$ and the shifted p-tuple of alphabets $\boldsymbol{D}^{(i)}$ respectively. In particular, we have $M_0 = M$ and $m_0 = m$. We define the *zero automaton* $\mathcal{Z}(\boldsymbol{\beta}, \boldsymbol{D})$ associated with the alternate base $\boldsymbol{\beta}$ and the p-tuple of alphabets $\boldsymbol{D}$ as the deterministic Büchi automaton $\mathcal{Z}(\boldsymbol{\beta}, \boldsymbol{D}) = (Q_{\boldsymbol{\beta}, \boldsymbol{D}}, (0, 0), Q_{\boldsymbol{\beta}, \boldsymbol{D}}, \cup_{i=0}^{p-1} D_i, \delta)$ where the set of states is

$$Q_{\boldsymbol{\beta}, \boldsymbol{D}} = \cup_{i=0}^{p-1} (\{i\} \times (X_i \cap [-M_i, -m_i]))$$

and the (partial) transition function $\delta \colon Q_{\boldsymbol{\beta}, \boldsymbol{D}} \times \cup_{i=0}^{p-1} D_i \to Q_{\boldsymbol{\beta}, \boldsymbol{D}}$ is defined as follows: for $(i, s) \in Q_{\boldsymbol{\beta}, \boldsymbol{D}}$ and $a \in D_i$, we have $\delta((i, s), a) = ((i+1) \bmod p, \beta_i s + a)$. Observe that since we have assumed that all the alphabets D_i contain the digit 0, the initial state $(0, 0)$ is indeed an element of $Q_{\boldsymbol{\beta}, \boldsymbol{D}}$. Moreover, if $s \in X_i$ and $a \in D_i$ then $\beta_i s + a \in X_{i+1}$ by Lemma 1.

Proposition 22. *The zero automaton $\mathcal{Z}(\boldsymbol{\beta}, \boldsymbol{D})$ accepts the set $Z(\boldsymbol{\beta}, \boldsymbol{D})$.*

Example 10. Consider the alternate base $\boldsymbol{\beta} = (\frac{1+\sqrt{13}}{2}, \frac{5+\sqrt{13}}{6})$ and the pair of alphabets $\boldsymbol{D} = (\{-2, -1, 0, 1, 2\}, \{-1, 0, 1\})$. Then $M_0 = -m_0 = \text{val}_{\beta^{(0)}}((21)^\omega) \simeq 1.67994$ and $M_1 = -m_1 = \text{val}_{\beta^{(1)}}((12)^\omega) \simeq 1.86852$. The zero automaton $\mathcal{Z}(\boldsymbol{\beta}, \boldsymbol{D})$ is depicted in Fig. 2 where the states with first components 0 and 1 are colored in pink and purple respectively, and where the edges labeled by $-2, -1, 0, 1$ and 2 are colored in dark blue, dark green, red, light green and light blue respectively. For instance, the sequences $1(\bar{1}0)^\omega$ and $(0\bar{1}21\bar{2}1)^\omega$ have value 0 in base $\boldsymbol{\beta}$ (where $\bar{1}$ and $\bar{2}$ designate the digits -1 and -2 respectively).

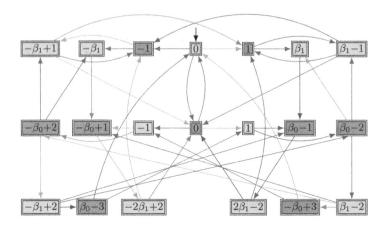

Fig. 2. The zero automaton $\mathcal{Z}(\boldsymbol{\beta}, \boldsymbol{D})$ for the alternate base $\boldsymbol{\beta} = (\frac{1+\sqrt{13}}{2}, \frac{5+\sqrt{13}}{6})$ and $\boldsymbol{D} = (\{-2, -1, 0, 1, 2\}, \{-1, 0, 1\})$. The used colors are described within Example 10.

In general, the zero automaton is infinite, i.e., it has infinitely many states. The following theorem, which generalizes a result from [19], describes when the zero automaton is actually finite in relation with some property of the spectrum.

Theorem 10 ([11]). *The following assertions are equivalent.*

1. *The set $Z(\boldsymbol{\beta}, \boldsymbol{D})$ is accepted by a finite Büchi automaton.*
2. *The spectrum $X^{\mathcal{D}}(\delta)$ has no accumulation point in $\mathbb{R}$.*
3. *The zero automaton $\mathcal{Z}(\boldsymbol{\beta}, \boldsymbol{D})$ is finite.*

Theorem 10 and Proposition 21 combined give us the following result.

Corollary 3. *If δ is a Pisot number and $\beta_i \in \mathbb{Q}(\delta)$ for all $i \in \{0, \ldots, p-1\}$ then the zero automaton $\mathcal{Z}(\boldsymbol{\beta}, \boldsymbol{D})$ is finite.*

5.2 Computing the Normalization

We show that the normalization in alternate bases is computable by finite automaton under certain conditions, in which case we construct such an automaton.

Following the same lines as in the real base case, we start by constructing a *converter* by using the zero automaton $\mathcal{Z}(\boldsymbol{\beta}, \boldsymbol{D})$) defined in Sect. 5.1. Consider two p-tuples of alphabets $\boldsymbol{D} = (D_0, \ldots, D_{p-1})$ and $\boldsymbol{D}' = (D'_0, \ldots, D'_{p-1})$. We denote the p-tuple of alphabets $(D_0 - D'_0, \ldots, D_{p-1} - D'_{p-1})$ by $\boldsymbol{D} - \boldsymbol{D}'$. The *converter from $\boldsymbol{D}$ to $\boldsymbol{D}'$* is the Büchi automaton

$$\mathcal{C}(\boldsymbol{\beta}, \boldsymbol{D}, \boldsymbol{D}') = (Q_{\boldsymbol{\beta}, \boldsymbol{D} - \boldsymbol{D}'}, (0, 0), Q_{\boldsymbol{\beta}, \boldsymbol{D} - \boldsymbol{D}'}, \cup_{i=0}^{p-1} (D_i \times D'_i), E)$$

where E is the set of transitions defined as follows: for $(i, s), (j, t) \in Q_{\boldsymbol{\beta}, \boldsymbol{D} - \boldsymbol{D}'}$ and for $(a, b) \in \cup_{i=0}^{p-1} (D_i \times D'_i)$, there is a transition

$$(i, s) \xrightarrow{(a, b)} (j, t)$$

if and only if $(a, b) \in D_i \times D_i'$ and there is a transition $(i, s) \xrightarrow{a-b} (j, t)$ in $\mathcal{Z}(\beta, D - D')$. Note that the converter is non-deterministic in general.

Proposition 23. *The converter* $\mathcal{C}(\beta, D, D')$ *accepts the set*

$$\left\{ (u, v) \in \prod_{n=0}^{\infty} (D_n \times D_n') : \mathrm{val}_\beta(u) = \mathrm{val}_\beta(v) \right\}.$$

In the case where β is a Parry alternate base, we consider a modification of the automaton $\mathcal{A}_\beta$ built in Sect. 3.3 in order to get a Büchi automaton accepting the set D_β. Without loss of generality, we suppose that $d^*_{\beta^{(i)}}(1)$ has a non-zero preperiod for all $i \in \{0, \dots, p-1\}$, i.e., in the case of a purely periodic expansion $(t_0 \cdots t_{n-1})^\omega$, we rather consider the writing $t_0(t_1 \cdots t_{n-1}t_0)^\omega$. Then we define the deterministic Büchi automaton $\mathcal{B}_\beta = (Q, I', F', A_\beta, \delta)$ where the states, the alphabet and the transitions are the same as those of the automaton $\mathcal{A}_\beta$, but now the sets of initial and final states are given by $I' = \{(0,0)\}$ and $F' = \{0, \dots, p-1\} \times \{0\}$.

Proposition 24. *If β is a Parry alternate base then the Büchi automaton $\mathcal{B}_\beta$ accepts the set D_β.*

Example 11. Consider again the alternate base $\beta = (\frac{1+\sqrt{13}}{2}, \frac{5+\sqrt{13}}{6})$. As explained above, since $d^*_{\beta^{(1)}}(1)$ is purely periodic, we consider the writing $1(01)^\omega$ instead of $(10)^\omega$. We obtain the Büchi automaton $\mathcal{B}_\beta$ depicted in Fig. 3.

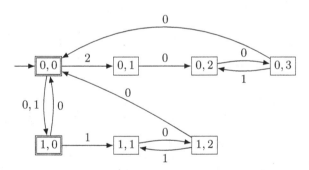

Fig. 3. A deterministic Büchi automaton accepting the set D_β for $\beta = (\frac{1+\sqrt{13}}{2}, \frac{5+\sqrt{13}}{6})$.

By combining the automata $\mathcal{B}_\beta$ and $\mathcal{C}(\beta, D, D')$ where D' is the p-tuple $(A_{\beta_0}, \dots, A_{\beta_{p-1}})$, we can prove the announced result on normalization.

Theorem 11 ([11]). *If δ is a Pisot number and $\beta_i \in \mathbb{Q}(\delta)$ for all $i \in \{0, \dots, p-1\}$, then the normalization function $\nu_{\beta, D}$ is computable by a finite automaton.*

6   Ergodic Properties of Alternate Base Expansions

In [9], we generalized the β-transformation for a real base β to the setting of alternate bases $\boldsymbol{\beta} = (\beta_0, \ldots, \beta_{p-1})$. As in the real base case, this new transformation, denoted $T_{\boldsymbol{\beta}}$, can be iterated in order to generate the digits of the $\boldsymbol{\beta}$-expansions of real numbers. The aim of this section is to report on the results describing the measure theoretical dynamical behavior of $T_{\boldsymbol{\beta}}$. The dynamical properties of a *lazy* β-transformations $L_{\boldsymbol{\beta}}$ are obtained by showing that the lazy dynamical system is isomorphic to the greedy one. We won't report on the lazy algorithm here, and refer the interested reader to [9,14].

6.1   The Alternate β-Transformation

A (measure preserving) dynamical system $(X, \mathcal{F}, \mu, T)$ is said to be *ergodic* if all $B \in \mathcal{F}$ such that $T^{-1}(B) = B$ satisfy $\mu(B) \in \{0, 1\}$, and it is said to be *exact* if $\cap_{n=0}^{\infty} \{T^{-n}(B) : B \in \mathcal{F}\}$ only contains sets of measure 0 or 1. Clearly, any exact dynamical system is ergodic.

For two measures μ and ν on the same σ-algebra $\mathcal{F}$, we say that μ is *absolutely continuous with respect to* ν if for all $B \in \mathcal{F}$, $\nu(B) = 0$ implies $\mu(B) = 0$, and we say that μ and ν are *equivalent* if they are absolutely continuous with respect to each other. In what follows, we will be concerned with the Borel σ-algebra $\mathcal{B}([0, 1))$.

A fundamental dynamical result of real base expansions is the following. This summarizes results from [29, 30, 32].

Theorem 12. *There exists a unique T_β-invariant probability measure μ_β on $\mathcal{B}([0, 1))$ that is absolutely continuous with respect to the Lebesgue measure λ restricted to $\mathcal{B}([0, 1))$. Furthermore, the measures μ_β and λ are equivalent on $\mathcal{B}([0, 1))$ and the dynamical system $([0, 1), \mathcal{B}([0, 1)), \mu_\beta, T_\beta)$ is ergodic and has entropy $\log(\beta)$.*

We now define the $\boldsymbol{\beta}$-*transformation* by

$$T_{\boldsymbol{\beta}} : \{0, \ldots, p-1\} \times [0, 1) \to \{0, \ldots, p-1\} \times [0, 1), \ (i, x) \mapsto ((i+1) \bmod p, T_{\beta_i}(x)).$$

The $\boldsymbol{\beta}$-transformation generates the digits d_n computed by the greedy algorithm as follows. For all $x \in [0, 1)$ and $n \geq 0$, we have $d_n = \lfloor \beta_n \pi_2(T_{\boldsymbol{\beta}}^n(0, x)) \rfloor$ where $\pi_2 \colon \mathbb{N} \times \mathbb{R} \to \mathbb{R}, \ (n, x) \mapsto x$.

6.2   Unique Absolutely Continuous $T_{\boldsymbol{\beta}}$-Invariant Measure

The following proposition provides us with the main tool for the construction of a $T_{\boldsymbol{\beta}}$-invariant measure. It is proved by using a result of Lasota and Yorke [24].

Proposition 25. *For all integers $n \geq 1$ and all real numbers $\beta_0, \ldots, \beta_{n-1} > 1$, there exists a unique $(T_{\beta_{n-1}} \circ \cdots \circ T_{\beta_0})$-invariant probability measure μ on $\mathcal{B}([0, 1))$ that is absolutely continuous with respect to the Lebesgue measure λ restricted to*

$\mathcal{B}([0,1))$. *Furthermore, the measure μ is equivalent to λ on $\mathcal{B}([0,1))$, its density function is bounded and decreasing, and the dynamical system*

$$([0,1), \mathcal{B}([0,1)), \mu, T_{\beta_{n-1}} \circ \cdots \circ T_{\beta_0})$$

is exact and has entropy $\log(\beta_0 \cdots \beta_{n-1})$.

For each $i \in \{0, \ldots, p-1\}$, we let $\mu_{\beta,i}$ denote the unique $(T_{\beta_{i+p-1}} \circ \cdots \circ T_{\beta_i})$-invariant absolutely continuous probability measure given by Proposition 25. Let us define a probability measure μ_β on the σ-algebra

$$\mathcal{T}_p = \left\{ \bigcup_{i=0}^{p-1} (\{i\} \times B_i) : \forall i \in \{0, \ldots, p-1\}, \ B_i \in \mathcal{B}([0,1)) \right\}$$

over $\{0, \ldots, p-1\} \times [0,1)$ as follows. For all $B_0, \ldots, B_{p-1} \in \mathcal{B}([0,1))$, we set

$$\mu_\beta \left(\bigcup_{i=0}^{p-1} (\{i\} \times B_i) \right) = \frac{1}{p} \sum_{i=0}^{p-1} \mu_{\beta,i}(B_i).$$

Let us also define another measure λ_p over the σ-algebra $\mathcal{T}_p$, which we call the *p-Lebesgue measure*. For all $B_0, \ldots, B_{p-1} \in \mathcal{B}([0,1))$, we set

$$\lambda_p \left(\bigcup_{i=0}^{p-1} (\{i\} \times B_i) \right) = \frac{1}{p} \sum_{i=0}^{p-1} \lambda(B_i). \tag{2}$$

We now state the announced generalization of Theorem 12 to alternate bases.

Theorem 13 ([9])**.** *The measure μ_β is the unique T_β-invariant probability measure on $\mathcal{T}_p$ that is absolutely continuous with respect to λ_p. Furthermore, μ_β is equivalent to λ_p on $\mathcal{T}_p$ and the dynamical system $(\{0, \ldots, p-1\} \times [0,1), \mathcal{T}_p, \mu_\beta, T_\beta)$ is ergodic and has entropy $\frac{1}{p} \log(\beta_0 \cdots \beta_{p-1})$.*

For $p \geq 2$, the dynamical system $(\{0, \ldots, p-1\} \times [0,1), \mathcal{T}_p, \mu_\beta, T_\beta)$ is not exact even though the dynamical systems $([0,1), \mathcal{B}([0,1)), \mu_{\beta,i}, T_{\beta_{i+p-1}} \circ \cdots \circ T_{\beta_i})$ are exact for all $i \in \{0, \ldots, p-1\}$. It suffices to note that the dynamical system $(\{0, \ldots, p-1\} \times [0,1), \mathcal{T}_p, \mu_\beta, T_\beta^p)$ is not ergodic for $p \geq 2$. Indeed, we have $T_\beta^{-p}(\{0\} \times [0,1)) = \{0\} \times [0,1)$ whereas $\mu_\beta(\{0\} \times [0,1)) = \frac{1}{p}$.

6.3   Density Functions

We obtain a formula for the density function $\frac{d\mu_\beta}{d\lambda_p}$ by using the density functions $\frac{d\mu_{\beta,i}}{d\lambda}$ for $i \in \{0, \ldots, p-1\}$.

Proposition 26. *The density function $\frac{d\mu_\beta}{d\lambda_p}$ of μ_β with respect to λ_p is*

$$\sum_{i=0}^{p-1} \left(\frac{d\mu_{\beta,i}}{d\lambda} \circ \pi_2 \right) \cdot \chi_{\{i\} \times [0,1)}.$$

6.4   Frequencies

We now turn to the frequencies of the digits in the β-expansions of real numbers in the interval $[0, 1)$. Recall that the frequency of a digit d occurring in the β-expansion $(a_n)_{n \geq 0}$ of a real number x in $[0, 1)$ is equal to

$$\lim_{n \to \infty} \frac{1}{n} \#\{0 \leq k < n : a_k = d\},$$

provided that this limit exists.

Proposition 27. *For λ-almost all $x \in [0, 1)$, the frequency of any digit d occurring in the β-expansion of x exists and is equal to*

$$\frac{1}{p} \sum_{i=0}^{p-1} \mu_{\beta,i}\left(\left[\frac{d}{\beta_i}, \frac{d+1}{\beta_i}\right) \cap [0, 1)\right).$$

Note that, when $p = 1$, the previous result gives back the classical formula $\mu_\beta([\frac{d}{\beta}, \frac{d+1}{\beta}) \cap [0, 1))$ where μ_β is the measure given in Theorem 12.

6.5   Isomorphism with the Dynamical Alternate β-Shift

The aim of this section is to generalize the isomorphism between the β-transformation and the β-shift to the framework of alternate bases. We start by providing some background of the real base case.

For an alphabet A, we let $\mathcal{C}_A$ denote the σ-algebra generated by the *cylinders*

$$C_A(a_0, \ldots, a_{\ell-1}) = \{(w_n)_{n \geq 0} \in A^{\mathbb{N}} : w_0 = a_0, \ldots, w_{\ell-1} = a_{\ell-1}\}$$

for all $\ell \geq 0$ and $a_0, \ldots, a_{\ell-1} \in A$. Consider the σ-algebra

$$\mathcal{G}_\beta = \left\{ \bigcup_{i=0}^{p-1} (\{i\} \times C_i) : \forall i \in \{0, \ldots, p-1\}, \; C_i \in \mathcal{C}_{A_\beta} \cap S_{\beta^{(i)}} \right\}$$

on $\bigcup_{i=0}^{p-1}(\{i\} \times S_{\beta^{(i)}})$. We define

$$\sigma_p \colon \bigcup_{i=0}^{p-1}(\{i\} \times S_{\beta^{(i)}}) \to \bigcup_{i=0}^{p-1}(\{i\} \times S_{\beta^{(i)}}), \; (i, w) \mapsto ((i+1) \bmod p, \sigma(w))$$

$$\psi_\beta \colon \{0, \ldots, p-1\} \times [0, 1) \to \bigcup_{i=0}^{p-1}(\{i\} \times S_{\beta^{(i)}}), \; (i, x) \mapsto (i, d_{\beta^{(i)}}(x)).$$

Note that the transformation σ_p is well defined by Proposition 16.

Proposition 28. *The map ψ_β defines an isomorphism between the dynamical systems $(\{0, \ldots, p-1\} \times [0, 1), \mathcal{T}_p, \mu_\beta, T_\beta)$ and $(\bigcup_{i=0}^{p-1}(\{i\} \times S_{\beta^{(i)}}), \mathcal{G}_\beta, \mu_\beta \circ \psi_\beta^{-1}, \sigma_p)$.*

However, although ψ_β is a continuous map, it does not define a topological isomorphism since it is not surjective.

In view of Proposition 28, the set $\cup_{i=0}^{p-1}(\{i\} \times S_{\beta^{(i)}})$ can be thought of as an *alternate β-shift*, that is, the generalization of the β-shift to alternate bases. However, in Sect. 3.3, what we called the alternate β-shift is the topological closure of the union $\cup_{i=0}^{p-1} S_{\beta^{(i)}}$. This definition was motivated by Theorem 5. So we can say that there are two ways to extend the notion of β-shift to alternate bases, depending on the way we look at it: either as a dynamical object or as a combinatorial object.

Thanks to Proposition 28, we obtain an analogue of Theorem 13 for the transformation σ_p.

Theorem 14 ([9]). *The measure $\mu_\beta \circ \psi_\beta^{-1}$ is the unique σ_p-invariant probability measure on $\mathcal{G}_\beta$ that is absolutely continuous with respect to $\lambda_p \circ \psi_\beta^{-1}$. Furthermore, $\mu_\beta \circ \psi_\beta^{-1}$ is equivalent to $\lambda_p \circ \psi_\beta^{-1}$ on $\mathcal{G}_\beta$ and the dynamical system $(\cup_{i=0}^{p-1}(\{i\} \times S_{\beta^{(i)}}), \mathcal{G}_\beta, \mu_\beta \circ \psi_\beta^{-1}, \sigma_p)$ is ergodic and has entropy $\frac{1}{p} \log(\beta_0 \cdots \beta_{p-1})$.*

References

1. Akiyama, S., Komornik, V.: Discrete spectra and Pisot numbers. J. Number Theory **133**(2), 375–390 (2013)
2. Bassino, F.: Beta-expansions for cubic pisot numbers. In: Rajsbaum, S. (ed.) LATIN 2002. LNCS, vol. 2286, pp. 141–152. Springer, Heidelberg (2002). https://doi.org/10.1007/3-540-45995-2_17
3. Bertrand-Mathis, A.: Développement en base θ; répartition modulo un de la suite $(x\theta^n)_{n\geq0}$; langages codés et θ-shift. Bull. Soc. Math. France **114**(3), 271–323 (1986)
4. Bertrand-Mathis, A.: Comment écrire les nombres entiers dans une base qui n'est pas entière. Acta Math. Hungar. **54**(3–4), 237–241 (1989)
5. Bruyère, V., Hansel, G.: Bertrand numeration systems and recognizability. Theoret. Comput. Sci. **181**(1), 17–43 (1997)
6. Caalima, J., Demegillo, S.: Beta cantor series expansion and admissible sequences. Acta Polytechnica **60**(3), 214–224 (2020)
7. Cantor, G.: Über die einfachen Zahlensysteme. Z. Math. Phys. **14**, 121–128 (1869)
8. Charlier, É., Cisternino, C.: Expansions in Cantor real bases. Monatshefte für Mathematik **195**(4), 585–610 (2021). https://doi.org/10.1007/s00605-021-01598-6
9. Charlier, É., Cisternino, C., Dajani, K.: Dynamical behavior of alternate base expansions. Ergodic Theory Dynam. Syst. **43**(3), 827–860 (2023)
10. Charlier, É., Cisternino, C., Kreczman, S.: On periodic alternate base expansions (2022). https://arxiv.org/abs/2206.01810
11. Charlier, É., Cisternino, C., Masáková, Z., Pelantová, E.: Spectrum, algebraicity and normalization in alternate bases. J. Number Theory **249**, 470–499 (2023)
12. Charlier, É., Cisternino, C., Stipulanti, M.: A full characterization of Bertrand numeration systems. In: Developments in language theory, vol. 13257, LNCS, pp. 102–114. Springer, Cham (2022). https://doi.org/10.1007/978-3-031-05578-2_8
13. Charlier, É., Rampersad, N., Rigo, M., Waxweiler, L.: The minimal automaton recognizing $m\mathbb{N}$ in a linear numeration system. Integers **11B**, Paper No. A4, 24 (2011)

14. Cisternino, C.: Combinatorial properties of lazy expansions in Cantor real bases (2021). https://arxiv.org/abs/2202.00437

15. Dajani, K., Kraaikamp, C.: Ergodic theory of numbers, Carus Mathematical Monographs, vol. 29. Mathematical Association of America, Washington, DC (2002)

16. Erdős, P., Rényi, A.: Some further statistical properties of the digits in Cantor's series. Acta Math. Acad. Sci. Hungar. **10**, 21–29 (1959)

17. Feng, D.J.: On the topology of polynomials with bounded integer coefficients. J. Eur. Math. Soc. **18**(1), 181–193 (2016)

18. Frougny, C.: Representations of numbers and finite automata. Math. Syst. Theory **25**(1), 37–60 (1992)

19. Frougny, Ch., Pelantová, E.: Two applications of the spectrum of numbers. Acta Math. Hungar. **156**(2), 391–407 (2018). https://doi.org/10.1007/s10474-018-0856-1

20. Galambos, J.: Representations of real numbers by infinite series, vol. 502. LNM. Springer-Verlag, Berlin-New York (1976). https://doi.org/10.1007/BFb0081642

21. Hollander, M.: Greedy numeration systems and regularity. Theory Comput. Syst. **31**(2), 111–133 (1998)

22. Kirschenhofer, P., Tichy, R.F.: On the distribution of digits in Cantor representations of integers. J. Number Theory **18**(1), 121–134 (1984)

23. Komornik, V., Lu, J., Zou, Y.: Expansions in multiple bases over general alphabets. Acta Math. Hungar. **166**(2) (2022)

24. Lasota, A., Yorke, J.A.: Exact dynamical systems and the Frobenius-Perron operator. Trans. Amer. Math. Soc. **273**(1), 375–384 (1982)

25. Li, Y.Q.: Expansions in multiple bases. Acta Math. Hungar. **163**(2), 576–600 (2021)

26. Loraud, N.: β-shift, systèmes de numération et automates. J. Théor. Nombres Bordeaux **7**(2), 473–498 (1995)

27. Lothaire, M.: Algebraic combinatorics on words, Encyclopedia of Mathematics and its Applications, vol. 90. Cambridge University Press, Cambridge (2002)

28. Neunhäuserer, J.: Non-uniform expansions of real numbers. Mediterr. J. Math. **18**(2), Paper No. 70, 8 (2021)

29. Parry, W.: On the β-expansions of real numbers. Acta Math. Acad. Sci. Hungar. **11**, 401–416 (1960)

30. Rényi, A.: Representations for real numbers and their ergodic properties. Acta Math. Acad. Sci. Hungar. **8**, 477–493 (1957)

31. Rényi, A.: On the distribution of the digits in Cantor's series. Mat. Lapok **7**, 77–100 (1956)

32. Rohlin, V.A.: Exact endomorphisms of a Lebesgue space. Izv. Akad. Nauk SSSR Ser. Mat. **25**, 499–530 (1961)

33. Schmidt, K.: On periodic expansions of Pisot numbers and Salem numbers. Bull. London Math. Soc. **12**(4), 269–278 (1980)

On the Number of Distinct Squares in Finite Sequences: Some Old and New Results

Srečko Brlek and Shuo Li[(⊠)]

Laboratoire de Combinatoire et d'Informatique Mathématique,
Université du Québec à Montréal,
CP 8888 Succ. Centre-ville, Montréal, QC H3C 3P8, Canada
{brlek.srecko,li.shuo}@uqam.ca

Abstract. A *square* is a word of the form uu, where u is a finite word. The problem of determining the number of distinct squares in a finite word was initially explored by Fraenkel and Simpson in 1998. They proved that the number of distinct squares, denoted as $\mathrm{Sq}(w)$, in a finite word w of length n is upper bounded by $2n$ and conjectured that $\mathrm{Sq}(w)$ is no larger than n. In this note, we review some old and new findings concerning the square-counting problem and prove that $\mathrm{Sq}(w) \leq n - \Theta(\log_2(n))$.

1 Introduction

A *square* is a word of the form uu for some finite word u. The study of squares, as well as other patterns in a word, is one of the fundamental topics in combinatorics on words. The problem of counting the number of distinct squares in a finite word was first investigated by Fraenkel and Simpson in [4] and has received noticeable attention.

In [4], Fraenkel and Simpson first proved that, for any finite word w of length n, the number of distinct squares in w is upper bounded by $2n$ and conjectured that this number is upper bounded by n. Their approach is based on the fact that, if we consider the last occurrence of each square factor in w, then at each position of w, there exist at most two different squares ending at this position. This particular structure is known as a double square. After some finer studies on the number of positions admitting a double square, Ilie [6] strengthened this bound to $2n - \Theta(\log(n))$; Lam [8] improved this result to $\frac{95}{48}n$; Deza, Franek and Thierry [3] achieved a bound of $\frac{11}{6}n$; Thierry [13] refined this bound to $\frac{3}{2}n$.

In a recent paper [2], Brlek and Li confirm Fraenkel and Simpson's conjecture by using an alternative approach. The strategy of the proof is twofold: One, they construct an injection from the set of distinct squares in w to a subset of independent circuits, defined as *small circuits* in [2], in the Rauzy graphs of w, and two, prove that the number of all small circuits in theses graphs is no larger than the length of w. In this note, we recall some details in their proof and show how can we develop their method to get better results. Particularly, we prove that

© The Author(s), under exclusive license to Springer Nature Switzerland AG 2023
A. Frid and R. Mercaş (Eds.): WORDS 2023, LNCS 13899, pp. 35–44, 2023.
https://doi.org/10.1007/978-3-031-33180-0_3

Theorem 1. *The number of distinct squares in a word* w *of length* n, *denoted as* Sq (w), *is bounded by*

$$\text{Sq}(w) \leq n - \Theta\Big(\log_2(n)\Big).$$

2   Preliminaries

We assume the reader familiar with the standard terminology: word, length, factor, prefix, suffix, lexicographic order $\prec$. Let w be a finite word. For any integer i satisfying $1 \leq i \leq |w|$, let $\text{Fac}_i(w)$ be the set of all length-i factors of w and let $|\text{Fac}_i(w)|$ denote the cardinality of $\text{Fac}_i(w)$. Then $\text{Fac}(w)$ is the set of all its factors. A word w is called *bordered* if there exists $u \neq w$ such that u is both a prefix and a suffix of w, and *unbordered* otherwise.

The conjugacy class $[w]$ is the set of the cyclic rotations of w.

For any natural number k, the k-*power* of a finite word u is the concatenation of k copies of u, and let it be denoted by u^k. Particularly, a *square* is a 2-power of some finite word u.

A word w is said to be *primitive* if it is not a power of another word. Then $\text{Prim}(w)$ is the set of primitive factors in w. A square is called a *primitive square* if it is of the form uu such that $u \in \text{Prim}(w)$.

For any word u and any rational number $\alpha = \frac{m}{|u|}$, the fractional α-power of u is defined to be $u^a u_0$ where u_0 is a prefix of u, $a = \lfloor \alpha \rfloor$ is the integer part of α, and $|u^a u_0| = m$. The α-power of u is denoted by u^α.

Definition 2. *For any finite word* w *and any positive integer* n, *let us define* $[w]_n = \left\{ u^{\frac{n}{|u|}} | u \in [w] \right\}$.

Example 3. *If we let* $u = aba$, *then* $[u]_1 = \left\{ v^{\frac{1}{3}} | v \in [u] \right\} = \{a, b\}$, $[u]_2 = \{aa, ab, ba\}$ *and* $[u]_5 = \{aabaa, abaab, baaba\}$. $\qquad\qquad\square$

Here we recall some elementary definitions and properties concerning graphs from Berge [1].

Let $G = (V, E)$ be a directed graph such that V is the set of its vertices and E is the set of its edges. By a *chain* we mean a sequence of edges $e_1, e_2, \ldots, e_k$, such that there exists a sequence of vertices $v_1, v_2, \ldots, v_{k+1}$ and that, for each i satisfying $1 \leq i \leq k$, e_i is either directed from v_i to v_{i+1} or from v_{i+1} to v_i. A *cycle* is a finite chain such that $v_{k+1} = v_1$. By a *path* we mean a sequence of edges $e_1, e_2, \ldots, e_k$, such that there exists a sequence of vertices $v_1, v_2, \ldots, v_{k+1}$ and that, for each i satisfying $1 \leq i \leq k$, e_i is directed from v_i to v_{i+1}. A *circuit* is a finite path such that $v_{k+1} = v_1$. From the definitions, any circuit is a cycle, but a cycle may not be a circuit.

A cycle or a circuit is called *elementary* if, apart from v_1 and v_{k+1}, every vertex which it meets is distinct, and consequently the number of its edges and vertices is equal. Each elementary cycle is represented by the sequence of its edges $C = (e_1, e_2, \ldots, e_l)$ to which corresponds an $|E|$-dimensional vector $\mu(C) = (c_1, c_2, \ldots, c_{|E|})$ in the E-dimensional vector space $\mathbb{R}^{|E|}$, where $c_i \in \{-1, 1, 0\}$ indicating in which direction the edge e_i is taken (± 1), or not belonging to the cycle (0). A set of cycles $C_1, C_2, \ldots, C_k, \ldots$ is called *independent* if their corresponding vectors are linearly independent.

A graph G is called *weakly connected*, if for any pair of distinct vertices, there exists a chain joining them.

Let $G = (V, E)$ be a weakly connected graph with l edges and s vertices, then the number $\chi(G) = l - s + 1$ is called the *cyclomatic number* of G.

Theorem 4 (Theorem 2, Chapter 4 in [1]). *The cyclomatic number of a graph is the maximum number of independent cycles in this graph.*

3   Rauzy Graph and Circuits

Let us first recall the construction of the Rauzy graphs for finite words.

Let w be a word of length k. For any integer n such that $1 \leq n \leq k$, the n-th Rauzy graph $\Gamma_n(w) = (\mathrm{Fac}_n(w), \mathrm{Fac}_{n+1}(w))$, is a directed graph where an edge $e \in \mathrm{Fac}_{n+1}(w)$ starts at the vertex u and ends at the vertex v, if and only if u is a prefix and v is a suffix of e. Observe that a circuit in a Rauzy graph is labelled by some word q obtained by taking the last letter of each edge, so that one can represent a circuit by $C = \left([q]_n, [q]_{n+1}\right)$ as well.

We remark that $\Gamma_n(w)$ is a weakly connected graph for any n.

For later use let $\Gamma(w) = \sqcup_{n=1}^{|w|} \Gamma_n(w)$ be the joint graph of all Rauzy graphs.

In this section, we are interested in enumerating the number of circuits in the Rauzy graphs. First let us give a standard notation for all these circuits. In [9], a standard notation is given for a subset of circuits in the Rauzy graphs, called *small circuits*, following Lemma 7 in [9]. Remarking that the argument given in Lemma 7 in [9] can be trivially generalized to all circuits, we then can generalize the notation to all circuits.

Lemma 5. (a generalization of Lemma 7 in [9]). *Let w be a finite word, and $n \leq |w|$. For any circuit C of length l in $\Gamma_n(w)$, there exists a unique primitive word q such that q is the smallest[1] in $[q]$, that $|q| = l$ and that $C = \left([q]_n, [q]_{n+1}\right)$.*

For sake of simplicity, we denote any circuit $C = \left([q]_n, [q]_{n+1}\right)$ in a Rauzy graph of w by $C_w(q, n)$.

Example 6. *Let $w = aabacabaabaaaaba$. Then $\Gamma_4(w)$ is the following:*

---

[1] A Lyndon word indeed.

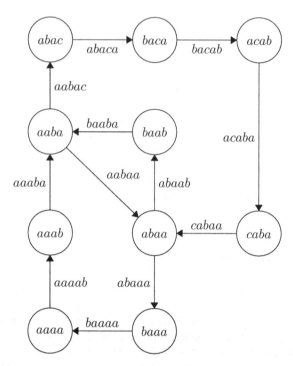

There are four elementary circuits in $\Gamma_4(w)$, and assuming that $a \prec b \prec c$ we have with the notation above

$$C_1 = \Big([aba]_4, [aba]_5 \Big) = C_w(aab, 4),$$

$$C_2 = \Big([aaaab]_4, [aaaab]_5 \Big) = C_w(aaaab, 4)$$

$$C_3 = \Big([cabaaba]_4, [cabaaba]_5 \Big) = C_w(aabacab, 4)$$

$$C_4 = \Big([cabaaaaba]_4, [cabaaaaba]_5 \Big) = C_w(aaaabacab, 4)$$

Here we recall some classes of circuits studied in previous work.

Definition 7. *From [2], a circuit $C_w(q,n)$ is called* small *if $|q| \le n$; from [9], a circuit $C_w(q,n)$ is called* quasi-small *if $|q| \le n+1$ and a circuit $C_w(q,n)$ is called* primitive *if $|q| = n+1$. Thus, a circuit is called quasi-small if it is either small or primitive.*

For a finite word w, let $S_n(w)$ (resp. $QS_n(w)$) denote the number of small circuit (resp. quasi-small circuits) in the n-th Rauzy graph of w, and let $S(w)$ (resp. $QS(w)$) denote the total number of small circuits (resp. quasi-small circuits) in $\Gamma(w)$.

Example 8. *Let us consider again the word w in Example 6. Among the four elementary circuits in $\Gamma_4(w)$, $C_1 = C_w(aab, 4)$ is a small circuit, $C_2 =*

$C_w(aaaab, 4)$ *is a primitive circuit, while* $C_3 = C_w(aabacab, 5)$ *and* $C_4 = C_w(aaaabacab, 5)$ *are neither small nor primitive. Thus,* $S_4(w) = 1$ *and* $QS_4(w) = 2.$

Lemma 9 (Theorem 2 in [9]). *Let* w *be a finite word, then for any positive integer* n *satisfying* $1 \leq n \leq |w|$, *we have*

$$S_n(w) \leq QS_n(w) \leq |\text{Fac}_{n+1}(w)| - |\text{Fac}_n(w)| + 1.$$

Moreover,

$$S(w) \leq QS(w) \leq |w| - |\text{Fac}_1(w)|.$$

The class of quasi-small circuits is a natural extension of the class of small circuits such that the cardinality of the number of quasi-small circuits on the Rauzy graphs of any finite word w is bounded by the length of w. Here we introduce another way to extend the class of small circuits.

Lemma 10. *Let* w *be a finite word and let* n *be an integer such that* $1 \leq n \leq |w|$. *If there exists a circuit* C *in* $\Gamma_n(w)$ *such that the edge set of* C *contains an unbordered element, then* C *is independent with all small circuits in* $\Gamma_n(w)$.

Proof. Let $C_w(p, n)$ be any small circuit in $\Gamma_n(w)$. From the definition, the edge set of $C_w(p, n)$ is $[p]_{n+1}$ with $n + 1 > |p|$. Thus, any element in the edge set of a small circuit is a bordered word. Consequently, the small circuits are independent with all circuits containing an unbordered edge.   □

Definition 11. *Let* w *be a finite word. For any integer* n *such that* $1 \leq n \leq |w|$, *a circuit* C *in* $\Gamma_n(w)$ *is called* special *if the edge set of* C *contains an unbordered element.*

Let $SpCircuit_n(w)$ be the special circuits characteristic function

$$SpCircuit_n(w) = \begin{cases} 1 \text{ if } \Gamma_n(w) \text{ contains a special circuit,} \\ 0 \text{ otherwise,} \end{cases}$$

and set

$$SpCircuit(w) = \sum_{1 \leq n \leq |w|} SpCircuit_n(w).$$

Example 12. *With the same word* w *and notation in Example 6,* C_2 *is special since it contains the unbordered edge* $baaaa$; C_3 *is special since it contains the unbordered edge* $aabac$; C_4 *is special since it contains the unbordered edge* $aabac$. *We remark here that a set of circuits containing two or more special circuits on the same Rauzy graph may not be independent. In this example,* C_1, C_2, C_3, C_4 *are not independent.*

The following lemma is a direct consequence of Lemma 10 and Theorem 4:

Lemma 13. *For any finite word* w,

$$S(w) + SpCircuit(w) \leq |w| - |\text{Fac}_1(w)|.$$

4   The Class of k-powers

Let w be a finite word. For any primitive factor p of w, let us define *the class of p* to be

$$\mathrm{Class}_p(w) = \{q^k | q \in [p], k \in \mathbf{N}^+, k \geq 2, q^k \in \mathrm{Fac}(w)\}.$$

Clearly, two classes $\mathrm{Class}_p(w)$ and $\mathrm{Class}_q(w)$ are equal if and only if p and q are conjugate. Let $\mathrm{Class}(w) = \{\mathrm{Class}_p(w) | p \in \mathrm{Prim}(w)\}$ and let $\mathrm{Class}^*(w) = \{\mathrm{Class}_p(w) | p \in \mathrm{Prim}(w), |p| > 1\}$.

Let $\mathrm{Sq}\,(w)$ be the number of distinct nonempty squares in w. For every class $\mathrm{Class}_p(w)$ of w, let us define

$$E_p(w) = \mathrm{Class}_p(w) \cap \{u^{2i} | i \in \mathbb{N}^+, u \in \mathrm{Prim}(w)\};$$
$$O_p(w) = \mathrm{Class}_p(w) \cap \{u^{2i+1} | i \in \mathbb{N}^+, u \in \mathrm{Prim}(w)\}.$$

It is obvious that $E_p(w) \cup O_p(w) = \mathrm{Class}_p(w)$ for any $p \in \mathrm{Prim}(w)$ and

$$\mathrm{Sq}\,(w) = \sum_{\mathrm{Class}_p(w) \in \mathrm{Class}(w)} |E_p(w)|.$$

Example 14. *Let $w = abababababcabc$. There are 4 distinct nonempty squares in $\mathrm{Fac}(w)$, namely $(ab)^2$, $(ba)^2$, $(ab)^4$ and $(abc)^2$. If we take $p = ab$, then*

$$\mathrm{Class}_p(w) = \{(ab)^2, (ba)^2, (ab)^3, (ba)^3, (ab)^4\};$$
$$E_p(w) = \{(ab)^2, (ba)^2, (ab)^4\};$$
$$O_p(w) = \{(ab)^3, (ba)^3\}.$$

Moreover, it is easy to check the following facts:

Lemma 15. *Let p be a primitive factor of w of length l. If $\mathrm{Class}_p(w) \neq \emptyset$, then one has*

$$|O_p(w)| \leq |E_p(w)| \leq |O_p(w)| + l \tag{1}$$

$$\frac{1 + |\mathrm{Class}_p(w)|}{|E_p(w)|} \geq 1 + \frac{1}{l}. \tag{2}$$

5   Injections from Squares to Circuits

In this section, we study the relations between the number of square factors in a word w and the number of small circuits in $\Gamma(w)$. It is first proved in [2] that there exists an injection from the set of square factors of w into the set of small circuits $Small(w)$ in w and later in [11] that there exists an injection from $\{p^k | p \in \mathrm{Prim}(w), p^k \in \mathrm{Fac}(w), k \in \mathbb{N}, k \geq 2\}$ into $Small(w)$. From these facts, one can prove the following theorem:

Theorem 16 (Theorem 1 in [11]). *Let k be an integer greater than 1. For any integer $n > 1$, let $N(n, k)$ denote the maximum number of different k-powers being factors of a word of length n. Then one has*

$$\frac{n}{k-1} - \Theta(\sqrt{n}) \leq N(n, k) \leq \frac{n-1}{k-1}.$$

The Fraenkel and Simpson's conjecture is a direct consequence of Theorem 16.

Moreover, inspired by the idea of studying the special circuits defined in the previous section, the second author studied the number of distinct squares in binary words. Using the fact that any circuit containing an edge of the form $0^k 1$ is a special circuit, thus independent with all small circuits, one can prove the following facts:

Theorem 17 (Theorem 3 in [10]). *Let w be a finite word over the alphabet $\{a, b\}$ such that $w = a^{r_1} b a^{r_2} b a^{r_3} b...a^{r_k} b a^{r_{k+1}}$, where $r_1, r_2, ..., r_{k+1}$ are nonnegative integers and $a^n = \underbrace{aa...a}_{n \ times}$. Let σ be a permutation of $1, 2, ..., k + 1$ such that $r_{\sigma(1)} \leq r_{\sigma(2)} \leq ... \leq r_{\sigma(k+1)}$, let $m(w)$ denote the number of distinct nonempty powers of exponent at least 2 in w and let $|w|$ denote the length of w. Then one has*

$$m(w) + r_{\sigma(k)} \leq |w| - 2.$$

This result helps to prove in particular a conjecture of Jonoska, Manea and Seki stated in [7]:

Theorem 18 (Theorem 2 in [10]). *Let w be a finite word over the alphabet $\{a, b\}$, let $\mathrm{Sq}(w)$ denote the number of distinct nonempty squares in w and let k denote the least of the number of a's and the number of b's in w. If $k \geq 2$, then one has*

$$\mathrm{Sq}(w) \leq \frac{2k - 1}{2k + 2}|w|.$$

6   Proof of the Main Theorem

In this section, we give a better estimation of $\mathrm{Sq}(w)$ for any finite word w by proving Theorem 1.

First, one can prove the following fact:

Lemma 19. *For any finite word w, there exists an injection from $\mathrm{Class}^*(w)$ to the set of primitive circuits in $\Gamma(w)$ by sending $\mathrm{Class}_p(w)$ to the primitive circuit $C_w(p, |p| - 1)$.*

Combining with Lemma 9 in [11], one has:

Lemma 20. *For any finite word w, there exists the following inequality:*

$$\sum_{\mathrm{Class}_q(w) \in \mathrm{Class}(w)} |\mathrm{Class}_q(w)| + 1 \leq |w|. \tag{3}$$

Using Lemma 15 and some elementary arithmetical argument, one can prove that:

Proposition 21. *Let w be a finite word of length n. If $|u^2| \leq 2\sqrt{n}$ for all primitive squares in $\mathrm{Fac}(w)$, then*

$$\mathrm{Sq}(w) \leq n - \frac{n}{\sqrt{n}+1}. \tag{4}$$

If there exists some primitive square $u^2 \in \mathrm{Fac}(w)$ such that $|u^2| > 2\sqrt{n}$, then one can estimate the number of special circuits in $\Gamma(w)$ by analyzing the number of unbordered factors in u in using the following lemma:

Lemma 22. *Let w be a finite word. Then there exist a finite sequence of unbordered words $v_1, v_2, \ldots, v_k$ and a finite sequence of positive integers $i_1, i_2, \ldots, i_k$ such that (i) $|v_1| = 1$, (ii) $|v_r| < |v_{r+1}| \leq 2|v_r|$ for any $1 \leq r \leq k-1$, (iii) $v_r^{i_r} \in \mathrm{Fac}(w)$ for any $1 \leq r \leq k$ and (iv) $\sum_{r=1}^{k} i_r |v_r| \geq \frac{|w|}{2}$.*

Lemma 22 is a consequence of Theorem 2 in [5]:

Lemma 23 (Theorem 2 in [5]). *Let w be an unbordered word and let v be a finite word such that $|w| = |v|$. If $w \neq v$, then there exists an unbordered word $u \in \mathrm{Fac}(wv)$ such that $|u| > |w|$.*

Example 24. *Let $w = abaabaabaababaabbb$ and let us compute the sequences $(v_r)_{r \geq 1}$ and $(i_r)_{r \geq 1}$ for w. Let us begin by setting $v_1 = w_1 = a$ and $j_1 = 1$. Since v_1^2 is not a prefix of w, let us set $i_1 = 1$ and $v_2 = w_1 w_2 = ab$. Since the first occurrence of the word ab occurs at the position 1 in w, let $j_2 = 1$. Remark that $w_{j_2} \cdots w_{j_2+|v_2|-1} = ab$ is a power of v_2 while $w_{j_1} \cdots w_{j_2+2|v_2|-1} = abaa$ is not. We have $i_2 = 1$. Further, from the recurrence relation, we can find an unbordered factor of $abaa$ with a length at least three. In fact, baa is such a word. Let $v_3 = baa$, $j_3 = 2$. we can check that $w_{j_3} \cdots w_{j_3+3|v_3|-1} = (baa)^3$ is a power of baa while $w_{j_3} \cdots w_{j_3+4|v_3|-1} = (baa)^3 bab$ is not. Thus, $i_3 = 3$. Similarly, we compute $v_4 = aabab$ with $i_4 = 1$ and $v_5 = aababaabbb$ with $i_5 = 1$. The algorithm stops here since $j_5 = 9$ and $j_5 + (i_5 + 1)|v_5| - 1 > |w|$. Thus,*

$$(v_r)_{1 \leq r \leq 5} = a, \ ab, \ baa, \ aabab, \ aababaabbb,$$

$$(i_r)_{1 \leq r \leq 5} = 1, \ 1, \ 3, \ 1, \ 1.$$

Using the sequence of unbordered words studied in Lemma 22, one can prove the following facts:

Proposition 25. *Let w be a finite word of length n. If w contains some primitive square u^2 such that $|u^2| \geq 2\sqrt{n}$, then*

$$\mathrm{Sq}(w) \leq n - \Theta\Big(\log_2(n)\Big). \tag{5}$$

Proof. Let u^2 be a primitive square contained in w and $|u| \geq \sqrt{n}$. Applying Lemma 22 on u, one has a sequence of unbordered words $(v_r)_{1 \leq r \leq k}$ and a sequence of integers $(i_r)_{1 \leq r \leq k}$ satisfying the conditions in Lemma 22 for some positive integer k.

If $k \geq \frac{\log_2(n)}{4}$, then, from Lemma 22, u^2 contains at least $\frac{\log_2(n)}{4} - 1$ distinct unbordered factors of different lengths but at least 2. Thus, $\text{Sq}(w) \leq n - \frac{1}{4}\log_2(n) + 1$.

If $k \leq \frac{\log_2(n)}{4}$, then, following the fact that $\sum_{r=1}^{k} i_r |v_r| \geq \frac{|u|}{2} \geq \frac{\sqrt{n}}{2}$, there exists some j such that $i_j |v_j| \geq \frac{2\sqrt{n}}{\log_2(n)}$. From the fact that $|v_k| < 2|v_{k-1}| < \dots < 2^k |v_1| = n^{\frac{1}{4}}$, $|v_j| \leq n^{\frac{1}{4}}$. Thus,

$$i_j \geq \frac{2\sqrt{n}}{\log_2(n)|v_j|} \geq \frac{2\sqrt{n}}{\log_2(n)n^{\frac{1}{4}}} = \frac{2n^{\frac{1}{4}}}{\log_2(n)}.$$

Consequently, $i_j > 3$ when n is large and, from Lemma 15, $O_{v_j}(v_j^{i_j}) \geq \frac{i_j |v_j|}{4} \geq \frac{\sqrt{n}}{2\log_2(n)}$ when n is large. From Lemma 20,

$$\text{Sq}(w) \leq n - \frac{\sqrt{n}}{2\log_2(n)},$$

when n is large.

In conclusion, in both cases, we have $sq(w) \leq n - \Theta(\log_2(n))$.

Proof. (of Theorem 1). It is a direct consequence of Proposition 21 and 25.

7 Discussion

Theorem 1 is an improvement of Theorem 1 in [2]. However, the upper bound obtained in Theorem 1 may still not be sharp. In fact, the maximum number of distinct nonempty squares in a binary string of length n has been computed for small n's and it is listed as A248958 in OEIS [12]. Motivated by this computation, we suggest a stronger upper bound for the number of distinct squares.

Conjecture 26. *With the same notation as above, for any finite word w, one has:*

$$\text{Sq}(w) \leq \lceil |w| + 1 - \sqrt{|w|} - \log_2(\sqrt{|w|}) \rceil,$$

where $\lceil x \rceil$ denotes the smallest integer larger than or equal to x.

Remark that the sequence A248958 is computed for binary words, if we let $(M(n))_{n \in \mathbb{N}}$ denote the sequence of the maximum number of distinct nonempty squares in any string of length n, there are the following open questions:

Question 27. *Is it true that if there exists some word w satisfying that $\text{Sq}(w) = M(|w|)$, then all square factors of w contain at most two distinct letters?*

Question 28. *Is it true that for any integer n, there exists a binary word w of length n such that $\text{Sq}(w) = M(n)$?*

Concerning the approach presented in this note, we remark that the upper bound obtained for words which only contain short primitive squares in Proposition 21 is quite close to the upper bound suggested in Conjecture 26. However, the upper bound obtained in Proposition 25 may still be improved. We believe that there are far more independent circuits including some unbordered edges than we estimate in Proposition 25 in the case that w contains some long primitive squares.

References

1. Berge, C.: The Theory of Graphs and Its Applications. Greenwood Press (1982)
2. Brlek, S., Li, S.: On the number of squares in a finite word. arXiv e-prints arXiv:2204.10204, April 2022
3. Deza, A., Franek, F., Thierry, A.: How many double squares can a string contain? Discret. Appl. Math. **180**, 52–69 (2015)
4. Fraenkel, A.S., Simpson, J.: How many squares can a string contain? J. Comb. Theory Ser. A **82**(1), 112–120 (1998)
5. Harju, T., Nowotka, D.: Periodicity and unbordered words. In: Diekert, V., Habib, M. (eds.) STACS 2004. LNCS, vol. 2996, pp. 294–304. Springer, Heidelberg (2004). https://doi.org/10.1007/978-3-540-24749-4_26
6. Ilie, L.: A note on the number of distinct squares in a word. In: Brlek, S., Reutenauer, C. (eds.) Proceedings Words 2005, 5-th International Conference on Words, vol. 36, pp. 289–294. Publications du LaCIM, Montreal, 13–17 September 2005
7. Jonoska, N., Manea, F., Seki, S.: A stronger square conjecture on binary words. In: Geffert, V., Preneel, B., Rovan, B., Štuller, J., Tjoa, A.M. (eds.) SOFSEM 2014. LNCS, vol. 8327, pp. 339–350. Springer, Cham (2014). https://doi.org/10.1007/978-3-319-04298-5_30
8. Lam, N.H.: On the number of squares in a string. AdvOL-Report 2 (2013)
9. Li, S.: A note on the Lie complexity and beyond. arXiv e-prints arXiv:2207.05859, July 2022
10. Li, S.: An upper bound of the number of distinct powers in binary words. arXiv e-prints arXiv:2209.06891, September 2022
11. Li, S., Pachocki, J., Radoszewski, J.: A note on the maximum number of k-powers in a finite word. arXiv e-prints arXiv:2205.10156, May 2022
12. OEIS Foundation Inc.: The On-Line Encyclopedia of Integer Sequences (2023). http://oeis.org
13. Thierry, A.: A proof that a word of length n has less than 1.5n distinct squares. arXiv e-prints arXiv:2001.02996, January 2020

Contributed Papers

Ranking and Unranking k-Subsequence Universal Words

Duncan Adamson[✉]

Leverhulme Research Centre for Functional Materials Design,
The University of Liverpool, Liverpool, UK
d.a.adamson@liverpool.ac.uk

Abstract. A subsequence of a word w is a word u such that $u = w[i_1]w[i_2], \ldots w[i_{|u|}]$, for some set of indices $1 \leq i_1 < i_2 < \cdots < i_k \leq |w|$. A word w is k-subsequence universal over an alphabet Σ if every word in Σ^k appears in w as a subsequence. In this paper, we provide new algorithms for k-subsequence universal words of fixed length n over the alphabet $\Sigma = \{1, 2, \ldots, \sigma\}$. Letting $\mathcal{U}(n, k, \sigma)$ denote the set of n-length k-subsequence universal words over Σ, we provide:

- an $O(nk\sigma)$ time algorithm for counting the size of $\mathcal{U}(n, k, \sigma)$;
- an $O(nk\sigma)$ time algorithm for ranking words in the set $\mathcal{U}(n, k, \sigma)$;
- an $O(nk\sigma)$ time algorithm for unranking words from the set $\mathcal{U}(n, k, \sigma)$;
- an algorithm for enumerating the set $\mathcal{U}(n, k, \sigma)$ with $O(n\sigma)$ delay after $O(nk\sigma)$ preprocessing.

1 Introduction

Words and subsequences are two fundamental combinatorial objects. Informally, a subsequence of a word w is a word u that can be found by deleting some subset of the symbols w. Subsequences are a heavily studied object within computer science [4,5,7,11,16,18,19,26–28] and beyond, with applications in a wide number of fields including bioinformatics [12,24], database theory [3], and modelling concurrency [23]. A recent survey of subsequence algorithms has been provided by Kosche et al. [17], highlighting major results for problems on funding subsequences in words.

This paper considers *k-subsequence universal* words. A word w is k-subsequence universal over an alphabet Σ if w contains every word of length k over Σ as a subsequence. These words were first defined by Karandikar and Schnoebelen [14,22] as *k-rich words*, however more recent work has used the term *k-subsequence universality* [4,5,16], which we will use here. The study of these words follows from work on *Simon's congruence* [25]. Informally, two words w, v are k-congruent if w and v share the same set of subsequences of length k. This relationship has been heavily studied [6,26–28], with a recent asymptotically optimal algorithm derived for testing if two words are k-congruent [9].

This work was completed at, and partially funded by the University of Göttingen.

© The Author(s), under exclusive license to Springer Nature Switzerland AG 2023
A. Frid and R. Mercaş (Eds.): WORDS 2023, LNCS 13899, pp. 47–59, 2023.
https://doi.org/10.1007/978-3-031-33180-0_4

Most relevant to this work are the papers by Barker et al. [4], and Day et al. [5], directly addressing k-subsequence universal words. In [4], the authors show that it is possible to determine, in linear time, if a word is k-subsequence universal or not, as well as the shortest k-subsequence universal prefix of a given word. Additionally, they provide results showing that the minimal set of ℓ-factors of a word w, $w_1 w_2 \ldots, w_\ell$ such that $w_1 w_2 \ldots w_\ell$ is k-subsequence universal, and the index i such that w^i is k-subsequence universal can be determined efficiently.

This is built on by [5], in which the authors provide a set of algorithmic results for minimising the number of edit operations to transform a word into a k-subsequence universal word, providing results on *insertions, deletions,* and *substitutions*. They show that the minimum number of *insertions* and *substitutions* needed to transform a word w into a k-subsequence universal word w' can be done in $O(nk)$ time, assuming that $k < n$. Additionally, they show that the number of *deletions* needed to reduce the universality index (the maximum k such that the word is k-subsequence universal) of a word to k can be determined in $O(nk)$ time.

This paper is interested in providing algorithms for some of the basic operations on classes of words, *counting, ranking, unranking,* and *enumerating* for the class of k-subsequence universal words. In providing these algorithms, we aim to expand the understanding of the space of k-subsequence universal words of a fixed length n. We use $\mathcal{U}(n, k, \sigma)$ to denote the set of k-subsequence universal words of length n over an alphabet of size σ (assumed to be the alphabet $\{1, 2, \ldots, \sigma\}$). The counting problem asks for the number of words in a given class. The ranking problem takes as input a word w and determines the number of words within the set which are lexicographically smaller than w. The unranking problem is the inverse of the ranking problem, taking a rank i and asking for the word in the set with the rank i. Finally, the enumeration problem asks for the explicit outputting of every word within the set in some fixed order. Each of these problems has been heavily studied for other classes of words, including cyclic words [1,2,8,10,15,21] and Gray codes [8,15,20].

Our Results. This paper builds upon the existing body of work on k-subsequence universal words to build a stronger understanding of the space of k-subsequence universal words of a fixed length. We provide a suite of algorithmic results for k-subsequence universal words of fixed length n. We denote the set of all k-subsequence universal words over the alphabet $1, 2, \ldots, \sigma$ with length n by $\mathcal{U}(n, k, \sigma)$. In Sect. 3, we provide an $O(nk\sigma)$ time algorithm for counting the size of $\mathcal{U}(n, k, \sigma)$. In Sect. 4, we use the observations from this counting algorithm to provide an $O(nk\sigma)$ time algorithm for ranking words in the set $\mathcal{U}(n, k, \sigma)$. Finally, in Sect. 5 we provide an $O(n\sigma)$ time algorithm for unranking within the set $\mathcal{U}(n, k, \sigma)$, with $O(nk\sigma)$ time preprocessing. We note this unranking algorithm directly provides an enumeration algorithm for the set $\mathcal{U}(n, k, \sigma)$ with $O(n\sigma)$ delay.

Computational Model. In this paper, we assume the unit cost RAM computational model, in this case, equivalent to the unit cost word-RAM with word size

$O(\log(N)\log(\sigma))$, where N is the larger of the input or the output. We note that this remains logarithmic relative to the number of k-subsequence universal words of length n, and thus the bits required to output the integer representation of the number of such words. All our complexities can be readjusted into the unit cost RAM computational model with word size $O(\log(n)\log(\sigma))$ where n is the size of the input, by applying a multiplicative factor of $O(n/\log(n))$ to the stated bounds. We avoid a factor of $O(n^2/\log^2(n))$ by noting that these algorithms only perform multiplications where at least one integer has size at most σ or addition between integers of size at most σ^n.

2 Preliminaries

We use the following notation. Given a pair of natural numbers $m, n \in \mathbb{N}$, the notation $[m, n]$ denotes the ordered set $\{m, m+1, \dots, n\}$, or the empty set if $m > n$. A word w is an ordered sequence of symbols over some alphabet Σ. The set of words of length n over the alphabet Σ is denoted Σ^n, and the set of all words over the alphabet Σ by Σ^*. The length of a word w is denoted $|w|$. The notation $w[i]$ is used to denote the ith symbol in the word w, and $w[i, j]$ is used to denote the contiguous sequence within w corresponding to the word $w[i]w[i+1]\dots w[j]$ (or the empty word ε if $i > j$). A word v is a *factor* of a word w if there exists some pair of indices i, j such that $v = w[i, j]$.

We assume the alphabet $\Sigma = [1, \sigma]$ for some natural number $\sigma \geq 1$. Given two words $w, v \in \Sigma^n$, w is *lexicographically smaller* than v if there exists some index $i \in [1, n]$ such that $w[1, i-1] = v[1, i-1]$ and $w[i] < v[i]$. Given two words $w, v \in \Sigma^*$, v is a subsequence of w if and only if there exists some series of indices $1 \leq i_1 < i_2 < \dots < i_{|v|} \leq |w|$ such that $v = w[i_1]w[i_2]\dots w[i_{|v|}]$.

Definition 1 (k-subsequence universality). *A word $w \in \Sigma^n$ is k-subsequence universal if and only if every word $v \in \Sigma^k$ is a subsequence of w. The set of words of length n that are k-subsequence universal over an alphabet of size σ is denoted $\mathcal{U}(n, k, \sigma)$.*

The *subsequence universality index* of a word w is the largest value k such that w is k-subsequence universal. In order to determine if a word is k-subsequence universal, we use *arch-factorisations*, first introduced by Hebard [13]. Informally, an *arch* of a word is a minimal length factor containing each symbol in the alphabet Σ at least once. For the remainder of this paper, we use the following formal definition:

Definition 2 (Arches). *An Arch over the alphabet $\Sigma = \{1, 2, \dots, \sigma\}$ is a word containing every symbol in Σ at least once. The universal subsequence of an arch v is the set of indices $(i_1, i_2, \dots, i_\sigma)$ satisfying the following:*

- *$v[i_1], v[i_2], \dots, v[i_\sigma]$ contains every symbol in Σ exactly once.*
- *The index i_j is the first position in v where the symbol $v[i_j]$ appears.*
- *The index i_σ is the last position in v.*

Any symbol not in the universal subsequence is called a free symbol.

We note that this definition of an arch corresponds to a 1-subsequence universal word where the last symbol is unique, i.e. it does not appear anywhere else in the word.

Definition 3 ([13], **Arch Factorisations**). *The arch-factorisation of a word* $w \in \Sigma^n$ *with a universality index of* k, *is a set of factors* $\{w_1, w_2, \ldots, w_k, v\}$, *denoted* $Arch(w)$, *such that* w_i *is a factor of* w *that is an arch for every* $i \in [1, k]$, v *is a suffix of* w *that does not contain any arch as a factor, and* $w_1 w_2 \ldots w_k v = w$.

$$w = 11234, 4321, 22314, 33214, 4$$
$$v = 12234, 323134, 11234, 4412$$

Fig. 1. An example of the arch-factorisation of two words $w, v \in \Sigma^{20}$ where $\Sigma = \{1, 2, 3, 4\}$. Each arch (or the ending suffix) is separated by a comma, and highlighted in a separate colour. Note that w is 4-subsequence universal while v is only 3-subsequence universal, despite sharing the same Parikh vector $(5, 5, 5, 5)$.

An example of this factorisation is given in Fig. 1. Day et al. [5] expanded upon Definition 3 to show that a word is k-subsequence universal if and only if there exists an arch-factorisation of w containing at least k-arches, and further, that such a factorisation can be computed in time linear to the length of the word.

Theorem 1 ([5]). *A word* $w \in \Sigma^n$ *is* k-subsequence universal over Σ if and only if $Arch(w)$ contains at least k arches. Further, $Arch(w)$ can be computed in $O(n)$ time.

In this paper, we use the following technical Lemma from [5].

Lemma 1 ([5]). *Let* $\Delta(w, i, j)$ *denote the number unique symbols in* $w[i, j]$ *for some* $w \in \Sigma^n$. *We can compute in* $O(n)$ *the values of* $\Delta(w, 1, j)$ *for every* $j \in [1, n]$.

We combine Theorem 1 and Lemma 1 to make the following observation.

Observation 1. *Let* $w \in \Sigma^n$ *be a* k-subsequence universal word with the arch-decomposition $w_1, w_2, \ldots, w_k, v$, and further let $\mathcal{A} = \{A_1, A_2, \ldots, A_k\}$ denote the set of indices where $A_\ell = 1 + \sum_{i \in [1, \ell-1]} |w_i|$, i.e. the set of indices in w corresponding to the first position of an arch in $Arch(w)$. Then, the values of $\Delta(w, A_\ell, i_\ell)$ can be computed in $O(n)$ time for every $\ell \in [k]$ and $i_\ell \in [A_\ell, A_{\ell+1} - 1]$, where $\Delta(w, A_\ell, i_\ell)$ denotes the number unique symbols in $w[A_\ell, i_\ell]$.

Using this notation, we provide a formal definition of the ranking and unranking problems as considered in this paper. The *rank* of a word w within an ordered set of words $\mathcal{S}$ is the number of words in $\mathcal{S}$ that are smaller than w under the ordering of the set. In this paper, we assume that the set of k-subsequence universal words is ordered lexicographically, and therefore the rank of a word w in the set $\mathcal{U}(n, k, \sigma)$ is the number of words in $\mathcal{U}(n, k, \sigma)$ that are lexicographically smaller than w. The *ranking* problem takes as input a word w integer triple $n, k, \sigma \in \mathbb{N}$ such that $n \geq k\sigma$, and returns the number of words in $\mathcal{U}(n, k, \sigma)$ lexicographically smaller than w. The *unranking* problem is conceptually the inverse of the ranking problem. Given an integer $i \in [1, |\mathcal{S}|]$, the unranking problem asks for the word in $\mathcal{S}$ with a rank of i. In this paper, the unranking problem takes as input a rank i and integer triple $n, k, \sigma \in \mathbb{N}$ such that $n \geq k\sigma$, and returns the word in $\mathcal{U}(n, k, \sigma)$ with a rank of i.

3   Counting Arches and k-Subsequence Universal Words

First, we present a tool for counting the size of $\mathcal{U}(n, k, \sigma)$, i.e. number of k-subsequence universal words of length n over the alphabet $\Sigma = \{1, 2, \ldots, \sigma\}$. As well as being an interesting result in and of itself, this provides the foundation for our tools for both ranking and unranking.

This section is split into two sections. First, we provide formulae for counting the number of arches, 0-subsequence universal words, and 1-universal words of length n over an alphabet of size σ. Second, we provide a recursive technique to count the number of k-subsequence universal words of length n over an alphabet of size σ.

3.1   Arches, 0-Subsequence Universal and 1-Subsequence Universal Words

We first consider how to count the number of arches, 0-subsequence universal and 1-subsequence universal word of length n. We note that these three special cases are closely interlinked. First, note that any word that is not 0-subsequence universal must be at least 1-subsequence universal. Therefore, the number of 1-subsequence universal words is equal to the number of words minus the number of 0-subsequence universal words. Similarly, the number of n-length arches is equal to the number of $(n - 1)$-length 1-subsequence universal words over an alphabet of size $\sigma - 1$, multiplied by σ. We start with 0-subsequence universal words.

Lemma 2. *The number of n-length 0-subsequence universal words over an alphabet $\Sigma = \{1, 2, \ldots, \sigma\}$ is given by:*

$$\sum_{i \in [1, \sigma]} (-1)^{i+1} \binom{\sigma}{i} (\sigma - i)^n.$$

Proof. Let $\varsigma \subseteq \Sigma$ be a i-length alphabet. Note first that the number of n-length words over ς is given by i^n, and further, as there are $\binom{\sigma}{i}$ such alphabets, the total number of words over *any* i-length alphabet is given by $\binom{\sigma}{i} i^n$. Observe that any string in ς^n is also in $(\varsigma \cup \{x\})^n$, for some $x \in \Sigma \setminus \varsigma$, and further, there are $\sigma - i$ such $i + 1$-length alphabets containing every symbol in ς. More generally, there are $\binom{\sigma-i}{j}$ alphabets of size $i + j$ containing every symbol in the i-length alphabet ς. Therefore, taking the sum of n-length words in all i-length alphabets, given by $\binom{\sigma}{i} (i)^n$, will also count every word in a $j < i$-length language $\binom{\sigma-j}{i-j}$ times. Combining this with the well-known binomial coefficient identities gives the equation for the total number of unique words in any alphabet in the set $\{\Sigma \setminus \{x\} \mid x \in \Sigma\}$ as:

$$\binom{\sigma}{1}(\sigma-1)^{n-1} - \binom{\sigma}{2}(\sigma-2)^{n-1} + \binom{\sigma}{3}(\sigma-3)^{n-1} \ldots (-1)^{\sigma+1}$$

$$= \sum_{i \in [1,\sigma]} (-1)^{i+1} \binom{\sigma}{i}(\sigma-i)^n$$

Using Lemma 2, the counting of n-length arches and 1-subsequence universal words follows directly.

Corollary 1. *The number of n-length 1-universal words over an alphabet $\Sigma = \{1, 2, \ldots, \sigma\}$ is given by:*

$$\sigma^n - \sum_{i \in [1,\sigma]} (-1)^{i+1} \binom{\sigma}{i}(\sigma-i)^n$$

Corollary 2. *The number of n-length Arches over an alphabet $\Sigma = \{1, 2, \ldots, \sigma\}$ is given by:*

$$\sigma(\sigma-1)^{n-1} - \sum_{i \in [2,\sigma]} (-1)^i i \binom{\sigma}{i}(\sigma-i)^{n-1}.$$

3.2   Counting k-Subsequence Universal Words

To count k-subsequence universal words with an arbitrary value of k, we employ a recursive approach. The high-level idea is to count the number of suffixes of k-subsequence universal words sharing a given prefix v. Let $\mathcal{S}(v)$ be the set of words of length $n- \mid v \mid$ such that for every word $u \in \mathcal{S}(v)$, the word vu is a k-subsequence universal word. Let $Arch(v) = v_1, v_2, \ldots, v_\ell, v'$ be the arch factorisation of v, or the set of the first k arches of v'. In order to count the size of $\mathcal{S}(v)$, we observe that every word $u \in \mathcal{S}(v)$ must contain a prefix u' such that $v'u'$ is an arch and the suffix $u[|u'| + 1, |u|]$ must contain $k - \ell - 1$ arches. Our recursive approach is based on the observation that the size of $\mathcal{S}(v)$ is equal to the size of $\bigcup_{x \in \Sigma} \mathcal{S}(vx)$. This leaves two major problems: determining whether or not the set $\mathcal{S}(vx)$ is empty, and ensuring that the total size of $\mathcal{S}(v)$ can be computed without having to explicitly check $\mathcal{S}(vw)$ for every suffix $w \in \Sigma^{n-|v|}$.

We solve these problems by introducing a new function, $\mathrm{CS}(q, m, c)$ (**Count Suffixes**) such that $|\mathcal{S}(v)| = \mathrm{CS}(q, m, c)$ where:

- q is the number of unique symbols in v'.
- m is the number of free symbols in every word $u \in \mathcal{S}(v)$, i.e. the number of symbols in u that do not belong to any universal subsequence of the first $k - \ell - 1$ arches of $u[|u'| + 1, |u|]$ or in the universal subsequence of u' in the word $v'u'$.
- c is the minimum number of arches in $v'u$, equal to $k - \ell$.

The value of $\mathrm{CS}(q, m, c)$ is determined in a recursive manner. We first provide the base cases. If $m = 0$, then every remaining symbol must be in the universal subsequence for one of the remaining c arches, giving $\mathrm{CS}(q, 0, c) = (\sigma - q)!(\sigma!)^{c-1}$. On the other hand, if $c = 0$, then the remaining symbols can be chosen arbitrarily from Σ, giving $\mathrm{CS}(q, m, 0) = \sigma^m$. Assuming both c and m are greater than 0, then the value of $\mathrm{CS}(q, m, k)$ is determined recursively. If $q = \sigma$, then the next symbol must be the first symbol of the $(k - c)^{th}$ arch of the word. As there are σ such possible symbols, followed by one of $\mathrm{CS}(1, m, c - 1)$ suffixes, the value of $\mathrm{CS}(\sigma, m, c)$ is $\sigma \, \mathrm{CS}(1, m, c - 1)$. Otherwise, the next symbol can either be one of the q symbols already in the universal subsequence of the current arch, or one of the $\sigma - q$ symbols not in the universal subsequence, giving $\mathrm{CS}(q, m, c) = (\sigma - q) \, \mathrm{CS}(q + 1, m, c) + q \, \mathrm{CS}(q, m - 1, c)$. Putting this together, the function $\mathrm{CS}(q, m, c)$ can be defined as:

$$\mathrm{CS}(q, m, c) = \begin{cases} (\sigma - q)!(\sigma!)^{c-1} & m = 0 \\ \sigma^m & c = 0 \\ \sigma \, \mathrm{CS}(1, m, c - 1) & q = \sigma, m > 0, c > 0 \\ (\sigma - q) \, \mathrm{CS}(q + 1, m, c) + q \, \mathrm{CS}(q, m - 1, c) & q > 0, m > 0, c > 0 \end{cases}$$

Lemma 3. *Let $\mathcal{S}(v)$ denote the set of suffixes such that for every $u \in \mathcal{S}(v)$, the word vu is an n-length k-subsequence word. Further, let ℓ be the number of complete arches in v, and let v' be the suffix of v such that $Arch(v) = v_1 v_2 \ldots v_{\ell-1} v_\ell v' = v$ where v_i is the ith arch in the arch factorisation for v. Then, the size of $\mathcal{S}(v)$ is equal to $\mathrm{CS}(q, m, k - \ell)$ where q is the number of unique symbols in v', and $m = n - (|v| + \sigma(k - \ell - 1))$.*

Proof. We will assume, for notational simplicity, that if $v' = \varepsilon$, then $q = \sigma$. We start with the base cases. If $m = 0$, every symbol in the suffixes of $\mathcal{S}(v)$ must be in the universal subsequence of one of the remaining arches. As there are q symbols in v', there are $(\sigma - q)!$ possible ways of extending v' to become an arch, and $\sigma!$ arches of length σ, the total number of suffixes in $\mathcal{S}(v)$ is $(\sigma - q)!(\sigma!)^{k-\ell-1} = (\sigma - q)(\sigma!)^{c-1}$. Alternatively, if $\ell \geq k$ (meaning that v is already a k-subsequence universal word), then every remaining symbol in the suffix is a free symbol, and as such there are no constraints on the contents of the suffix. Therefore in this case, there are σ^m suffixes in $\mathcal{S}(v)$. Note that in the case $m = 0$ and $\ell \geq k$, both of these formulae return 0, corresponding to the empty word.

In the general case, assume that the size of $\mathcal{S}(vx)$ is equal to $\text{CS}(q', m', c')$, where q' is the number of unique symbols in $v'x$, m' is equal to $n - (|v| + 1) - (q + \sigma(k - \ell))$, and c' is $k - \ell$. Observe that the total number of suffixes in $\mathcal{S}(v)$ is equal to $\sum_{x \in \Sigma} |\mathcal{S}(vx)|$. If $q = \sigma$, then the next symbol must belong to the universal subsequence of the next arch, equal to the value of $\text{CS}(1, m, k - \ell - 1)$. If x is one of the q symbols that have already appeared in v', then the size of $\mathcal{S}(vx)$ is $\text{CS}(q, m - 1, c)$. Otherwise, the size of $\mathcal{S}(vx)$ is $\text{CS}(q + 1, m, c)$. As there are q unique symbols in v', the number of suffixes is given by the sum $q\,\text{CS}(q, m - 1, c) + (\sigma - q)\,\text{CS}(q + 1, m, c)$. Hence the size of $\mathcal{S}(v)$ is given by $\text{CS}(q, m, k - \ell)$.

Lemma 4. *The values of* $\text{CS}(q, m, c)$ *can be computed for every* $q \in [1, \sigma], m \in [0, n], c \in [0, k]$ *in* $O(nk\sigma)$ *time.*

Proof. The correctness follows for the arguments above. We assume that the values of $(\sigma - q)!(\sigma!)^c$ have been precomputed for every $q \in [1, \sigma], c \in [0, k]$, requiring $O(k\sigma)$ time, and the values of σ^m have been precomputed for every $m \in [0, n]$ requiring $O(n)$ time. To determine the time complexity, note that the value of $\text{CS}(q, m, c)$ can be computed in constant time assuming that the values of $\text{CS}(q+1, m, c)$, $\text{CS}(1, m, c-1)$, and $\text{CS}(q, m-1, c)$ have already been computed. As the base cases of $c = 0$, and $m = 0$ can be computed in constant time, and for every other case the values of m and $(\sigma - q) + \sigma c$ are monotonically decreasing, the values of $\text{CS}(q, m, c)$ can be computed for every $q \in [1, \sigma], m \in [0, n], c \in [0, k]$ in a dynamic manner, starting with the base cases, and proceeding in increasing value of m, $\sigma - q$ and c. We note that the order in which m, q and c are incremented is irrelevant provided $\text{CS}(q + 1, m, c)$, $\text{CS}(1, m, c - 1)$, and $\text{CS}(q, m - 1, c)$ are computed before $\text{CS}(q, m, c)$. Therefore, the total time complexity of computing the values of $\text{CS}(q, m, c)$ for every $q \in [1, \sigma], m \in [0, n], c \in [0, k]$, is $O(nk\sigma)$.

Note that the number of k-subsequence universal words is equal to the number of words in the set $\mathcal{S}(\varepsilon)$, i.e. the number of n-length words with an empty prefix. Therefore, from Lemma 4, it follows that the size of $\mathcal{U}(n, k, \sigma)$ can be computed by counting the size of $\mathcal{S}(\varepsilon)$, equivalent to evaluating $\sigma\,\text{CS}(1, n - (k\sigma), k)$. Theorem 2 follows from this observation.

Theorem 2. *The size of* $\mathcal{U}(n, k, \sigma)$ *can be computed in* $O(nk\sigma)$ *time.*

4   Ranking

Using the counting techniques outlined in Sect. 3, we can now rank a given word $w \in \Sigma^n$ amongst the set of n-length k-subsequence universal words. This is done in an iterative manner. For each $i \in [1, n]$, we count the number of words of n-length k-subsequence universal words with the prefix $w[1, i-1]x$, where x is some symbol lexicographically smaller than $w[i]$. Taking the sum of such words for every $i \in [1, n]$ gives the total number of n-length k-subsequence universal words that are lexicographically smaller than w. By taking the sum of such words for each prefix, the total number of n-length k-subsequence universal words that are lexicographically smaller than w can be computed.

Let $Arch(w) = w_1, w_2, \ldots, w_m v$. The first key observation is that given any word $u = w_1 w_2 \ldots w_i u'$, for some $i \leq k$, u is k-subsequence universal if and only if u' is $(k-i)$-subsequence universal. Secondly, given a word $s = w_1 w_2 \ldots w_i[1:j]s'$, s is k-subsequence universal if and only if $w_i[1:j]s'$ is $(k-i+1)$-subsequence universal.

Preprocessing. In order to make our ranking algorithm more efficient, we first provide an overview of the preprocessing that is performed before the main ranking algorithm. Let $Arch(w) = w_1, w_2, \ldots, w_k, v$. Using the notation from Observation 1, let $A_1, A_2, \ldots, A_k$ be the indices such that A_ℓ corresponds to the first position in w at which the arch w_ℓ appears in w, i.e. $w_\ell = w[A_\ell, A_{\ell+1} - 1]$. Further, let $\Delta(w, A_\ell, i)$ be the number of unique symbols in the i-length prefix of w_ℓ. We assume that the values of $\Delta(w, A_\ell, i_\ell)$ have been computed for every $i \in [A_\ell, A_{\ell+1} - 1]$, $\ell \in [0, k]$.

In order to count the number of free symbols within each suffix of w, let m be an n-length array such that $m[i]$ contains the number of free symbols in the i-length suffix of w. The values of $m[i]$ are computed by starting with $i = n$, and working in decreasing value of i. Note that the value of $m[n]$ is equal to 0. In the general case, the value of $m[i]$ is either $m[i + 1]$, if $w[i]$ belongs to the universal subsequence of some arch in the arch decomposition, of $m[i + 1] + 1$ otherwise. Letting ℓ be the index such that $A_\ell \leq i < A_{\ell+1}$, note that if $w[i]$ is in the universal subsequence of w_ℓ, then $\Delta(w, A_\ell, i) = \Delta(w, A_\ell, i - 1) + 1$, otherwise $\Delta(w, A_\ell, i) = \Delta(w, A_\ell, i - 1)$. Hence using the previous computation, the values of $m[i]$ can be determined in $O(n)$ time for every $i \in [1, n]$.

In order to determine the number of symbols smaller than $w[i]$, an additional n-length array l such that $l[i]$ contains the set of symbols that appear between A_ℓ and i in w, where ℓ is the index such that $A_\ell \leq i < A_{\ell+1}$. This complements m by ensuring providing a quick method of checking if a given symbol x has already been used by $w_\ell[1, i + 1 - A_\ell]$. The array l is computed in $O(n\sigma)$ time as follows. For each $i \in [1, n]$, note that the value of $l[i]$ is either $l[i - 1] \cup \{w[i]\}$, if $i \neq A_\ell$ for every $\ell \in [0, k]$, or $\{w[i]\}$ otherwise. By storing each array as a σ-length binary vector, requiring at most $O(n\sigma)$ time to initialise, the values of $l[i]$ can be computed for every $i \in [1, n]$ in $O(n)$ time. Finally, we assume that the value of $CS(q, m, c)$ has been precomputed for every $q \in [1, \sigma], m \in [n]$ and $c \in [k]$.

Ranking. We now have the tools we need to rank the input word $w \in \Sigma^n$. We note that w does not have to be a k-subsequence universal word, allowing this tool to be used in a more general setting. At a high level, our approach is to take each prefix of w, $w[1, i]$, and count the number of words in $\mathcal{U}(k, n, \sigma)$ that are lexicographically smaller than w with the prefix $w[1, i]x$, where $x < w[i + 1]$.

By taking the sum of such words for each prefix of w, the total number of words smaller than w can be determined.

Let $\mathcal{R}(i) = \{u \in \mathcal{U}(k, n, \sigma) \mid u < w, u[1, i] = w[1, i], u[i + 1] < w[i + 1]\}$, and let ℓ be the index such that $A_\ell \leq i < A_{\ell+1}$. Note that the number of possible values for the symbol $u[i + 1]$ is equal to $w[i + 1] - 1$. Further, the number of words in $\mathcal{R}(i)$ with the prefix $w[1, i]x$ for some fixed $x < w[i+1]$ is equal to either $\text{CS}(\Delta(w, A_\ell, i) + 1, m[i], k + 1 - \ell)$, if x is in the universal subsequence of w_ℓ or $\text{CS}(\Delta(w, A_\ell, i), m[i] - 1, k + 1 - \ell)$ otherwise. Recall that the array l contains at position i the set of unique symbols in the factor of w $w[A_\ell, i]$. Therefore, the size of $\mathcal{R}(i)$ can be computed with this following sum:

$$|\mathcal{R}(i)| = \sum_{x \in [1, w[i+1]-1]} \begin{cases} \text{CS}(\Delta(w, A_\ell, i) + 1, m[i], k + 1 - \ell) & x \in l[i] \\ \text{CS}(\Delta(w, A_\ell, i), m[i] - 1, k + 1 - \ell) & x \notin l[i] \end{cases}$$

Using $\mathcal{R}(i)$, rank of w in the set $\mathcal{U}(n, k, \sigma)$, denoted $rank(w)$, is given by:

$$rank(w) = \sum_{i \in [0, n-1]} |\mathcal{R}(i)|$$

Theorem 3. *The rank of a given word $w \in \Sigma^n$ can be determined in $O(nk\sigma)$ time.*

Proof. Observe that for any word of the form $w[i]xv$ to be k-subsequence universal, the suffix v must belong to $\mathcal{S}(w[i]x)$. Let ℓ be the index such that $A_\ell \leq i < A_{\ell+1}$, q be the number of unique symbols in $w[A_\ell, i]$ and m the number of free symbols following $w[1, i]x$. Note that the number of possible values of v is either $\text{CS}(q + 1, m, k - \ell)$, if x is not in the universal subsequence of $w[A_\ell, i]$, or $\text{CS}(q, m - 1, k - \ell)$ if x has already appeared in $w[A_\ell, i]$. Using the list l, it can be determined in constant time if the symbol x appears in $w[A_\ell, i]$. By extension, the total number of n-length k-subsequence universal words with the prefix $w[i]x$ can be computed in constant time, assuming that the values of $\text{CS}(q, m, k - \ell)$ has been precomputed, and hence the value of $\mathcal{R}(i)$ can be computed in $O(\sigma)$ time. As there are n possible prefixes of w, the total rank of w within $\mathcal{U}(n, k, \sigma)$ can be computed in $O(n\sigma)$ time after $O(n\sigma k)$ preprocessing.

5   Unranking

We complement our counting and ranking techniques by showing how to unrank n-length k-subsequence universal words. Note that an efficient unranking technique may be used as an effective tool to enumerate the set of all k-subsequence universal words. We assume that the values of $\text{CS}(q, m, k)$ have been precomputed for every $q \in [1, \sigma], m \in [0, n]$, and $c \in [0, k]$.

Our unranking processes operates in an iterative manner. Let w be the word of rank i that is being unranked. Starting with $j = 1$, the value of $w[j]$ is computed by counting the number of n-length k-subsequence universal words with the prefix $w[1, j-1]x$, for $x \in \Sigma$ starting with $x = 1$. The value of x is increased until the number of words with a prefix smaller than or equal to $w[1, j-1]x$ is greater than i. Once this value of x has been computed, $w[j]$ is set to $x-1$, and the algorithm proceeds to compute the value of $w[j+1]$.

Theorem 4. *The k-subsequence universal word w of length n with a rank of i can be determined in $O(n\sigma + nk\sigma)$ time.*

Proof. Starting with $w[1]$, note that the number of words with the prefix x, for any $x \in \Sigma$, is given by $CS(1, n-(k\sigma), k)$. Further, any word with the first symbol x has a rank in the range $(x-1)CS(1, n-(k\sigma), k)+1$ to $xCS(1, n-(k\sigma), k)$. Therefore the value of $w[1]$ is the value of x such that $(x-1)CS(1, n-(k\sigma), k) < i \leq xCS(1, n-(k\sigma), k)$.

More generally, let $t(j)$ be the smallest rank of words with the prefix $w[1, j]$, determined by the sum:

$$t(j) = \sum_{\ell \in [1,j]} \sum_{x \in [1, w[\ell]-1]} |S(w[1, \ell-1]x)| = t(j-1) + \sum_{x \in [1, w[j]-1]} |S(w[1, j-1]x)|.$$

Note that the value of $t(j)$ can therefore be computed in $O(\sigma)$ time using $t(j-1)$ and the values of $CS(q, m, c)$. The value of $w[j+1]$ is, therefore, the symbol x such that $t(j) + \sum_{y \in [1, x-1]} |S(w[1, j]y)| < i \leq t(j) + \sum_{y \in [1, x]} |S(w[1, j]y)|$, and further can be computed in $O(\sigma)$ time, giving the total time complexity of the unranking of w as $O(n\sigma)$ after $O(n\sigma k)$ preprocessing.

Corollary 3. *The set of k-subsequence universal words of length n can be output explicitly with $O(n\sigma)$ delay after $O(nk\sigma)$ preprocessing.*

Proof. Following Theorem 4, each index $i \in [1, |\mathcal{U}(n, k, \sigma)|]$ can be unranked in $O(n\sigma)$ time after at most $O(n\sigma k)$ preprocessing. Hence the set $\mathcal{U}(n, \sigma, k)$ can be enumerated with $O(n \cdot \sigma)$ delay after $O(nk\sigma)$ preprocessing.

6   Conclusion

In this paper, we provided new tools for understanding the space of k-subsequence universal words. Notably, we have shown how to count, rank, unrank, and enumerate these words with efficient algorithms for words of fixed length. We note that all of these algorithms can be extended to the setting of words of length at most n. We see two key open questions asked in this paper. First, if there is a general formula for counting the number of n-length k-subsequence universal words. Indeed, such a formula may allow for a speed up for the preprocessing of the ranking, unranking, and enumeration algorithms, if it can be extended to count the size of $S(v)$ efficiently. Secondly, if there is an enumeration algorithm outputting every word in $\mathcal{U}(n, k, \sigma)$ with at most $O(n)$ delay after polynomial-time preprocessing.

The author thanks the Leverhulme Trust for funding this research via the Leverhulme Research Centre for Functional Materials Design. Further, the author would like to thank the reviewers for their helpful comments that have improved the readability of this paper.

References

1. Adamson, D.: Ranking binary unlabelled necklaces in polynomial time. In: Descriptional Complexity of Formal Systems: 24th IFIP WG 1.02 International Conference, DCFS 2022, Debrecen, Hungary, 29–31 August 2022, Proceedings, pp. 15–29. Springer, Cham (2022). https://doi.org/10.1007/978-3-031-13257-5_2
2. Adamson, D., Deligkas, A., Gusev, V.V., Potapov, I.: Ranking bracelets in polynomial time. In: 32nd Annual Symposium on Combinatorial Pattern Matching, pp. 4–17 (2021)
3. Artikis, A., Margara, A., Ugarte, M., Vansummeren, S., Weidlich, M.: Complex event recognition languages: tutorial. In: Proceedings of the 11th ACM International Conference on Distributed and Event-based Systems, pp. 7–10 (2017)
4. Barker, L., Fleischmann, P., Harwardt, K., Manea, F., Nowotka, D.: Scattered factor-universality of words. In: Jonoska, N., Savchuk, D. (eds.) DLT 2020. LNCS, vol. 12086, pp. 14–28. Springer, Cham (2020). https://doi.org/10.1007/978-3-030-48516-0_2
5. Day, J.D., Fleischmann, P., Kosche, M., Koß, T., Manea, F., Siemer, S.: The edit distance to k-subsequence universality. In: Bläser, M., Monmege, B. (eds.) 38th International Symposium on Theoretical Aspects of Computer Science (STACS 2021). Leibniz International Proceedings in Informatics (LIPIcs), vol. 187, pp. 25:1–25:19. Schloss Dagstuhl - Leibniz-Zentrum für Informatik, Dagstuhl (2021)
6. Fleischer, L., Kufleitner, M.: Testing Simon's congruence. In: 43rd International Symposium on Mathematical Foundations of Computer Science (MFCS 2018). Schloss Dagstuhl-Leibniz-Zentrum fuer Informatik (2018)
7. Fleischmann, P., Haschke, L., Huch, A., Mayrock, A., Nowotka, D.: Nearly k-universal words-investigating a part of Simon's congruence. In: Han, Y.S., Vaszil, G. (eds.) Descriptional Complexity of Formal Systems: 24th IFIP WG 1.02 International Conference, DCFS 2022, Debrecen, Hungary, 29–31 August 2022, Proceedings, pp. 57–71. Springer, Cham (2022). https://doi.org/10.1007/978-3-031-13257-5_5
8. Fredricksen, H., Maiorana, J.: Necklaces of beads in k colors and k-ary de Bruijn sequences. Discret. Math. 23(3), 207–210 (1978)
9. Gawrychowski, P., Kosche, M., Koß, T., Manea, F., Siemer, S.: Efficiently testing Simon's congruence. In: Bläser, M., Monmege, B. (eds.) 38th International Symposium on Theoretical Aspects of Computer Science (STACS 2021). Leibniz International Proceedings in Informatics (LIPIcs), vol. 187, pp. 34:1–34:18. Schloss Dagstuhl - Leibniz-Zentrum für Informatik, Dagstuhl (2021)
10. Gilbert, E.N., Riordan, J.: Symmetry types of periodic sequences. Ill. J. Math. 5(4), 657–665 (1961)
11. Halfon, S., Schnoebelen, P., Zetzsche, G.: Decidability, complexity, and expressiveness of first-order logic over the subword ordering. In: 2017 32nd Annual ACM/IEEE Symposium on Logic in Computer Science (LICS), pp. 1–12. IEEE (2017)
12. Han, R., Wang, S., Gao, X.: Novel algorithms for efficient subsequence searching and mapping in nanopore raw signals towards targeted sequencing. Bioinformatics 36(5), 1333–1343 (2020)
13. Hebrard, J.-J.: An algorithm for distinguishing efficiently bit-strings by their subsequences. Theoret. Comput. Sci. 82(1), 35–49 (1991)

14. Karandikar, P., Schnoebelen, P.: The height of piecewise-testable languages with applications in logical complexity. In: 25th EACSL Annual Conference on Computer Science Logic (CSL 2016). Schloss Dagstuhl-Leibniz-Zentrum fuer Informatik (2016)

15. Kociumaka, T., Radoszewski, J., Rytter, W.: Computing k-th Lyndon word and decoding lexicographically minimal de Bruijn sequence. In: Kulikov, A.S., Kuznetsov, S.O., Pevzner, P. (eds.) CPM 2014. LNCS, vol. 8486, pp. 202–211. Springer, Cham (2014). https://doi.org/10.1007/978-3-319-07566-2_21

16. Kosche, M., Koß, T., Manea, F., Siemer, S.: Absent subsequences in words. In: Bell, P.C., Totzke, P., Potapov, I. (eds.) RP 2021. LNCS, vol. 13035, pp. 115–131. Springer, Cham (2021). https://doi.org/10.1007/978-3-030-89716-1_8

17. Kosche, M., Koß, T., Manea, F., Siemer, S.: Combinatorial algorithms for subsequence matching: a survey. In: Bordihn, H., Horváth, G., Vaszil, G. (eds.) Proceedings 12th International Workshop on Non-Classical Models of Automata and Applications, Debrecen, Hungary, 26–27 August 2022. Electronic Proceedings in Theoretical Computer Science, vol. 367, pp. 11–27. Open Publishing Association (2022)

18. Lothaire, M.: Combinatorics on Words, vol. 17. Cambridge University Press (1997)

19. Mateescu, A., Salomaa, A., Sheng, Yu.: Subword histories and Parikh matrices. J. Comput. Syst. Sci. **68**(1), 1–21 (2004)

20. Savage, C.: A survey of combinatorial gray codes. SIAM Rev. **39**(4), 605–629 (1997)

21. Sawada, J., Williams, A.: Practical algorithms to rank necklaces, Lyndon words, and de Bruijn sequences. J. Discrete Algorithms **43**, 95–110 (2017)

22. Schnoebelen, P., Karandikar, P.: The height of piecewise-testable languages and the complexity of the logic of subwords. Logical Methods Comput. Sci. **15** (2019)

23. Shaw, A.C.: Software descriptions with flow expressions. IEEE Trans. Software Eng. **3**, 242–254 (1978)

24. Shikder, R., Thulasiraman, P., Irani, P., Pingzhao, H.: An OpenMP-based tool for finding longest common subsequence in bioinformatics. BMC. Res. Notes **12**, 1–6 (2019)

25. Simon, I.: Piecewise testable events. In: Brakhage, H. (ed.) GI-Fachtagung 1975. LNCS, vol. 33, pp. 214–222. Springer, Heidelberg (1975). https://doi.org/10.1007/3-540-07407-4_23

26. Simon, I.: Words distinguished by their subwords. Proc. WORDS **2003**(27), 6–13 (2003)

27. Troniček, Z.: Common subsequence automaton. In: Champarnaud, J.-M., Maurel, D. (eds.) CIAA 2002. LNCS, vol. 2608, pp. 270–275. Springer, Heidelberg (2003). https://doi.org/10.1007/3-540-44977-9_28

28. Zetzsche, G.: The complexity of downward closure comparisons. In: Chatzigiannakis, I., Mitzenmacher, M., Rabani, Y., Sangiorgi, D. (eds.) 43rd International Colloquium on Automata, Languages, and Programming (ICALP 2016). Leibniz International Proceedings in Informatics (LIPIcs), vol. 55, pp. 123:1–123:14. Schloss Dagstuhl-Leibniz-Zentrum fuer Informatik, Dagstuhl (2016)

Longest Common Subsequence with Gap Constraints

Duncan Adamson, Maria Kosche, Tore Koß[(✉)], Florin Manea,
and Stefan Siemer

Department of Computer Science, University of Göttingen, Göttingen, Germany
tore.koss@cs.uni-goettingen.de

Abstract. We consider the longest common subsequence problem in the context of subsequences with gap constraints. In particular, following Day et al. 2022, we consider the setting when the distance (i. e., the gap) between two consecutive symbols of the subsequence has to be between a lower and an upper bound (which may depend on the position of those symbols in the subsequence or on the symbols bordering the gap) as well as the case where the entire subsequence is found in a bounded range (defined by a single upper bound), considered by Kosche et al. 2022. In all these cases, we present efficient algorithms for determining the length of the longest common constrained subsequence between two given strings.

1 Introduction

A *subsequence* of a string $w = w[1]w[2]\dots w[n]$, where $w[i]$ is a symbol from a finite alphabet Σ for $i \in \{1, \dots, n\}$, is a string $v = w[i_1]w[i_2]\dots w[i_k]$, with $k \leq n$ and $1 \leq i_1 < i_2 \leq \dots < i_k \leq n$. The positions $i_1, i_2, \dots, i_k$ on which the symbols of v appear in w are said to define the embedding of v in w.

In general, the concept of subsequences appears and plays important roles in many different areas of theoretical computer science such as: formal languages and logics (e. g., in connection to piecewise testable languages [34–36,57,58], or in connection to subword-order and downward-closures [29,42,43,61]); combinatorics on words (e. g., in connection to binomial equivalence, binomial complexity, or to subword histories [20,44,45,50,53–55]); the design and complexity of algorithms. To this end, we mention some classical algorithmic problems such as the computation of *longest common subsequences* or of the *shortest common supersequences* [6,9,15,31,32,48,49,51], the testing of the Simon congruence of strings and the computation of the arch-factorisation and universality of words [7,16,17,19,26,27,30,40,59,60]; see also [41] for a survey of some combinatorial algorithmic problems related to subsequence matching. Moreover, these problems and some other closely related ones have recently regained interest in the context of fine-grained complexity (see [1,3,11,12]). Nevertheless, subsequences appear also in more applied settings: for modelling concurrency [13,52,56], in database theory (especially *event stream processing* [5,28,62]), in data mining [46,47], or in problems related to bioinformatics [10].

© The Author(s), under exclusive license to Springer Nature Switzerland AG 2023
A. Frid and R. Mercaş (Eds.): WORDS 2023, LNCS 13899, pp. 60–76, 2023.
https://doi.org/10.1007/978-3-031-33180-0_5

Most problems related to subsequences are usually considered in the setting where the embedding of subsequences in words are arbitrary. However, in [18], a novel setting is considered, based on the intuition that, in practical scenarios, some properties with respect to the *gaps* that are induced by the embeddings can be inferred. As such, [18] introduces the notion of *subsequences with gap constraints*: these are strings v which can be embedded by some mapping e in a word w in such a way that the *gaps* of the embedding, i. e., the factors between the images of the mapping, satisfy certain properties. The main motivation of introducing and studying this model of subsequences in [18] comes from database theory [37,38], and the properties which have to be satisfied by the gaps are specified either in the form of length constraints (i. e., bounds on the length of the gap) or regular-language constraints. We refer the reader to [18] for a detailed presentation of subsequences with gap constraints and their motivations, as well as a discussion of the various related models. The main results of [18] are related to the complexity of the matching problem: given two strings w and v, decide if there is an embedding e of v as a subsequence of w, such that the gaps induced by e fulfil some given length and regular gap constraints. A series of other complexity results related to analysis problems for the set of subsequences of a word, which fulfil a given set of gap constraints, were obtained. The results of [18] are further extended in [39], where the authors consider *subsequences in bounded ranges*: these are strings v which can be embedded by some mapping e in a word w in such a way that the range in which all the symbols of v are embedded has length at most B, for some given integer B. This investigation was motivated in the context of sliding window algorithms [21–25], and the obtained results are again related to the complexity of matching and analysis problems.

One of the most studied algorithmic problem for subsequences is the problem of finding the length of the longest common subsequence of two strings (for short LCS), see, e. g., [6,15,31,32,48,49,51] or the survey [9]. In this problem, we are given two strings v and w, of length m and n, respectively, over an alphabet of size σ, and want to find the largest k for which there exists a string of length k which can be embedded as a subsequence in both v and w. The results on LCS are efficient algorithms (in most of the papers cited above) but also conditional complexity lower bounds [1–3]. In particular, there is a folklore algorithm solving LCS in $O(N)$ time, for $N = mn$, and, interestingly, the existence of an algorithm whose complexity is $O(N^{1-c})$, for some $c > 0$, would refute the Strong Exponential Time Hypothesis (SETH), see [1].

Our Contributions. In this paper, we investigate the LCS problem in the context of subsequences with gap-length constraints, which seem to have a strong motivation and many application (see [18,39] and the references therein). Clearly, in the model considered by [18], the gap constraints depend on the length of the subsequence, while in LCS this is not known, and we actually need to compute this length. So, the model of [18] needs to be adapted to the setting of LCS. One way to do this is to consider that all the gaps of the common subsequence we search for are restricted by the same pair of lower and upper bounds; in this case, the length of the common subsequence plays no role anymore. We extend this initial idea significantly. On the one hand, we consider the case when there is a constant number of different length constraints which restrict the gaps (and they are given alongside the words). A further extension is the case when we are given, alongside the input words, an arbitrarily long tuple of gap-length constraints: the ith gap constraint in this tuple refers to the gap between the ith and $(i + 1)^{th}$ symbol of the common subsequence we try to find (and it plays some role only if that subsequence has length at least $i + 1$); clearly, the longest common subsequence can be as long as the input words, so it has at most length $\min\{m, n\}$. We also consider the case when the gap constraints are given by the actual letters bounding the gap. All these extensions of LCS refer to models of constrained subsequences, where the constraints are local, i.e., they depend on the embedding of the symbols bounding the gap. Finally, we also consider LCS in the case of subsequences in bounded ranges, where the upper bound B on the size of the ranges in which we look for subsequences is given as input; in this case, we have a global constraint on the embedding of the subsequence.

After defining these variants of the LCS, which seem interesting and well motivated to us, we propose efficient algorithms for each of them. In most cases, these algorithms are non-trivial extensions of the standard dynamic programming algorithm solving LCS. A quick overview of our results for variants of LCS, when the gaps are local: if all gap constraints are identical or we have a constant number of different gap constraints (and the sequence of gap constraints fulfils some additional synchronization condition) we obtain algorithms running in $O(N)$ time; if we have arbitrarily many different constraints, we obtain an algorithm running in $O(Nk)$, where k is the length of the longest common constrained subsequence of the input words; if, moreover, the sequence of constraints is increasing, then the problem can be solved in $O(N \operatorname{polylog} N)$; if the constraints on the gaps are defined according to both letters bounding them, we obtain an algorithm running in $O(\min\{N\sigma \log N, N\sigma^2\})$; if the constraints on the gaps are defined according only to the letter coming after (respectively, before) them, we obtain an algorithm running in $O(\min\{N \log N, N\sigma\})$. In the case of subsequences in bounded range, we show an algorithm which runs in $O(NB^{o(1)})$ time for the respective extension of LCS (i.e., it runs in $O(NB^d)$, for some $0 < d < 1$), as well as an $\frac{1}{3}$-approximation algorithm running in $O(N)$ time.

Related Work. With respect to algorithms, the results of [33] cover the case when all gap constraints are identical. In particular, [33] considers a variant of LCS where the lengths of the gaps induced by the embeddings of the common subsequence in the two input strings are all constrained by the same lower and upper bounds and, additionally, there is an upper bound on the absolute value of the difference between the lengths of the ith gap induced in w and the ith gap induced in v, for all i. The authors of that paper propose a quadratic-time algorithm for this problem and then derive more efficient algorithms in some particular cases. To the best of our knowledge, the case of multiple gap-length constraints was not addressed so far in the literature. In [1], the authors consider LCS for subsequences in a bounded range, called there LOCAL-2-LCS, as an intermediate step in showing complexity lower bounds for LCS; they only mention the trivial $O(NB^2)$ algorithm solving it. The results of [14] lead to an $O(N^{1+o(1)})$ solution for this problem; our solution builds on that approach.

With respect to lower bounds, LCS is a particular case for all the problems we consider in our paper, as we can simply take all the length constraints to be trivial: $(0, n)$ in the case of gap constraints or $B = n$ in the case of subsequences in bounded ranges. Therefore, for each of our problems, the existence of an algorithm whose worst case complexity is $O(N^{1-c})$, with $c > 0$, would refute SETH. Thus, if $\sigma \in O(1)$ (i.e., when the input is over alphabets of constant size), most of our algorithms solving LCS with gap-length constraints are optimal (unless SETH is false) up to polylog-factors; the exceptions are the two cases when we do not impose any monotonicity or synchronization condition on the tuple of gap-constraints. If σ is not constant, the previous claim also does not hold anymore for the case when the constraints on the gaps are defined according to both letters bounding them. In the case of LCS for subsequences in a bounded range, [1] shows quadratic lower bounds even for B being polylogarithmic in n. Thus, unless $B \in O(\text{polylog } N)$, there is a super-logarithmic mismatch between the upper bound provided by our algorithm and the existing lower bound.

It is natural to ask what happens when we have non-trivial constraints, such as, e.g., constraints of the form (a, b) with $a, b \in O(1)$. In [18], it is shown that deciding whether there exists an embedding of a string v as a subsequence of another string w, such that this embedding satisfies a sequence of $|v|$ constraints of the form (a, b) with $a, b \leq 6$, cannot be done in $O(N^{1-c})$ time, with $c > 0$, unless SETH is false; moreover, the respective reduction can be modified so that the embedding fulfills our synchronization property for the case of $O(1)$ distinct gap constraints. The respective decision problem can also be solved by checking whether the longest common constrained subsequence of v and w, where the gap-length constraints for the common subsequence are exactly those defined for the embedding of v in w, has length $|v|$. So, the same lower bound from [18] (which coincides with the lower bound for the classical LCS problem) holds for the constrained LCS problem, even when we have a constant number of constant gap-length constraints, fulfilling, on top, the aforementioned synchronization property also. Due to page limitations some proofs are omitted in this version; see the full version [4].

2   Preliminaries

Let $\mathbb{N} = \{1, 2, \ldots\}$ be the set of natural numbers, $[n] = \{1, \ldots, n\}$, and $[m : n] = [n] \setminus [m - 1]$, for $m, n \in \mathbb{N}$. For $(a, b), (c, d) \in \mathbb{N}^2$, we write $(a, b) \subseteq (c, d)$ if and only if $c \leq a$ and $b \leq d$. All logarithms used in this paper are in base 2.

For a finite alphabet Σ, Σ^+ denotes the set of non-empty words (or strings) over Σ and $\Sigma^* = \Sigma^+ \cup \{\varepsilon\}$ (where ε is the empty word). For a word $w \in \Sigma^*$, $|w|$ denotes its length (in particular, $|\varepsilon| = 0$). We set $w^0 = \varepsilon$ and $w^k = ww^{k-1}$ for every $k \geq 1$. For a word w of length n and some $i \in [n]$, we denote by $w[i]$ the letter on the ith position of w, so $w = w[1]w[2] \cdots w[n]$. For every $i, j \in [|w|]$, we define $w[i : j] = w[i]w[i+1]\ldots w[j]$ if $i \leq j$, and $w[i : j] = \varepsilon$, if $i > j$. For $w \in \Sigma^*$, we define $\mathsf{alph}(w) = \{b \in \Sigma \mid b \text{ occurs at least once in } w\}$. The strings

$w[i : j]$ are called *factors* of the string w; if $i = 1$ (respectively, $j = |w|$), then $w[i : j]$ is called a *prefix* (respectively, *suffix*) of w. For simplicity, for a word w and two natural numbers $m \leq n$, we write $w \in [m, n]$ if $m \leq |w| \leq n$.

For an $m \times n$ matrix $M = (M[i, j])_{i \in [m], j \in [n]}$ and two sets $I \subset [m], J \subset [n]$, let $M[I, J]$ be the submatrix $(M[i, j])_{i \in I, j \in J}$ consisting in the elements which are at the intersection of row $M[i, \cdot]$ and column $M[\cdot, j]$ for $i \in I$ and $j \in J$.

Further, we define the notions of subsequence and subsequence with gap-length constraints, following [18]. Our definitions are based on the notion of *embedding*. For a string w, of length n, and a natural number $k \in [n]$, an embedding is a function $e \colon [k] \rightarrow [n]$ such that $i < j$ implies $e(i) < e(j)$ for all $i, j \in [k]$. We say e is a *matching* embedding if $e(k) = n$. For strings $u, w \in \Sigma^*$ with $|u| \leq |w|$, an embedding $e \colon [|u|] \rightarrow [|w|]$ is an *embedding of u into w* if $u = w[e(1)]w[e(2)] \ldots w[e(k)]$, then u is called a *subsequence of w*.

For an embedding $e \colon [k] \rightarrow [|w|]$ and every $j \in [k - 1]$, the jth gap of w induced by e is the string $\mathsf{gap}_e(w, j) = w[e(j) + 1 : e(j + 1) - 1]$. A *$t$-tuple of gap-length constraints* is a t-tuple $gc = (C_1, C_2, \ldots, C_t)$ with $C_i = (\ell_i, u_i)$ and $0 \leq \ell_i \leq u_i \leq n$ for every $i \in [t]$. We set $gc[i] = C_i$ for every $i \in [t]$, and $gc[1 : i] = (C_1, C_2, \ldots, C_i)$. We say that an embedding e *satisfies a $(k - 1)$-tuple of gap-length constraints gc with respect to a string w* if it has the form $e \colon [k] \rightarrow [|w|]$, and, for every $i \in [k - 1]$, $\ell_i \leq |\mathsf{gap}_e(w, i)| \leq u_i$ (that is, $\mathsf{gap}_e(w, i) \in C_i$).

If there is an embedding e of u into w satisfying the gap constraints gc, we denote this by $u \preceq_{gc} w$. For a $(k-1)$-tuple gc of gap constraints, let $SubSeq(gc, w)$ be the set of all subsequences of w induced by embeddings satisfying gc, i.e., $SubSeq(gc, w) = \{u \mid u \preceq_{gc} w\}$. The elements of $SubSeq(gc, w)$ are also called the *gc-subsequences of w*. For more details see [18].

We are interested in defining and investigating the longest common subsequence problem (LCS for short) in the context of subsequences with gap constraints. However, in the framework introduced in [18], the gap constraints depend on (the length of) the subsequence, and this is not known for the LCS problem. As such, we propose a series of problems where we introduce variants of LCS accommodating gap-length constraints. In all our problems, we have two input strings v and w, with $|v| = m$ and $|w| = n$ and $m \leq n$, and these strings are over an alphabet $\Sigma = \{1, 2, \ldots, \sigma\}$, with $\sigma \leq m$. For the rest of this paper, let $N = mn$. We also consider w.l.o.g. that, when the input contains gap-length constraints, every individual constraint $C = (\ell, u)$ fulfils $0 \leq \ell \leq u \leq n$.

First, some additional definitions. Let $v, w \in \Sigma^*$ be two words; a word s is a common subsequence of v and w if s is a subsequence of both v and w. Let gc be a $(k - 1)$-tuple of gap-length constraints. A word s of length k is a *common gc-subsequence of v and w* if both $s \preceq_{gc} w$ and $s \preceq_{gc} v$ hold. Let $ComSubSeq(gc, v, w)$ denote $SubSeq(gc, v) \cap SubSeq(gc, w)$.

A $(k - 1)$-tuple gc of gap-length constraints is called *increasing* if $gc[i] \subseteq gc[i + 1]$, for all $i \in [k - 2]$. Let gc be an increasing $(k - 1)$-tuple of gap-length constraints and let $i \in [k - 2]$. Assume s is a $gc[1 : i]$-subsequence embedded in $w[1 : i']$, such that the last position of s is mapped to i', and t is a $gc[1 : j]$-subsequence embedded in $w[1 : i']$ as well, such that the last position of t is mapped to i', and $j > i$. If there exists $a \in \Sigma$ such that the embedding of s

in $w[1 : i']$ can be extended to an embedding of sa in $w[1 : i'']$, which satisfies $gc[1 : i+1]$, for some $i'' > i'$, then the embedding of t in $w[1 : i']$ can be extended as well to an embedding of ta in $w[1 : i'']$ which satisfies $gc[1 : i+1]$.

A $(k-1)$-tuple gc of gap-length constraints is called *synchronized* when it satisfies the property that for all $i, j \in [k-1]$, if $gc[i] = gc[j]$ and $i \leq j$ then $gc[i+e] \subseteq gc[j+e]$ for all $e \geq 0$ such that $i + e \leq j + e \leq m - 1$; for example, the tuple $((0, 5)(0, 1)(0, 2)(0, 3)(0, 1)(0, 5)(0, 3)(0, 4))$ is synchronized. Let gc be a synchronized $(k-1)$-tuple of gap-length constraints and let $i \in [k-2]$. Assume s is a $gc[1 : i]$-subsequence embedded in $w[1 : i']$, such that the last position of s is mapped to i', and t is a $gc[1 : j]$-subsequence embedded in $w[1 : i']$ as well, such that the last position of t is mapped to i', and $j > i$ and $gc[i+1] = gc[j+1]$. Now, if there exists a letter a such that the embedding of s in $w[1 : i']$ can be extended to an embedding of sa in $w[1 : i'']$ which satisfies $gc[1 : i + 1]$, for some $i'' > i'$, then the embedding of t in $w[1 : i']$ can be extended as well to an embedding of ta in $w[1 : i'']$ which satisfies $gc[1 : i + 1]$.

The Longest Common Subsequence Problem (LCS) is defined as follows.

Problem 1 (LCS). Given v, w, compute the largest $k \in [m]$ such that there exists a common subsequence s of both v and w with $|s| = k$.

We now extend LCS to the case of subsequences with gap constraints (for a more detailed discussion on variants of this problem, see the full version [4]). Firstly we consider the case when the constraints are *local*, as in [18]: they concern only the gaps occurring between two consecutive symbols of the subsequence.

Problem 2 (LCS-MC). Given $v, w \in \Sigma^*$ and an $(m-1)$-tuple of gap-length constraints gc, compute the largest $k \in \mathbb{N}$ such that there exists a common $gc[1 : k-1]$-subsequence s of v and w, with $|s| = k$. That is, find the largest k for which $ComSubSeq(gc[1 : k-1], v, w)$ is non-empty.

Clearly, LCS is a particular case of LCS-MC, where $gc = ((0, n), \ldots, (0, n))$.

In LCS-MC the input tuple gap-length constraints contains arbitrarily many constraints (therefore the acronym MC in the name of the problem), as many as the maximum amount of gaps that a common subsequence of v and w may have (that is, $m - 1$). We also consider LCS-MC for increasing tuples of gap-length constraints gc; this variant is called LCS-MC-INC.

We consider two special cases of Problem LCS-MC, where all these constraints are either identical (i.e., LCS with one constraint) or drawn from a set of constant size (i.e., LCS with $O(1)$ constraints), which seem interesting to us.

Problem 3 (LCS-1C). Given $v, w \in \Sigma^*$ and an $(m-1)$-tuple of identical gap-length constraints $gc = ((\ell, u), \ldots, (\ell, u))$, compute the largest $k \in \mathbb{N}$ such that there exists a common $gc[1 : k-1]$-subsequence s of v and w, with $|s| = k$.

Problem 4 (LCS-O(1)C). Given $v, w \in \Sigma^*$ and an $(m-1)$-tuple of gap-length constraints $gc = ((\ell_1, u_1), \ldots, (\ell_{m-1}, u_{m-1}))$, where $|\{(\ell_i, u_i) \mid i \in [m-1]\}|$ (the number of distinct constraints of gc) is in $O(1)$, compute the largest $k \in \mathbb{N}$ such that there exists a common $gc[1 : k-1]$-subsequence s of v and w, with $|s| = k$.

The general results obtained for LCS-MC are improved for LCS-O(1)C, by considering the latter problem in the restricted setting of synchronized gap-length constraints only. The resulting problem is called LCS-O(1)C-SYNC.

In the problems introduced so far, the gap between two consecutive symbols in the searched subsequence depends on the positions of these symbols inside the respective subsequence (i. e., the gap between the ith and $i + 1^{th}$ symbols is always the same). That is, the actual symbols of the subsequence play no role in defining the constraints; it is only the length of the subsequence which is important. For the next problems, the constraints on a gap between consecutive symbols are determined by one or both symbols bounding the respective gap, and do not depend on the position of the gap inside the subsequence.

For this, we first need to modify our setting. Let $left : \Sigma \rightarrow [n] \times [n]$ and $right : \Sigma \rightarrow [n] \times [n]$ be two functions, defining the gap constraints. For an embedding $e \colon [k] \rightarrow [n]$, we say that e *satisfies the gap constraints defined by* $(left, right)$ *with respect to a string* x if for every $i \in [k-1]$ we have that $\mathsf{gap}_e(x, i) \in left(x[e(i)]) \cap right(x[e(i+1)])$; in other words, $\mathsf{gap}_e(x, i)$ has to simultaneously fulfil the constraints $left(x[e(i)])$ and $right(x[e(i+1)])$, defined by the symbols bounding that gap. If there is an embedding e of a string y into x satisfying the gap constraints $(left, right)$, we denote this by $y \preceq_{left,right} x$ and call y a $(left, right)$-*subsequence of* x. In the following algorithmic problems, functions $g : \Sigma \rightarrow [n] \times [n]$ are given as sequences of σ tuples $(a, g(a))_{a \in \Sigma}$.

Problem 5 (LCS-Σ). Given two words $v, w \in \Sigma^*$ and two functions $left : \Sigma \rightarrow [n] \times [n]$ and $right : \Sigma \rightarrow [n] \times [n]$, compute the largest number $k \in \mathbb{N}$ such that there exists a common $(left, right)$-subsequence s of v and w, with $|s| = k$.

When $left(a) = (0, n)$ for all $a \in \Sigma$ (respectively, $right(a) = (0, n)$ for all $a \in \Sigma$), the gap constraints are defined only by the function $right$ (respectively, $left$), and the problem LCS-Σ is denoted LCS-ΣR (respectively, LCS-ΣL).

In the problems introduced so far, the constraints were local (in the sense that they were defined by consecutive problems in the subsequence). In the last problem we introduce, we build on the works [1,39], and consider subsequences which occur inside factors of bounded length of the input words. In particular, for a given integer B, a word s is a B-subsequence of w if there exists a factor $w[i+1 : i+B]$ of w containing s as subsequence, and we look for the largest common B-subsequence of two input words. This problem was called Local-2-Longest Common Subsequence in [1], but as the constraint acts now globally on the subsequence, we prefer to call it LCS-BR (LCS in bounded range),

Problem 6 (LCS-BR). Given $v, w \in \Sigma^*$ and $B \in [n]$, compute the largest $k \in \mathbb{N}$ such that there exists a common B-subsequence s of v and w, with $|s| = k$.

Note that the *computational model* used to describe our algorithms is the standard unit-cost RAM with logarithmic word size (see the full version [4]).

3   LCS with Local Gap Constraints

An initial approach for LCS-MC. For all variants of LCS where the constraints are local (i. e., they depend on the position of the gap in the subsequence, or on the letters bounding it), the sets of subsequences which are candidates for s can be, in the worst case, of exponential size in N. Therefore, computing the respective sets for both input words, their intersection, and then finding the longest string in this intersection would result in an exponential time algorithm. LCS can be, however, solved by a dynamic programming approach (considered folklore) in $O(N)$ time. Similarly, LCS-MC (and its particular cases LCS-1C and LCS-$O(1)$C, as well as LCS-Σ) can also be solved by a dynamic programming approach, running in polynomial time. We describe this general idea for LCS-MC only (as it can be easily adapted to all other problems). This idea reflects, to a certain extent, a less efficient implementation of the folklore algorithm for LCS.

For input strings v, w and constraints $gc = (C_1, \ldots, C_{m-1})$ we define, for each $p \in [m]$, a matrix $M_p \in \{0, 1\}^{m \times n}$, where $M_p[i, j] = 1$ if and only if there exists a string s_p with $|s_p| = p$ and matching embeddings e_v, e_w, respectively into $v[1 : i]$ and $w[1 : j]$, satisfying $gc[1 : p-1]$. We compute M_1 by setting $M_1[i, j] = 1$ if and only if $v[i] = w[j]$. Then we compute M_p recursively by dynamic programming: let $C_{p-1} = (\ell, u)$ and note that $M_p[i, j] = 1$ if and only if $v[i] = w[j]$ and there are positions i' with $\ell \leq i - i' - 1 \leq u$ and j' with $\ell \leq j - j' - 1 \leq u$ such that there is a string s_{p-1} of length $p - 1$ with matching embeddings into $v[1 : i']$ and $w[1 : j']$, respectively, satisfying $gc[1 : p - 2]$. That is $M_p[i, j] = 1$ if and only if there is a 1 in the submatrix $M_{p-1}[I, J]$ with $I = [i - u - 1 : i - \ell - 1]$ and $J = [j - u - 1 : j - \ell - 1]$. In the end, the length k of the longest common subsequence of v and w satisfying gc equals the largest p such that M_p is not the 0-matrix. A naïve implementation of this approach runs in $O(N^2 k)$ time.

A more efficient implementation is given in the following.

Lemma 1. LCS-MC *can be solved in* $O(Nk)$ *time, where* k *is the largest number for which there exists a common* $gc[1 : k - 1]$-*subsequence* s *of* v *and* w.

Proof. As mentioned above, we can compute M_1 in $O(N)$ time. So, let $2 \leq p \leq m$, $C_{p-1} = (\ell, u)$, and $d = |C_{p-1}| = u - \ell + 1$. We want to compute the elements of M_p and assume that M_{p-1} was already computed. For convenience we treat $M_{p-1}[i, j] = 0$ when either $i < 1$ or $j < 1$.

We use a pair of $m \times n$ matrices A and B, where $A[i, j]$ stores the sum of (or equivalently the amount of 1s in) d consecutive entries $M_{p-1}[i, j - d + 1], \ldots, M_{p-1}[i, j]$ in the rows of M_{p-1}. Then $A[i, 1] = M_{p-1}[i, 1]$ and $A[i, j] = A[i, j - 1] - M_{p-1}[i, j - d] + M_{p-1}[i, j]$ for all $i \in [m]$ and $j \in [2 : n]$. The entry $B[i, j]$ stores the sum of all entries $M_{p-1}[i', j']$ with $0 \leq i - i' < d$ and $0 \leq j - j' < d$. Again, for convenience, we treat all entries $A[i, j]$ as 0 if either $i < 1$ or $j < 1$. Then we compute $B[i, j]$ as follows. We set $B[1, 1] = M_{p-1}[1, 1]$, $B[1, j] = B[1, j - 1] - M_{p-1}[1, j - d] + M_{p-1}[1, j]$, and $B[i, j] = B[i - 1, j] - A[i - d, j] + A[i, j]$ for all $i \in [2 : m]$ and $j \in [2 : n]$.

Since the computation of each entry in A or B takes $O(1)$ time, we can compute the matrices A and B in time $O(N)$. Now $M_p[i, j] = 1$ if and only if there is a 1 in the submatrix $M_{p-1}[I, J]$, which is true if $B[i - \ell - 1, j - \ell - 1] > 0$. Hence we can compute $M_p[i, j]$ in constant time, M_p in $O(N)$ time, and the sequence $M_1, \ldots, M_{k+1}$ in time $O(Nk)$. □

An $O(N \log^2 N)$ time algorithm for LCS-MC-INC. We now consider LCS-MC-INC, a variant of LCS-MC where the tuple gc is increasing. We begin with a lemma describing a data structure, which is then used to solve LCS-MC-INC.

Lemma 2. *Given an $m \times n$ matrix M with all elements initially equal to 0, we can maintain a data structure (two dimensional segment tree) T for M, so that we can execute the following operations efficiently:*

- *$update_T(i', i'', j', j'', x)$: set $M[i,j] = \max\{M[i,j], x\}$, for all $i \in [i' : i'']$ and $j \in [j' : j'']$; here, x is a natural number. Time: $O(\log n \log m)$.*
- *$query_T(i,j)$: return $M[i,j]$. Time: $O(\log m)$.*

Lemma 3. LCS-MC-INC *can be solved in $O(N \log^2 N)$.*

Proof. Main idea. Our algorithm computes, one by one, the elements of an $m \times n$ matrix M (whose elements are initially set to 0). The approach is to define $M[i,j]$, for each pair of positions $(i,j) \in [m] \times [n]$ such that $v[i] = w[j]$, to equal the length p of the longest string s_p which has matching embeddings into $v[1:i]$ and $w[1:j]$, respectively, satisfying $gc[1:p-1]$. Because gc is increasing (for all $i \in [m-2]$, $gc[i] \subseteq gc[i+1]$), p can be determined as follows. It is enough to extend with the symbol $a = v[i] = w[j]$ (mapped to position i of v and position j of w) the longest subsequence $s_{p'}$, with $|s_{p'}| = p'$, such that the embeddings of $s_{p'}$ in v and w end on positions i' and j', respectively, where the gap between i' and i and the gap between j' and j fulfil the gap constraint $gc[p' + 1]$. Indeed, this is enough: as gc is increasing, this longest subsequence can be extended in exactly the same way as any other shorter subsequence with the same properties. Then, to obtain $s_{p'}$, it is enough to set $p = p' + 1$ and extend $s_{p'}$ with the letter a, mapped to $v[i]$ and $w[j]$ in the two embeddings, respectively.

However, when considering the position (i,j), we do not know the value of p', and, as such, we do not know the range where we need to look for i' and j'. Therefore, we need to find a way around this.

Dynamic Programming Approach. We now show how to compute the elements of M. In the case of the dynamic programming algorithms solving LCS, the element $M[i,j]$ of the matrix M is computed by looking at some elements $M[i',j']$, with $i' \le i, j' \le j, (i,j) \ne (i',j')$. By the arguments presented above, such an approach does not seem to work directly for LCS-MC-INC. However, if we know the value p of some entry $M[i,j]$, we can be sure that $M[i'',j''] \ge p+1$ for all i'' and j'' such that $i + \ell + 1 \le i'' \le i + u + 1$ and $j + \ell + 1 \le j'' \le j + u + 1$, where $gc[p+1] = (\ell, u)$; we store this information. Moreover, if we know already all the values $M[i,j]$, with $i \le i', j \le j', (i,j) \ne (i',j')$, then we have already seen (and stored) all possible values for $M[i',j']$ (or, in other words, all possible subsequences that we can extend in order to get $M[i,j]$), so we simply set $M[i',j']$ to the largest such possible value.

So, we compute the elements $M[i,j]$ one by one, by traversing the elements of M for i from 1 to m, for j from 1 to n, and proceed as follows. When we reach an element $M[i,j]$ in our traversal of M, we simply set it permanently to its current value. Then, if we set $M[i,j]$ to some value p, and $gc[p] = (\ell, u)$, we update each element $M[i', j']$ of submatrix $M[I, J]$, where $I = [i+\ell+1 : i+u+1]$ and $J = [j+\ell+1 : j+u+1]$, to be $M[i', j'] = \max\{M[i', j'], p+1\}$.

The Algorithm. First we define the matrix M, and initialize all its entries with 0. Then, we build the data structure $\mathcal{T}$ from Lemma 2 for M. In an initial step, we set all values $M[1, j] = 1$, where $v[1] = w[j]$, and $M[i, 1] = 1$, where $v[i] = w[1]$; this is done by using update-operations on $\mathcal{T}$ (to set the entry $M[i, j] = x$, for some $x > 0$, given that $M[i, j]$ was equal to 0, it is enough to execute update(i, i, j, j, x)). Further, we execute the following procedure.

1: for $i = 2$ to m do
2:    for $j = 2$ to n do
3:       Set $M[i, j]$ =query$_{\mathcal{T}}(i, j) = p$; $M[i, j]$ remains equal to p permanently;
4:       For $(\ell, u) = gc[p]$ update$_{\mathcal{T}}(i+\ell+1, i+u+1, j+\ell+1, j+u+1, p+1)$.

The solution to LCS-MC-INC is the maximum element of M.

Conclusion. The correctness of our algorithm follows from the arguments presented above. The time complexity of the algorithm is $O(N \log^2 N)$, as we need $O(N \log^2 N)$ time for the preprocessing (setting up $\mathcal{T}$ and doing the initial updates on it). Then the 4-step procedure described above takes also $O(N \log^2 N)$ time, as in each iteration of the inner loop we perform as the most time consuming operation an update on $\mathcal{T}$. Our claim follows.   □

Summing up, we have shown the following theorem regarding LCS-MC.

Theorem 1. LCS-MC *can be solved in* $O(Nk)$ *time, where* k *is the largest number for which there exists a common* $gc[1 : k-1]$*-subsequence* s *of* v *and* w. LCS-MC-INC *can be solved in* $O(N \log^2 N)$.

$O(N)$ *Solutions for* LCS-1C *and* LCS-O(1)C. While this problem was already solved in [33], we also briefly describe our solution for it. Our approach is based on the following data-structures lemma, which are also used to solve some of the other problems we discuss here.

Lemma 4. *Let* $\Psi : [m] \times [n] \to \{0, 1\}$ *be a predefined function, such that* $\Psi(i, j)$ *can be retrieved in* $O(1)$ *time. Given an* $m \times n$ *matrix* M, *with all elements initially equal to 0, and four positive integers* $\ell_1 \leq u_1, \ell_2 \leq u_2$, *we can maintain a data structure* $\mathcal{D}$ *for* M, *so that the following process runs in* $O(N)$ *time:*

1: for $i = 1$ to m do
2:    update $\mathcal{D}$ (set up for processing line i);
3:    for $j = 1$ to n do
4:       update $\mathcal{D}$ (set up for computing $M[i, j]$);
5:       use $\mathcal{D}$ to retrieve $\mathfrak{m}$, the maximum of the submatrix $M[I, J]$
          where $I = [i - u_1 : i - \ell_1]$ and $J = [j - u_2 : j - \ell_2]$;
       $\mathfrak{m}$ is set to be 0 when I or J are empty.
6:       if $\Psi(i, j) = 1$ then set $M[i, j] = \mathfrak{m} + 1$.

We can now show immediately the following result.

Theorem 2. LCS-1C *can be solved in* $O(N)$ *time.*

Proof. Let (ℓ, u) be the single gap-length constraint appearing in gc. We define the $m \times n$ matrix M, where $M[i,j] = p$ if and only if p is the greatest number for which there exists a subsequence s_p, with $|s_p| = p$, such that there are two matching embeddings e_v and e_w of s_p into $v[1 : i]$ and $w[1 : j]$, respectively, both satisfying $gc[1 : p-1]$. We have that $M[i,j] = p$ if and only if $v[i] = w[j]$ and $p-1$ is the greatest number for which there exist i' and j' with $i - u - 1 \le i' \le i - \ell - 1$ and $j - u - 1 \le j' \le j - \ell - 1$ and $M[i', j'] = p - 1$. Hence, the entries of M can be computed using Lemma 4, for $u_1 = u_2 = u + 1$, $\ell_1 = \ell_2 = \ell + 1$ and $\Psi(i,j) = 1$ if and only if $v[i] = w[j]$.   □

This result can be extended to the case of LCS-$O(1)$C-SYNC. It is, however, open if a similar result holds for the unrestricted problem LCS-$O(1)$C.

Theorem 3. LCS-$O(1)$C-SYNC *can be solved in* $O(N)$ *time, where the constant hidden by the O-notation depends linearly on the number h of distinct gap-length constraints of gc.*

The result of Theorem 3 holds, in fact, for a larger family of constraints, namely constraints whose elements can be partitioned in $h \in O(1)$ classes, such that, for all $1 \le i < j \le k-1$, if i, j are in the same class of the partition then $gc[i] = gc[j]$ and $gc[i + e] \subseteq gc[j + e]$ for all $e \ge 0$ such that $i + e \le j + e \le m - 1$.

Solutions for LCS-ΣR, LCS-ΣL, *and* LCS-Σ. In general, the solutions to all these problems are based on a dynamic programming approach. We compute an $m \times n$ matrix M such that $M[i,j] = p$ if and only if p is the largest number for which there exists a *(right)*-subsequence (respectively, *(left)*- or *(left, right)*-subsequence) s_p of $v[1 : i]$ and $w[1 : j]$ such that there are two matching embeddings e_v and e_w of s_p into $v[1 : i]$ and $w[1 : j]$, respectively, both fulfilling the constraints. The elements of M can be computed in two ways. On the one hand, we can use Lemma 4 and get a time complexity of $O(N\sigma)$ (or $O(N\sigma^2)$ for LCS-Σ). On the other hand, we can use a dynamic version of 2D Range Maximum Queries structures, extending the Sparse Tables from [8], and obtain a time complexity of $O(N \log N)$ (or $O(N\sigma \log N)$ for LCS-Σ). In all cases, the answer to the considered problems is the maximum element of M. For space reasons, our results are described in detail in the full version [4].

4   LCS with Global Constraints

In this section, we present our solution to LCS-BR. First, we note the naïve solution: we consider every pair $(v[i + 1 : i + B], w[j + 1 : j + B])$ of factors of length B of the two input words, respectively, and find their longest common subsequence, using the folklore dynamic programming algorithm for LCS. As each word of length n has $n - B + 1$ factors of length B, this approach requires solving LCS for $O((m - B)(n - B)) \subseteq O(N)$ words, with each such LCS-computation requiring $O(B^2)$ time. This yields a total time complexity of $O(NB^2)$.

Here, we improve this by providing an $O(NB^{o(1)})$ time algorithm via the *alignment oracles* provided by Charalampopoulos et al. [14]. Each such oracle is built for a pair of words v and w, with $|v| = m$, $|w| = n$ and $N = mn$, and is able to return the answer to queries asking for the length of the LCS between two factors $v[i : i']$ and $w[j : j']$. One of the results of [14] is the following theorem.

Theorem 4 ([14]). *We can construct in $N^{1+o(1)}$ time an alignment oracle for the words v and w, with $\log^{2+o(1)} N$ query time.*

Theorem 4 can be used directly to build an alignment oracle $\mathcal{A}$ for the input words w and v in $O(N^{1+o(1)})$ time. Using this oracle, we can solve LCS-BR by making $O(N)$ queries to $\mathcal{A}$, for every pair of indices $i \leq m, j \leq n$, each requiring $O(\log^{2+o(1)} N)$ time. The total time complexity of this direct approach is therefore $O(N^{1+o(1)})$. We improve this approach by creating a set of smaller oracles, allowing us to avoid the extra work required to answer LCS queries beyond the bounded range. For simplicity, assume w.l.o.g. that m and n are multiples of $2B$. Let, for all i, $w_i = w[iB + 1 : (i+2)B]$, or $w_i = w[iB + 1 : n]$ if $(i+2)B > n$ and $v_i = v[iB + 1 : (i+2)B]$, or $v_i = v[iB + 1 : n]$ if $(i+2)B > m$. Observe that every factor of length B of w appears in at least one subword w_i and every factor of length B of v appears in at least one subword v_i. Therefore, solving LCS-BR with input v_i, w_j for every $i \in [0, m/B], j \in [0, n/B]$ would also give us the solution to LCS-BR for input words v, w.

To find the solution to LCS-BR with input v_i, w_j, we use Theorem 4 to construct an oracle $\mathcal{A}_{i,j}$ for v_i and w_j. As $|v_i| = |w_j| = 2B$, the oracle $\mathcal{A}_{i,j}$ can be constructed in $O(B^{2+o(1)})$ time. The solution of LCS-BR for the input words v_i, w_j can be then determined by making $O(B^2)$ queries to $\mathcal{A}_{i,j}$, each requiring $O(\log^{2+o(1)} B)$ time. So, the total time complexity of solving LCS-BR for input words v_i, w_j is $O(B^{2+o(1)})$. As there are $O(N/B^2)$ pairs of factors v_i, w_j, the time complexity of solving LCS-BR for words v, w is $O\left(\frac{N}{B^2}B^{2+o(1)}\right) = O(NB^{o(1)})$.

Theorem 5. LCS-BR *can be solved in $O(NB^{o(1)})$ time.*

An Approximation Algorithm. We complement our $O(NB^{o(1)})$ exact algorithm with an $O(N)$ time 1/3-approximation algorithm (for details see the version on arXiv). With $LCS(v_i, w_j)$ denoting the length of the longest common subsequence of v_i and w_j, the key insight is that the length of the longest common B-subsequence s of v and w fulfils $\max_{i,j \in [n/B]} LCS(v_i, w_j)/3 \leq |s| \leq \max_{i,j \in [n/B]} LCS(v_i, w_j)$.

5   Future Work

A series of problems remain open from our work. Can our results regarding LCS-MC be improved, at least in its particular case LCS-O(1)C? If not, can one show tight complexity lower bounds for these problems? Can the dependency of Σ from the solutions to LCS-Σ and its variants be removed? We were not focused on shaving polylog factors from the time complexity of our algorithms, but it would be also interesting to see if this is achievable. Nevertheless, it would be interesting to address also the problem of efficiently computing the actual longest common constrained subsequences in the case of all addressed problems.

References

1. Abboud, A., Backurs, A., Williams, V.V.: Tight hardness results for LCS and other sequence similarity measures. In: Guruswami, V. (ed.) IEEE 56th Annual Symposium on Foundations of Computer Science, FOCS 2015, Berkeley, CA, USA, 17–20 October 2015, pp. 59–78. IEEE Computer Society (2015). https://doi.org/10.1109/FOCS.2015.14
2. Abboud, A., Rubinstein, A.: Fast and deterministic constant factor approximation algorithms for LCS imply new circuit lower bounds. In: 9th Innovations in Theoretical Computer Science Conference, ITCS 2018, 11–14 January 2018, Cambridge, MA, USA, pp. 35:1–35:14 (2018). https://doi.org/10.4230/LIPIcs.ITCS.2018.35
3. Abboud, A., Williams, V.V., Weimann, O.: Consequences of faster alignment of sequences. In: Automata, Languages, and Programming - 41st International Colloquium, ICALP 2014, Copenhagen, Denmark, 8–11 July 2014, Proceedings, Part I, pp. 39–51 (2014). https://doi.org/10.1007/978-3-662-43948-7_4
4. Adamson, D., Kosche, M., Koß, T., Manea, F., Siemer, S.: Longest common subsequence with gap constraints. arXiv e-prints arXiv:2304.05270 (2023)
5. Artikis, A., Margara, A., Ugarte, M., Vansummeren, S., Weidlich, M.: Complex event recognition languages: Tutorial. In: Proceedings of the 11th ACM International Conference on Distributed and Event-based Systems, DEBS 2017, Barcelona, Spain, 19–23 June 2017, pp. 7–10 (2017). https://doi.org/10.1145/3093742.3095106
6. Baeza-Yates, R.A.: Searching subsequences. Theor. Comput. Sci. **78**(2), 363–376 (1991)
7. Barker, L., Fleischmann, P., Harwardt, K., Manea, F., Nowotka, D.: Scattered factor-universality of words. In: Jonoska, N., Savchuk, D. (eds.) DLT 2020. LNCS, vol. 12086, pp. 14–28. Springer, Cham (2020). https://doi.org/10.1007/978-3-030-48516-0_2
8. Bender, M.A., Farach-Colton, M.: The LCA problem revisited. In: Gonnet, G.H., Viola, A. (eds.) LATIN 2000. LNCS, vol. 1776, pp. 88–94. Springer, Heidelberg (2000). https://doi.org/10.1007/10719839_9
9. Bergroth, L., Hakonen, H., Raita, T.: A survey of longest common subsequence algorithms. In: de la Fuente, P. (ed.) Seventh International Symposium on String Processing and Information Retrieval, SPIRE 2000, A Coruña, Spain, 27–29 September 2000, pp. 39–48. IEEE Computer Society (2000). https://doi.org/10.1109/SPIRE.2000.878178
10. Bille, P., Gørtz, I.L., Vildhøj, H.W., Wind, D.K.: String matching with variable length gaps. Theor. Comput. Sci. **443**, 25–34 (2012). https://doi.org/10.1016/j.tcs.2012.03.029
11. Bringmann, K., Chaudhury, B.R.: Sketching, streaming, and fine-grained complexity of (weighted) LCS. In: Proceedings FSTTCS 2018. LIPIcs, vol. 122, pp. 40:1–40:16 (2018)
12. Bringmann, K., Künnemann, M.: Multivariate fine-grained complexity of longest common subsequence. In: Proceedings of SODA 2018, pp. 1216–1235 (2018)
13. Buss, S., Soltys, M.: Unshuffling a square is NP-hard. J. Comput. Syst. Sci. **80**(4), 766–776 (2014). https://doi.org/10.1016/j.jcss.2013.11.002

14. Charalampopoulos, P., Gawrychowski, P., Mozes, S., Weimann, O.: An almost optimal edit distance oracle. In: Bansal, N., Merelli, E., Worrell, J. (eds.) 48th International Colloquium on Automata, Languages, and Programming, ICALP 2021, 12–16 July 2021, Glasgow, Scotland (Virtual Conference). LIPIcs, vol. 198, pp. 48:1–48:20. Schloss Dagstuhl - Leibniz-Zentrum für Informatik (2021). https://doi.org/10.4230/LIPIcs.ICALP.2021.48

15. Chvátal, V., Sankoff, D.: Longest common subsequences of two random sequences. J. Appli. Probability **12**(2), 306–315 (1975). http://www.jstor.org/stable/3212444

16. Crochemore, M., Melichar, B., Tronícek, Z.: Directed acyclic subsequence graph – overview. J. Discrete Algorithms **1**(3–4), 255–280 (2003)

17. Day, J.D., Fleischmann, P., Kosche, M., Koß, T., Manea, F., Siemer, S.: The edit distance to k-subsequence universality. In: 38th International Symposium on Theoretical Aspects of Computer Science, STACS 2021, 16–19 March 2021, Saarbrücken, Germany (Virtual Conference), pp. 25:1–25:19 (2021). https://doi.org/10.4230/LIPIcs.STACS.2021.25

18. Day, J.D., Kosche, M., Manea, F., Schmid, M.L.: Subsequences with gap constraints: Complexity bounds for matching and analysis problems. In: Bae, S.W., Park, H. (eds.) 33rd International Symposium on Algorithms and Computation, ISAAC 2022, 19–21 December 2022, Seoul, Korea. LIPIcs, vol. 248, pp. 64:1–64:18. Schloss Dagstuhl - Leibniz-Zentrum für Informatik (2022). https://doi.org/10.4230/LIPIcs.ISAAC.2022.64

19. Fleischer, L., Kufleitner, M.: Testing Simon's congruence. In: Proceedings of MFCS 2018. LIPIcs, vol. 117, pp. 62:1–62:13 (2018)

20. Freydenberger, D.D., Gawrychowski, P., Karhumäki, J., Manea, F., Rytter, W.: Testing k-binomial equivalence. In: Multidisciplinary Creativity, a collection of papers dedicated to G. Păun 65th birthday, pp. 239–248, available in CoRR abs/arXiv: 1509.00622 (2015)

21. Ganardi, M., Hucke, D., König, D., Lohrey, M., Mamouras, K.: Automata theory on sliding windows. In: STACS. LIPIcs, vol. 96, pp. 31:1–31:14. Schloss Dagstuhl - Leibniz-Zentrum für Informatik (2018)

22. Ganardi, M., Hucke, D., Lohrey, M.: Querying regular languages over sliding windows. In: FSTTCS. LIPIcs, , vol. 65, pp. 18:1–18:14. Schloss Dagstuhl - Leibniz-Zentrum für Informatik (2016)

23. Ganardi, M., Hucke, D., Lohrey, M.: Randomized sliding window algorithms for regular languages. In: ICALP. LIPIcs, vol. 107, pp. 127:1–127:13. Schloss Dagstuhl - Leibniz-Zentrum für Informatik (2018)

24. Ganardi, M., Hucke, D., Lohrey, M.: Sliding window algorithms for regular languages. In: Klein, S.T., Martín-Vide, C., Shapira, D. (eds.) LATA 2018. LNCS, vol. 10792, pp. 26–35. Springer, Cham (2018). https://doi.org/10.1007/978-3-319-77313-1_2

25. Ganardi, M., Hucke, D., Lohrey, M., Starikovskaya, T.: Sliding window property testing for regular languages. In: ISAAC. LIPIcs, vol. 149, pp. 6:1–6:13. Schloss Dagstuhl - Leibniz-Zentrum für Informatik (2019)

26. Garel, E.: Minimal separators of two words. In: Apostolico, A., Crochemore, M., Galil, Z., Manber, U. (eds.) CPM 1993. LNCS, vol. 684, pp. 35–53. Springer, Heidelberg (1993). https://doi.org/10.1007/BFb0029795

27. Gawrychowski, P., Kosche, M., Koß, T., Manea, F., Siemer, S.: Efficiently testing Simon's congruence. In: 38th International Symposium on Theoretical Aspects of Computer Science, STACS 2021, 16–19 March 2021, Saarbrücken, Germany (Virtual Conference), pp. 34:1–34:18 (2021). https://doi.org/10.4230/LIPIcs.STACS.2021.34

28. Giatrakos, N., Alevizos, E., Artikis, A., Deligiannakis, A., Garofalakis, M.: Complex event recognition in the Big Data era: a survey. VLDB J. **29**(1), 313–352 (2019). https://doi.org/10.1007/s00778-019-00557-w

29. Halfon, S., Schnoebelen, P., Zetzsche, G.: Decidability, complexity, and expressiveness of first-order logic over the subword ordering. In: Proceedings LICS 2017, pp. 1–12 (2017)

30. Hebrard, J.J.: An algorithm for distinguishing efficiently bit-strings by their subsequences. Theor. Comput. Sci. **82**(1), 35–49 (1991)

31. Hirschberg, D.S.: Algorithms for the longest common subsequence problem. J. ACM **24**(4), 664–675 (1977). https://doi.org/10.1145/322033.322044

32. Hunt, J.W., Szymanski, T.G.: A fast algorithm for computing longest subsequences. Commun. ACM **20**(5), 350–353 (1977). https://doi.org/10.1145/359581.359603

33. Iliopoulos, C.S., Kubica, M., Rahman, M.S., Waleń, T.: Algorithms for computing the longest parameterized common subsequence. In: Ma, B., Zhang, K. (eds.) CPM 2007. LNCS, vol. 4580, pp. 265–273. Springer, Heidelberg (2007). https://doi.org/10.1007/978-3-540-73437-6_27

34. Karandikar, P., Kufleitner, M., Schnoebelen, P.: On the index of Simon's congruence for piecewise testability. Inf. Process. Lett. **115**(4), 515–519 (2015)

35. Karandikar, P., Schnoebelen, P.: The height of piecewise-testable languages with applications in logical complexity. In: Proceedings CSL 2016. LIPIcs, vol. 62, pp. 37:1–37:22 (2016)

36. Karandikar, P., Schnoebelen, P.: The height of piecewise-testable languages and the complexity of the logic of subwords. Log. Methods Comput. Sci. **15**(2) (2019)

37. Kleest-Meißner, S., Sattler, R., Schmid, M.L., Schweikardt, N., Weidlich, M.: Discovering event queries from traces: Laying foundations for subsequence-queries with wildcards and gap-size constraints. In: 25th International Conference on Database Theory, ICDT 2022. LIPIcs, vol. 220, pp. 18:1–18:21. Schloss Dagstuhl - Leibniz-Zentrum für Informatik (2022). https://doi.org/10.4230/LIPIcs.ICDT.2022.18

38. Kleest-Meißner, S., Sattler, R., Schmid, M.L., Schweikardt, N., Weidlich, M.: Discovering multi-dimensional subsequence queries from traces - from theory to practice. In: König-Ries, B., Scherzinger, S., Lehner, W., Vossen, G. (eds.) Datenbanksysteme für Business, Technologie und Web (BTW 2023), 20. Fachtagung des GI-Fachbereichs, Datenbanken und Informationssysteme" (DBIS), 06.-10, März 2023, Dresden, Germany, Proceedings. LNI, vol. P-331, pp. 511–533. Gesellschaft für Informatik e.V. (2023). https://doi.org/10.18420/BTW2023-24

39. Kosche, M., Koß, T., Manea, F., Pak, V.: Subsequences in bounded ranges: Matching and analysis problems. In: Lin, A.W., Zetzsche, G., Potapov, I. (eds.) Reachability Problems - 16th International Conference, RP 2022, Kaiserslautern, Germany, 17–21 October 2022, Proceedings. LNCS, vol. 13608, pp. 140–159. Springer (2022). https://doi.org/10.1007/978-3-031-19135-0_10

40. Kosche, M., Koß, T., Manea, F., Siemer, S.: Absent subsequences in words. In: Bell, P.C., Totzke, P., Potapov, I. (eds.) RP 2021. LNCS, vol. 13035, pp. 115–131. Springer, Cham (2021). https://doi.org/10.1007/978-3-030-89716-1_8

41. Kosche, M., Koß, T., Manea, F., Siemer, S.: Combinatorial algorithms for subsequence matching: A survey. In: Bordihn, H., Horváth, G., Vaszil, G. (eds.) Proceedings 12th International Workshop on Non-Classical Models of Automata and Applications, NCMA 2022, Debrecen, Hungary, 26–27 August 2022. EPTCS, vol. 367, pp. 11–27 (2022). https://doi.org/10.4204/EPTCS.367.2

42. Kuske, D.: The subtrace order and counting first-order logic. In: Fernau, H. (ed.) CSR 2020. LNCS, vol. 12159, pp. 289–302. Springer, Cham (2020). https://doi.org/10.1007/978-3-030-50026-9_21

43. Kuske, D., Zetzsche, G.: Languages ordered by the subword order. In: Bojańczyk, M., Simpson, A. (eds.) FoSSaCS 2019. LNCS, vol. 11425, pp. 348–364. Springer, Cham (2019). https://doi.org/10.1007/978-3-030-17127-8_20

44. Lejeune, M., Leroy, J., Rigo, M.: Computing the k-binomial complexity of the thue–morse word. In: Hofman, P., Skrzypczak, M. (eds.) DLT 2019. LNCS, vol. 11647, pp. 278–291. Springer, Cham (2019). https://doi.org/10.1007/978-3-030-24886-4_21

45. Leroy, J., Rigo, M., Stipulanti, M.: Generalized Pascal triangle for binomial coefficients of words. Electron. J. Combin. **24**(1.44), 36 (2017)

46. Li, C., Wang, J.: Efficiently mining closed subsequences with gap constraints. In: SDM, pp. 313–322. SIAM (2008)

47. Li, C., Yang, Q., Wang, J., Li, M.: Efficient mining of gap-constrained subsequences and its various applications. ACM Trans. Knowl. Discov. Data **6**(1), 2:1–2:39 (2012)

48. Maier, D.: The complexity of some problems on subsequences and supersequences. J. ACM **25**(2), 322–336 (1978)

49. Masek, W.J., Paterson, M.: A faster algorithm computing string edit distances. J. Comput. Syst. Sci. **20**(1), 18–31 (1980). https://doi.org/10.1016/0022-0000(80)90002-1

50. Mateescu, A., Salomaa, A., Yu, S.: Subword histories and Parikh matrices. J. Comput. Syst. Sci. **68**(1), 1–21 (2004)

51. Nakatsu, N., Kambayashi, Y., Yajima, S.: A longest common subsequence algorithm suitable for similar text strings. Acta Informatica **18**, 171–179 (1982). https://doi.org/10.1007/BF00264437

52. Riddle, W.E.: An approach to software system modelling and analysis. Comput. Lang. **4**(1), 49–66 (1979). https://doi.org/10.1016/0096-0551(79)90009-2

53. Rigo, M., Salimov, P.: Another generalization of abelian equivalence: Binomial complexity of infinite words. Theor. Comput. Sci. **601**, 47–57 (2015)

54. Salomaa, A.: Connections between subwords and certain matrix mappings. Theoret. Comput. Sci. **340**(2), 188–203 (2005)

55. Seki, S.: Absoluteness of subword inequality is undecidable. Theor. Comput. Sci. **418**, 116–120 (2012)

56. Shaw, A.C.: Software descriptions with flow expressions. IEEE Trans. Software Eng. **4**(3), 242–254 (1978). https://doi.org/10.1109/TSE.1978.231501

57. Simon, I.: Hierarchies of events with dot-depth one – Ph.D. thesis. University of Waterloo (1972)

58. Simon, I.: Piecewise testable events. In: Brakhage, H. (ed.) GI-Fachtagung 1975. LNCS, vol. 33, pp. 214–222. Springer, Heidelberg (1975). https://doi.org/10.1007/3-540-07407-4_23

59. Simon, I.: Words distinguished by their subwords (extended abstract). In: Proceedings of WORDS 2003, vol. 27, pp. 6–13. TUCS General Publication (2003)

60. Troníček, Z.: Common subsequence automaton. In: Champarnaud, J.-M., Maurel, D. (eds.) CIAA 2002. LNCS, vol. 2608, pp. 270–275. Springer, Heidelberg (2003). https://doi.org/10.1007/3-540-44977-9_28

61. Zetzsche, G.: The complexity of downward closure comparisons. In: Proceedings of ICALP 2016. LIPIcs, vol. 55, pp. 123:1–123:14 (2016)
62. Zhang, H., Diao, Y., Immerman, N.: On complexity and optimization of expensive queries in complex event processing. In: International Conference on Management of Data, SIGMOD 2014, Snowbird, UT, USA, 22–27 June 2014, pp. 217–228 (2014). https://doi.org/10.1145/2588555.2593671

On Substitutions Preserving Their Return Sets

Valérie Berthé and Herman Goulet-Ouellet[(⊠)]

IRIF, Université Paris Cité, 75013 Paris, France
{berthe,hgoulet}@irif.fr

Abstract. We consider the question of whether or not a given primitive substitution preserves its sets of return words—or *return sets* for short. More precisely, we study the property asking that the image of the return set to a word equals the return set to the image of that word. We show that, for bifix encodings (where images of letters form a bifix code), this property holds for all but finitely many words. On the other hand, we also show that every conjugacy class of Sturmian substitutions contains a member for which the property fails infinitely often. Various applications and examples of these results are presented, including a description of the subgroups generated by the return sets in the shift of the Thue–Morse substitution. Up to conjugacy, these subgroups can be sorted into strictly decreasing chains of isomorphic subgroups weaving together a simple pattern. This is in stark contrast with the Sturmian case, and more generally with the dendric case (including in particular the Arnoux–Rauzy case), where it is known that all return sets generate the free group over the underlying alphabet.

Keywords: Substitutions · Return words · Sturmian substitutions

Since the seminal work of Durand at the end of the 90s—including a generalization of Cobham's theorem [13], a characterization of sequences defined by primitive substitutions [14], and with Host and Skau, a description of the dimension groups of substitution dynamical systems via sequences of Kakutani–Rohlin partitions [15]—return words have proved to be a useful tool in combinatorics on words and symbolic dynamics. They have been used, among other things, to characterize Sturmian words [31], study maximal bifix codes [6,11], or else study the *Schützenberger group*, a topological/algebraic invariant of minimal shift spaces which takes the form of a projective profinite group [1–3].

A *return set* is a set of (first) return words to a given word in a given set of words, for instance the language of a substitution. The *Return Theorem*, a striking result by a group of authors which includes the first author of this paper, states that in a dendric shift space (a common generalization of episturmian shifts and codings of interval exchanges), every return set is a basis for the free group over the alphabet of the shift [10]. This kind of stable behavior for subgroups generated by return sets was observed in the more general

This work was supported by the Agence Nationale de la Recherche through the project "Codys" (ANR-18-CE40-0007).

© The Author(s), under exclusive license to Springer Nature Switzerland AG 2023
A. Frid and R. Mercaş (Eds.): WORDS 2023, LNCS 13899, pp. 77–90, 2023.
https://doi.org/10.1007/978-3-031-33180-0_6

class of suffix-connected shift spaces, introduced by the second author [19]. The behavior of return sets in a given shift space has also been linked, by Almeida and Costa, to the Schützenberger group; in particular, global stability of the subgroups generated by the return sets entails freeness of the Schützenberger group [3]. Together with results from [12,18], this can be used to show that dendric shift spaces are not topologically conjugate to shift spaces of primitive uniform substitutions—though the same conclusion can also be reached using other means [9].

In addition to the algebraic regularity from the Return Theorem, let us recall that Durand proved in [14] a combinatorial regularity property for return sets of substitutive shift spaces. His result uses *derived sequences*, which are obtained by recoding an infinite word with respect to some return set. Durand showed that having a finite number of derived sequences precisely characterizes substitutive sequences. Coming from a dynamical point of view, this might set some expectation that return sets in substitutive shift spaces should also show some algebraic regularity. However, looking at concrete examples quickly reveals that, outside of the well-charted dendric case, the subgroups generated by return sets can exhibit a more complicated behavior—and in fact, little is known about how these subgroups behave in general.

In a roundabout way, this was the initial motivation for the work presented in this paper, which started as a study of the subgroups generated by the return sets in the Thue–Morse shift. In this case the subgroups, far from all being equal, form instead several decreasing chains of subgroups, crossing and weaving into each other. Moreover, these chains can be completely described—details are given in Proposition 1 and in § 5—thanks to a simple preservation property which sparked our curiosity: with finite exceptions, the Thue–Morse substitution maps its return sets to other return sets. Our first main result (Theorem 1) states that this property holds for every primitive aperiodic *bifix* substitution (and by bifix, we mean that the letter images form a bifix code). Our second main result (Theorem 2) states that, on the other hand, it fails for primitive *Sturmian* substitutions, at least up to conjugacy.

1   Preliminaries

Throughout this paper, A is a finite alphabet, A^* is the set of words over A equipped with concatenation, and ε is the empty word. For a subset $B \subseteq A^*$, let B^* denote the submonoid of A^* generated by B. We use $u \cdot v$ as a shorthand for the pair (u, v), in particular when u and v are in A^*. We say that $s \cdot t$ *refines* $u \cdot v$ if u is a suffix of s and v is a prefix of t, i.e. $s \cdot t = s'u \cdot vt'$ for some $s', t' \in A^*$; we then write $u \cdot v \preceq s \cdot t$.

Let $A^{\mathbb{Z}}$ be the set of two-sided infinite word over A. Recall that a *shift space* is a subset $X \subseteq A^{\mathbb{Z}}$ invariant under the shift map $(x_n)_{n \in \mathbb{Z}} \mapsto (x_{n+1})_{n \in \mathbb{Z}}$ and its inverse, and closed for the product topology of $A^{\mathbb{Z}}$ (A being equipped with the discrete topology). The *language of X*, denoted $\mathcal{L}(X)$, is the set of all finite words occurring as factors (i.e. consecutive blocks) in the elements $x \in X$. A *left* *extension* of $u \in \mathcal{L}(X)$ is a word $x \in A^*$ such that $xu \in \mathcal{L}(X)$. Likewise, y is a *right extension* of u if $uy \in \mathcal{L}(X)$. We say that u is *left special* if it admits at least two left extensions of length 1, and *right special* if those are instead right extensions. We say that u is *bispecial* if it is both left and right special.

Given a pair $u \cdot v$ with $uv \in \mathcal{L}(X)$, a word $r \in A^*$ is a *return word* (sometimes called *first return word*) to $u \cdot v$ in X if $urv \in \mathcal{L}(X) \cap uvA^* \cap A^*uv$ and contains exactly two occurrences of uv. We denote by $\mathcal{R}_{u \cdot v}$ the set of return words to $u \cdot v$ in X (the shift space is easily inferred from context). Those are called the *return sets* of X. Note that this definition differs from the one—perhaps more common in the literature—where return words are *one-sided*; in our notation, this would correspond to return sets of the form $\mathcal{R}_{\varepsilon \cdot v}$ or $\mathcal{R}_{u \cdot \varepsilon}$. Concrete instances of return sets are given, for instance, in Example 3. The lemma below collects some standard properties of return sets that will be useful later.

Lemma 1 ([2,14,15]). *Let X be a shift space and $u \cdot v$ be a pair with $uv \in \mathcal{L}(X)$.*

(i) The set $\mathcal{R}_{u \cdot v}$ generates freely the submonoid $\mathcal{R}^*_{u \cdot v}$ of A^*, i.e. it is a code.
(ii) The submonoid $\mathcal{R}^*_{u \cdot v}$ contains $\{w \in \mathcal{L}(X) : uwv \in \mathcal{L}(X) \cap uvA^* \cap A^*uv\}$.
(iii) $\mathcal{R}_{u \cdot v} = u^{-1}\mathcal{R}_{\varepsilon \cdot uv}u = v\mathcal{R}_{uv \cdot \varepsilon}v^{-1}$.

The item (iii) above says that we can pass from two-sided to one-sided return sets and back simply by conjugating. This however should not be construed as saying that two-sided return sets are useless: there are circumstances in which the two-sided version plays an important role, as in [2,3,15]. In the present paper, it is helpful in handling some aspects of the proof of our first main result (Theorem 1).

Let $\sigma \colon A^* \to A^*$ be a primitive substitution. We denote by X_σ the two-sided shift space associated with σ (as per the standard definition; see e.g. [28, § 5] or [17, § 1.4]) and by $\mathcal{L}(\sigma)$ the language of X_σ. By primitivity, all the return sets in X_σ are finite. We say that σ is *aperiodic* if the shift space X_σ does not consist of a single finite orbit of the shift map. We say that σ is an *encoding* if σ is injective on A^*. In particular, encodings preserve codes [7, Corollary 2.1.6]. We say that σ is a *bifix encoding*, or simply *bifix*, if for all distinct letters $a \neq b \in A$, $\sigma(a)$ is neither prefix nor suffix of $\sigma(b)$. Every bifix substitution is an encoding [7, Proposition 2.1.9]. Finally, we say that σ is *left proper* (or *right proper*) if there is a letter $a_0 \in A$ such that $\sigma(a) \in a_0A^*$ (or $\sigma(a) \in A^*a_0$) for all $a \in A$.

A *Sturmian substitution* is a binary substitution, here defined on the alphabet $\{0, 1\}$, which preserves *Sturmian sequences*—aperiodic sequences in $\{0,1\}^{\mathbb{N}}$ of minimal factor complexity (see [23]). By [8, Theorem 5], this is the same as being *weakly Sturmian* (the image of *some* Sturmian sequence is Sturmian); in the primitive case, this is also the same as generating a Sturmian shift space. In particular, primitive Sturmian substitutions are aperiodic.

In the next example, we recall two important substitutions which will reappear several times in this paper.

Example 1. Consider the following primitive binary substitutions:

$$\tau \colon 0 \mapsto 01, 1 \mapsto 10 \quad \text{and} \quad \phi \colon 0 \mapsto 01, 1 \mapsto 0.$$

The former is known as the *Thue–Morse* substitution, and it is bifix and aperiodic; it is neither left nor right proper. The latter, the *Fibonacci* substitution, is Sturmian, hence aperiodic, and left proper; it is neither bifix nor right proper.

Let $F(A)$ be the free group over A. It consists, we recall, of the words over the alphabet $A \cup A^{-1}$ (where $A^{-1} = \{a^{-1} : a \in A\}$ is a disjoint copy of A) which are irreducible under the rewriting rules $aa^{-1} \to \varepsilon$ for all $a \in A$. There is a natural embedding $A^* \hookrightarrow F(A)$ which allows us to view words as elements of $F(A)$ and substitutions as endomorphisms of $F(A)$—and we do so whenever convenient. The endomorphisms of $F(A)$ that are also substitutions are called *positive*; the positive automorphisms of $F(\{0,1\})$ are precisely the Sturmian substitutions [26,32].

For a given subset $B \subseteq F(A)$, we let $\langle B \rangle$ denote the subgroup of $F(A)$ generated by B. Similar to codes in A^*, we call B *free* if the inclusion $B \hookrightarrow F(A)$ extends to an isomorphism $F(B) \cong \langle B \rangle$. We also say that B is a *basis* of the subgroup $\langle B \rangle$. In contrast with free monoids, every subgroup of the free group admits a basis (this is the celebrated Nielsen–Schreier theorem; see [25, Proposition 2.11]), and in fact infinitely many bases whenever it is not cyclic. We will be interested in the subgroups generated by the return sets of a shift space, dubbed the *return groups* of the shift space.

Example 2. If $X \subseteq A^{\mathbb{Z}}$ is a minimal dendric shift space, then all return groups are equal to $F(A)$, and moreover every return set viewed as a subset of $F(A)$ is free; this is the aforementioned Return Theorem [10, Theorem 4.5]. This applies in particular to Sturmian shift spaces, as these are all dendric. More generally, if X is suffix-connected, then it has only finitely many return groups which all belong to the same conjugacy class [19, Theorem 1.1].

2   Preservation of Return Words

Let σ be a primitive substitution over A. Given a pair $u \cdot v$ with $uv \in \mathcal{L}(\sigma)$, we consider the following preservation property:

$$\sigma(\mathcal{R}_{u \cdot v}) = \mathcal{R}_{\sigma(u) \cdot \sigma(v)}. \tag{P}$$

Let us state our two main results.

Theorem 1. *Let σ be a primitive aperiodic bifix substitution. There exists a constant $K > 0$ such that the property* (P) *holds for every pair $u \cdot v$ with $uv \in \mathcal{L}(X)$ and $|uv| \geq K$.*

Theorem 2. *Let σ be a primitive Sturmian substitution. There is a substitution in the conjugacy class of σ for which the property* (P) *fails for infinitely many pairs $u \cdot v$ with $uv \in \mathcal{L}(\sigma)$.*

The proofs are given later, in § 3 and § 4 respectively. The first main result will be used to describe the return groups in the Thue–Morse shift.

Proposition 1. *Up to conjugacy, all return groups in the shift space of the Thue–Morse substitution are equal to one of the following subgroups of $F(\{0, 1\})$:*

$$\langle \tau^n(0), \tau^n(1) \rangle, \quad \langle \tau^n(1), \tau^n(00), \tau^n(0110) \rangle, \quad \langle \tau^n(0), \tau^n(11), \tau^n(1001) \rangle,$$

with $n \geq 0$. Moreover, the given generating sets are bases of these subgroups.

The proof of Proposition 1 will be given in § 5. The remainder of this section explores various aspects of the property (P), starting with an example.

Example 3. For the Thue–Morse substitution $\tau: 0 \mapsto 01, 1 \mapsto 10$, the pair $0 \cdot \varepsilon$ fails the property (P) since

$$\mathcal{R}_{0 \cdot \varepsilon} = \{0, 10, 110\} \quad \text{and} \quad \mathcal{R}_{01 \cdot \varepsilon} = \{01, 001, 101, 1001\}.$$

On the other hand, $\mathcal{R}_{0 \cdot 1} = \{10, 100, 110, 1100\}$ and

$$\tau(\mathcal{R}_{0 \cdot 1}) = \{1001, 100101, 101001, 10100101\} = \mathcal{R}_{01 \cdot 10}.$$

In fact, every pair $u \cdot v$ with $uv \in \mathcal{L}(\tau)$ and $|uv| \geq 2$ satisfies the property (P). This is a consequence of the proof of Theorem 1 detailed in Proposition 4.

Next is a simple lemma that will be useful for the proof of Theorem 1.

Lemma 2. *Let σ be a primitive substitution which is also an encoding. A pair $u \cdot v$ with $uv \in \mathcal{L}(\sigma)$ satisfies the property (P) if and only if $\mathcal{R}_{\sigma(u) \cdot \sigma(v)} \subseteq \sigma(\mathcal{R}_{u \cdot v}^*)$.*

Proof. The inclusion $\sigma(\mathcal{R}_{u \cdot v}) \subseteq \mathcal{R}_{\sigma(u) \cdot \sigma(v)}^*$ always holds by Lemma 1 (ii). Since $\sigma(\mathcal{R}_{u \cdot v})$ and $\mathcal{R}_{\sigma(u) \cdot \sigma(v)}$ are both codes (by Lemma 1 (i)), they are equal if and only if they generate the same submonoid (by [7, Corollary 2.2.4]). □

Remark 1. In the previous lemma, we can replace the assumption that σ is an encoding by the local condition $\text{Card}\, \sigma(\mathcal{R}_{u \cdot v}) = \text{Card}\, \mathcal{R}_{\sigma(u) \cdot \sigma(v)}$.

The lemma below shows that the property (P) depends only on the word uv, rather than on the pair $u \cdot v$.

Lemma 3. *Let σ be a primitive substitution. A pair $u \cdot v$ with $uv \in \mathcal{L}(\sigma)$ satisfies the property (P) if and only if the pair $uv \cdot \varepsilon$ does.*

Proof. This is straightforward by Lemma 1 (iii); for instance, assuming that $u \cdot v$ satisfies the property (P), we get

$$\sigma(\mathcal{R}_{uv \cdot \varepsilon}) = \sigma(v)^{-1} \sigma(\mathcal{R}_{u \cdot v}) \sigma(v) = \sigma(v)^{-1} \mathcal{R}_{\sigma(u) \cdot \sigma(v)} \sigma(v) = \mathcal{R}_{\sigma(uv) \cdot \varepsilon}. \qquad \square$$

The next proposition gives a *sufficient* condition for a pair of words to fail the property (P). It is used to prove Theorem 2.

Proposition 2. *Let σ be a primitive substitution which is also left proper. Let $u \cdot v$ be a pair with $uv \in \mathcal{L}(\sigma)$. If $\sigma(uv)$ is right special, then $u \cdot v$ fails the property (P).*

Proof. Thanks to the previous lemma, we may assume that $v = \varepsilon$. Since σ is left proper, there is a letter $a_0 \in A$ such that $\sigma(\mathcal{R}_{u \cdot \varepsilon}) \subseteq a_0 A^*$. As $\sigma(u)$ is right special, it has a right extension $b \in A$ with $b \neq a_0$. Then, there is a return word $r \in \mathcal{R}_{\sigma(u) \cdot \varepsilon}$ such that $r \in bA^*$, and clearly $r \notin \sigma(\mathcal{R}_{u \cdot v})$. $\qquad\square$

Remark 2. In Proposition 2, the assumption that σ is left proper can be replaced by the local condition that uv is not right special.

Of course, similar statements hold with the left special property for right proper substitutions or under the local assumption that uv is not left special. We thus deduce the following result as a straightforward consequence of Theorem 1; to the best of our knowledge, this is new.

Corollary 1. *Let σ be a primitive aperiodic bifix substitution. There is a constant $K > 0$ such that, for every $u \in \mathcal{L}(\sigma)$ with $|u| \geq K$, u is left special, right special or bispecial whenever $\sigma(u)$ is. If σ is left proper (or right proper), then $\mathrm{Im}(\sigma) \cap \mathcal{L}(\sigma)$ contains only finitely many right special words (or left special words).*

Finally, we give an example illustrating failure of the property (P); it is a special case of Theorem 2.

Example 4. Consider the Fibonacci substitution $\phi \colon 0 \mapsto 01, 1 \mapsto 0$. Let $u \in \mathcal{L}(\phi)$ be right special and consider the word $v = u1$. We claim that $\phi(v)$ is right special. Indeed, we have $u0$ and $v0 \in \mathcal{L}(\phi)$, hence $\phi(u0) = \phi(v)1$ and $\phi(v1) = \phi(v)0$ both belong to $\mathcal{L}(\phi)$ as well. Since ϕ is left proper, it follows from Proposition 2 that ϕ fails the property (P) for all such words v. Take for instance $u = 0010$. Then, $v = 00101$ and $\sigma(v) = 01010010$ have for return sets

$$\mathcal{R}_{v \cdot \varepsilon} = \{00100101, 00101\} \quad \text{and} \quad \mathcal{R}_{\sigma(v) \cdot \varepsilon} = \{01010010, 10010\},$$

confirming that $v \cdot \varepsilon$ fails the property (P).

Remark 3. Example 3 shows that the property (P) may fail for reasons other than Proposition 2 and Remark 2, as τ is neither left nor right proper and 0 is bispecial in X_τ.

3   The Bifix Case

This section presents the proof of Theorem 1, which relies on the following notions and subsequent lemma. Let $\sigma \colon A^* \to A^*$ be a primitive substitution. A *parse* of $u \cdot v$ (under σ) is a pair $x \cdot y$ such that $xy \in \mathcal{L}(\sigma)$ and $\sigma(x) \cdot \sigma(y) \succeq u \cdot v$; we call *parsable* a pair that admits a parse. On the other hand, an *interpretation* of $u \in \mathcal{L}(\sigma)$ is a triple (s, w, t) such that $w \in \mathcal{L}(\sigma)$ and $\sigma(w) = sut$. A pair $u \cdot v$ is called *synchronizing* if, for every interpretation (s, w, t) of uv, there exists a pair $x' \cdot y'$ such that $w = x'y'$ and $\sigma(x') \cdot \sigma(y') = su \cdot vt$.

The next lemma is a reformulation of Mossé's celebrated recognizability theorem [27] which is due to Kyriakoglou [22, Proposition 3.3.20] (see also [17, Proposition 1.4.38]).

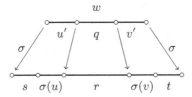

Fig. 1. An illustration of the proof of Theorem 1.

Lemma 4 ([22]). *Let σ be a primitive aperiodic substitution. There exists a constant $L > 0$ such that every parsable pair $u \cdot v$ with $|u|$ and $|v| \geq L$ is synchronizing.*

Proof (Theorem 1). Let $L > 0$ be the constant given by the previous lemma. Let us write

$$|\sigma| = \max\{|\sigma(a)| : a \in A\} \quad \text{and} \quad \langle\sigma\rangle = \min\{|\sigma(a)| : a \in A\}.$$

Let $M > 0$ be such that $\min\{|r| : r \in \mathcal{R}_{u \cdot v}\} \geq |\sigma|\lceil L/\langle\sigma\rangle\rceil$ for all pairs $u \cdot v$ with $uv \in \mathcal{L}(\sigma)$ and $|uv| \geq M$. Such a constant M exists by primitivity [14, Lemma 3.2]. Finally, let $K = \max\left(M, 2\lceil L/\langle\sigma\rangle\rceil\right)$.

Fix a pair $u \cdot v$ with $uv \in \mathcal{L}(\sigma)$ and $|uv| \geq K$. By Lemma 2, it suffices to show that the inclusion $\mathcal{R}_{\sigma(u) \cdot \sigma(v)} \subseteq \sigma(\mathcal{R}^*_{u \cdot v})$ holds. Fix a return word $r \in \mathcal{R}_{\sigma(u) \cdot \sigma(v)}$; we wish to show that $r \in \sigma(\mathcal{R}^*_{u \cdot v})$. Thanks to Lemma 3, we may assume that $|u|$ and $|v|$ are both $\geq \lfloor K/2 \rfloor$.

We start by showing that r admits a preimage under σ. Let (s, w, t) be an interpretation of $\sigma(u)r\sigma(v)$. Note that $\sigma(u) \cdot \sigma(v)$ is parsable with $|\sigma(u)|$ and $|\sigma(v)| \geq L$, therefore it is synchronizing. Since moreover $\sigma(u)r \cdot \sigma(v) \succeq \sigma(u) \cdot \sigma(v)$, there is a factorization $w = w'v'$ such that $\sigma(w') \cdot \sigma(v') = s\sigma(u)r \cdot \sigma(v)t$. Let p be the prefix of v of length $\lceil L/\langle\sigma\rangle\rceil$. Then $|\sigma(p)| \geq L$, and since $|r| \geq |\sigma|\lceil L/\langle\sigma\rangle\rceil$, $\sigma(p)$ is a prefix of r. The pair $\sigma(u) \cdot \sigma(p)$ is synchronizing, being parsable with $|\sigma(u)|$ and $|\sigma(p)| \geq L$. Since $\sigma(u) \cdot r \succeq \sigma(u) \cdot \sigma(p)$, there exists a factorization $w' = u'q$ such that $\sigma(u') = s\sigma(u)$ and $\sigma(q) = r$ (see Fig. 1).

We finish by showing that $q \in \mathcal{R}^*_{u \cdot v}$. For this purpose, recall that bifix codes generate submonoids that are biunitary [7, Proposition 2.2.7]. In the case at hand, it means that the two following properties hold:

$$x, xy \in \mathrm{Im}(\sigma) \implies y \in \mathrm{Im}(\sigma) \quad \text{and} \quad x, yx \in \mathrm{Im}(\sigma) \implies y \in \mathrm{Im}(\sigma).$$

Applying these properties to $\sigma(u') = s\sigma(u)$ and $\sigma(v') = \sigma(v)t$ respectively yields words s', t' such that $u' = s'u$ and $v' = vt'$. In particular, $uqv \in \mathcal{L}(\sigma)$. Since $r \in \mathcal{R}_{\sigma(u) \cdot \sigma(v)}$, there are words x, y that $\sigma(u)r\sigma(v) = \sigma(uv)x = y\sigma(uv)$. Applying the above properties once again, this time to $\sigma(uqv) = \sigma(uv)x = y\sigma(uv)$, produces words x', y' such that $uqv = uvx' = y'uv$. By Lemma 1 (ii), $q \in \mathcal{R}^*_{u \cdot v}$. $\square$

For instance, in the case of the Thue–Morse substitution, the constant K from the proof above equals 2 (see Proposition 4). Using a computability result of Durand and Leroy, we also deduce the following.

Proposition 3. *The constant K of Theorem 1 is computable.*

Proof. The constant L of Lemma 4 is related in a straightforward way to the recognizability constant (see the proof of [22, Proposition 3.3.20]), which is computable by a result of Durand and Leroy [16]. As for the constant M, note that it can be set equal to any $m \in \mathbb{N}$ such that $\min\{|r| : r \in \mathcal{R}_{w \cdot \varepsilon}\} \geq |\sigma| \lceil L/\langle \sigma \rangle \rceil$ for all w in the finite set $\mathcal{L}(\sigma) \cap A^m$. Since the language of a primitive substitution and its return sets are computable, this property is decidable for each natural number, and therefore the constant M is computable.   □

4   The Sturmian Case

In preparation for the proof of Theorem 2, let us set up some notation and recall a handful of facts about Sturmian substitutions. Given a pair $u \cdot v$ of words in $\{0,1\}^*$, we let $[u \cdot v]$ be the substitution $0 \mapsto u, 1 \mapsto v$. We say that two substitutions σ and ρ defined on the same alphabet A are *rotationally conjugate*, or simply *conjugate*, if there is a word $w \in A^*$ such that either $\sigma(a)w = w\rho(a)$ for all $a \in A$, or $w\sigma(a) = \rho(a)w$ for all $a \in A$. Then, we write respectively $\sigma = w^{-1}\rho w$ or $\sigma = w\rho w^{-1}$.

Consider the set of *standard pairs*: it is the smallest subset of $\{0,1\}^* \times \{0,1\}^*$ which contains $\{0 \cdot 1, \ 1 \cdot 0\}$ and which is closed under $u \cdot v \mapsto u \cdot uv$ and $u \cdot v \mapsto vu \cdot v$. Substitutions of the form $[u \cdot v]$ with $u \cdot v$ standard are called *standard substitutions*. Every standard (binary) substitution is Sturmian [5, Proposition 2.6]. If σ is a Sturmian substitution, then its conjugacy class contains $|\sigma(01)| - 1$ substitutions, all of which are Sturmian and one of which is standard [29, Propositions 9 and 10]. Moreover, all the primitive Sturmian substitutions in the same conjugacy class share the same shift space [29, Lemma 8].

Proof (Theorem 2). Let $\sigma = [x \cdot y]$ be a primitive standard substitution whose conjugacy class is given by

$$\sigma_0 = \sigma, \quad \sigma_1 = a_0^{-1}\sigma_0 a_0, \quad \sigma_2 = a_1^{-1}\sigma_1 a_1, \quad \cdots \quad \sigma_n = a_{n-1}^{-1}\sigma_{n-1}a_{n-1},$$

with $n = |x| + |y| - 1$, and where $\sigma_i(0)$ and $\sigma_i(1)$ both start with the letter a_i for $0 \leq i < n$. Let $\sigma_i = [x_i \cdot y_i]$. Let E be the automorphism of $\{0,1\}^*$ exchanging 0 and 1 and consider the substitutions $\sigma_i^E = E \circ \sigma_i \circ E$. They also form a Sturmian conjugacy class where $\mathcal{L}(\sigma_i^E) = E(\mathcal{L}(\sigma_i))$. Thus, up to replacing σ by σ^E if needed, we may assume that $x = ys$ for some word s, which is non-empty since σ is not periodic. Next, note that x_n and y_n start with distinct letters by [29, Lemma 5]; in particular, y_n is not a prefix of x_n. Since y_0 is a prefix of x_0, we may let j be the largest index satisfying $0 \leq j < n$ such that y_j is a prefix of x_j but y_{j+1} is not a prefix of x_{j+1}. Write $x_j = y_j s_j$ and let b_j be the first letter of s_j; the choice of j implies that $a_j \neq b_j$.

Next, observe that there exists $\ell \geq 0$ such that $01^\ell 0$ and $01^{\ell+1}0 \in \mathcal{L}(\sigma)$ with these being the only two factors of the form $01^k 0$ in $\mathcal{L}(\sigma)$. (Note that ℓ may be equal to 0, i.e. the letter 1 does not need to be the most frequent letter.)

In particular, 1^ℓ is right special and admits 0 and 10 as right extensions. Let u be a right special factor in X_σ. Since the right special factors in X_σ are all suffixes of one another, either $u = 1^k$ with $k < \ell$ or 1^ℓ is a suffix of u. Either way, it follows that $u10 = v0 \in \mathcal{L}(\sigma)$. Since σ_j is left proper, by Proposition 2, we are done if we can show that $\sigma_j(v)$ is right special. But note that

$$\sigma_j(u0) = \sigma_j(u)y_j s_j = \sigma_j(v)s_j \in \mathcal{L}(\sigma) \quad \text{and} \quad \sigma_j(v0) = \sigma_j(v)x_j \in \mathcal{L}(\sigma).$$

Hence, $\sigma_j(v)a_j$ and $\sigma_j(v)b_j \in \mathcal{L}(\sigma)$. Since $a_j \neq b_j$, this shows that $\sigma_j(v)$ is indeed right special.   □

The next example illustrates the construction in the above proof.

Example 5. Let $\sigma = [1101 \cdot 110]$. It is a primitive standard substitution with conjugacy class

$$\sigma_0 = [1101 \cdot 110], \quad \sigma_1 = [1011 \cdot 101], \quad \sigma_2 = [0111 \cdot 011],$$
$$\sigma_3 = [1110 \cdot 110], \quad \sigma_4 = [1101 \cdot 101], \quad \sigma_5 = [1011 \cdot 011].$$

Here, σ_3 is the first member of the conjugacy class whose second component is not a prefix of the first, so $j = 2$. Moreover, 0110 and 01110 $\in \mathcal{L}(\sigma)$, so $\ell = 2$.

According to the argument above, if u is right special, then $v = u1$ is such that $\sigma_2(v)$ is also right special. And indeed, $\sigma_2(v0) = \sigma_2(v)011$ and $\sigma_2(u01) = \sigma_2(v)1011$ both belong to $\mathcal{L}(\sigma)$. Since σ_2 is left proper, we can apply Proposition 2 to conclude that v fails the property (P) with respect to σ_2.

For a concrete example, take $u = 11$. In this case, $v = 111, \sigma_2(v) = 011011011$, and the fact that v fails the property (P) is apparent from the equalities

$$\mathcal{R}_{v \cdot \varepsilon} = \{0110110111, 0110110110111\}, \quad \mathcal{R}_{\sigma_2(v) \cdot \varepsilon} = \{011, 1011011011\}.$$

5   Return Words of Thue–Morse Substitution

In this section, we give precise formulas for the return sets in the shift of the Thue–Morse substitution $\tau: 0 \mapsto 01, 1 \mapsto 10$ and proceed to give the proof of Proposition 1. Our starting point is the following simple proposition, which is a consequence of the proof of Theorem 2.

Proposition 4. *Every pair $u \cdot v$ with $uv \in \mathcal{L}(\tau)$ and $|uv| \geq 2$ satisfies the property (P) with respect to τ.*

Proof. First, we claim that the constant L from Lemma 4 can be set to $L = 2$. To establish this, it suffices to show that every parsable pair $u \cdot v$ with $|u| = |v| = 2$ is synchronizing for τ. There are four such pairs, namely, $01 \cdot 10$, $10 \cdot 01$, $01 \cdot 01$ and $10 \cdot 10$. That the first two pairs are synchronizing is obvious: otherwise, 11 and 00 would be images of letters, and this is not the case. If, say, $01 \cdot 01$ was not synchronizing, then there would be a pair $x \cdot y$ such that $010 \cdot 1 \preceq x \cdot y$, $xy \in \mathcal{L}(\tau)$ and $x, y \in \operatorname{Im}(\tau)$. This would then imply $1010 \cdot 10 \preceq x \cdot y$, and therefore $101010 \in \mathcal{L}(\tau)$, a contradiction. The fact that $10 \cdot 10$ is synchronizing is proved similarly. This proves the claim. Finally, since $L = |\tau| = \langle \tau \rangle = 2$, we find that $R(|\tau|\lceil L/\langle \tau \rangle \rceil) = R(2) = 2$, so $K = \max(2, 2) = 2$.   □

Table 1. Return sets in X_τ.

uv	$u \cdot v$	$\mathcal{R}_{u \cdot v}$
w_n	$w_{n-1} \cdot \overline{w}_{n-1}$	$\{\tau^n(10),\ \tau^n(110),\ \tau^n(100),\ \tau^n(1100)\}$
$\overline{w}_n$	$\overline{w}_{n-1} \cdot w_{n-1}$	$\{\tau^n(01),\ \tau^n(001),\ \tau^n(011),\ \tau^n(0011)\}$
z_n	$w_n \cdot w_{n-1}$	$\{\tau^n(001),\ \tau^n(01101),\ \tau^n(0011001),\ \tau^n(011001101)\}$
$\overline{z}_n$	$\overline{w}_n \cdot \overline{w}_{n-1}$	$\{\tau^n(110),\ \tau^n(10010),\ \tau^n(1100110),\ \tau^n(100110010)\}$

As noted in Balková et al. [4, § 3.1], every return set is in fact determined by a bispecial pair—a pair $u \cdot v$ such that uv is bispecial. To see why, fix a minimal aperiodic shift space $X \subseteq A^{\mathbb{Z}}$ and consider a pair $u \cdot v$ with $uv \in \mathcal{L}(X)$. If $u \cdot v$ is not left special, then we may replace it by $au \cdot v$ where $a \in A$ is the unique left extension of uv of length 1 in $\mathcal{L}(X)$, and this does not change the return set. A similar phenomenon occurs with the unique right extension $b \in A$ of uv of length 1 when uv is not right special. We can therefore consider the minimal refinement $s \cdot t \succeq u \cdot v$ which is bispecial; let us call this the *bispecial closure* of $u \cdot v$. What Balková et al. observed is stated (with an extra maximality property) in the next lemma.

Lemma 5. *Let $X \subseteq A^{\mathbb{Z}}$ be a minimal aperiodic shift space and $u \cdot v$ be a pair with $uv \in \mathcal{L}(X)$. The bispecial closure of $u \cdot v$ is the greatest refinement $s \cdot t \succeq u \cdot v$ with $st \in \mathcal{L}(X)$ such that $\mathcal{R}_{s \cdot t} = \mathcal{R}_{u \cdot v}$.*

Next, we recall the classification of the bispecial factors in the Thue–Morse shift given by de Luca and Mione in 1994 [24]. Let us mention in passing that the idea behind de Luca and Mione's result has been significantly generalized by Klouda, who gave an algorithm for computing similar classifications for any primitive aperiodic substitution [21].

Proposition 5 ([24]). *Every bispecial factor of length ≥ 2 in X_τ is equal to one of the following words, for some $n \geq 0$:*

$$w_n = \tau^n(01), \quad \overline{w}_n = \tau^n(10), \quad z_n = \tau^n(010), \quad \overline{z}_n = \tau^n(101).$$

Let also $w_{-1} = 0$ and $\overline{w}_{-1} = 1$. Note that $w_n = \tau(w_{n-1})$, $z_n = \tau(z_{n-1})$, and so on. Note moreover the following factorizations for $n \geq 0$:

$$w_n = w_{n-1}\overline{w}_{n-1}, \quad z_n = w_n w_{n-1}, \quad \overline{w}_n = \overline{w}_{n-1}w_{n-1}, \quad \overline{z}_n = \overline{w}_n \overline{w}_{n-1}.$$

Thanks to Proposition 4, computing the return sets to the bispecial factors is as simple as computing them for w_0, z_0, $\overline{w}_0$ and $\overline{z}_0$, and taking direct images by τ^n. Table 1 gives the return sets to each of the pairs given by the factorizations above.

Next, we proceed to describe the return groups as stated in Proposition 1, again noting that we only need to consider those that are determined by bispecial pairs. For this purpose, it is useful to have in mind the following lemma.

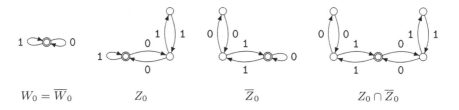

Fig. 2. Stallings automata of the subgroups W_0, $\overline{W}_0$, Z_0, $\overline{Z}_0$ and $Z_0 \cap \overline{Z}_0$.

Lemma 6. *The Thue–Morse substitution $\tau\colon 0 \mapsto 01, 1 \mapsto 10$ viewed as an endomorphism of $F(\{0,1\})$ is injective.*

A quick and easy way to establish this is to show that $\{01, 10\}$ is a free subset of $F(\{0,1\})$, which is indeed equivalent to τ being injective. This can be done using *Stallings' algorithm*, a powerful algorithm that we also use in the proof of Proposition 1 below. This algorithm, given a finitely generated subgroup of a free group, produces an automaton, known as the *Stallings automaton* of the subgroup, which allows to easily test for membership. It can also be used to test whether a subset is a basis of the subgroup, and has numerous other applications beyond (calculate subgroup indexes, test for conjugacy, compute intersections, etc.). See the paper by Kapovich and Myasnikov [20] for a detailed description of Stallings' algorithm and many of its applications; see also the paper by Touikan [30] for an efficient implementation of the algorithm.

Proof (Proposition 1). Consider the following return groups in X_τ:

$$W_n = \langle \mathcal{R}_{w_{n-1} \cdot \overline{w}_{n-1}} \rangle, \quad Z_n = \langle \mathcal{R}_{w_n \cdot w_{n-1}} \rangle,$$
$$\overline{W}_n = \langle \mathcal{R}_{\overline{w}_{n-1} \cdot w_{n-1}} \rangle, \quad \overline{Z}_n = \langle \mathcal{R}_{\overline{w}_n \cdot \overline{w}_{n-1}} \rangle.$$

By Lemma 5 and Proposition 5, every return set $\mathcal{R}_{u \cdot v}$ with $uv \in \mathcal{L}(\tau)$ and $|uv| \geq 2$ is conjugate to one of these. It remains to show that these subgroups have the bases stated in the proposition. First, we claim that, for all $n \geq 0$, $\{\tau^n(0), \tau^n(1)\}$ forms a basis of W_n and $\{\tau^n(1), \tau^n(00), \tau^n(0110)\}$ forms a basis of Z_n. This can be checked directly in the case $n = 0$ using Stallings' algorithm (the relevant Stallings automata are found in Fig. 2) and it then holds for all $n \geq 0$ thanks to Proposition 5 and Lemma 6.

To finish, let E be the automorphism exchanging 0 and 1, viewed as an automorphism of $F(\{0,1\})$. On the one hand, it is clear that $W_n = E(W_n) = \overline{W}_n$, which shows that the return groups $\overline{W}_n$ are redundant. On the other hand, $E(Z_n) = \overline{Z}_n$ and thus $\overline{Z}_n$ have the required basis for all $n \geq 0$. This concludes the proof. $\qquad\square$

Consider the following sequences of subgroups:

$$\boldsymbol{W} = (W_i)_{i \in \mathbb{N}} = (\overline{W}_i)_{i \in \mathbb{N}}, \quad \boldsymbol{Z}_{\text{even}} = (Z_{2i})_{i \in \mathbb{N}}, \quad \boldsymbol{Z}_{\text{odd}} = (Z_{2i+1})_{i \in \mathbb{N}},$$
$$\overline{\boldsymbol{Z}}_{\text{even}} = (\overline{Z}_{2i})_{i \in \mathbb{N}}, \quad \overline{\boldsymbol{Z}}_{\text{odd}} = (\overline{Z}_{2i+1})_{i \in \mathbb{N}}.$$

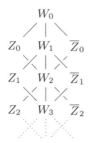

Fig. 3. Hasse diagram of return groups of X_τ ordered by inclusion.

Up to conjugacy, every return group in the Thue–Morse shift occurs in one (and only one) of these sequences. It is not hard to see that they are all strictly decreasing with respect to inclusion. The next proposition clarifies how these sequences are related to one another. The proof is a straightforward application of Stallings algorithm; all relevant Stallings automata are found in Fig. 2.

Proposition 6. *For all $n \geq 0$, $W_n = \langle Z_n \cup \overline{Z}_n \rangle > Z_n \cap \overline{Z}_n > W_{n+2}$.*

The Hasse diagram of the poset formed by the return groups $W_n = \overline{W}_n$, Z_n and $\overline{Z}_n$ ordered by inclusion is depicted in Fig. 3.

6   Conclusion

The statement of Theorem 2 is somewhat unsatisfactory, in the sense that we are left wondering what happens with the other Sturmian substitutions within the conjugacy class. Based on examples, we expect all Sturmian substitutions to satisfy the conclusion of Theorem 2. It is not clear at the moment how generalizations, like dendric substitutions, might behave.

On the other hand, Theorem 1 shows that being bifix is a sufficient condition for having the property (P) for all but finitely many pairs, but we are not sure whether or not it is necessary. We would like to know if other natural classes of primitive substitutions satisfy the property (P) for all but finitely many pairs (clearly, by Theorem 2, such classes cannot contain the class of primitive Sturmian substitutions). It might also be interesting to investigate the property (P) for non-primitive substitutions.

Finally, we wonder whether or not the sophisticated pattern formed by the return groups of the Thue–Morse shift is common among primitive aperiodic bifix substitutions. While classifications of bispecial factors can always be obtained in the primitive case thanks to the aforementioned algorithm of Klouda, the return preservation property might not suffice to determine all return groups, since the general form of Klouda's algorithm involves taking more than direct images—it uses what Klouda calls f_B-*images*, which in the case of the Thue–Morse substitution reduces to the direct image. In this regard, the Thue–Morse substitution seems to be quite special.

References

1. Almeida, J.: Profinite groups associated with weakly primitive substitutions. J. Math. Sci. **144**(2), 3881–3903 (2007). https://doi.org/10.1007/s10958-007-0242-y, translated from Fundam. Prikl. Mat., **11**(3), 13–48 (2005)

2. Almeida, J., Costa, A.: Presentations of Schützenberger groups of minimal subshifts. Israel J. Math. **196**(1), 1–31 (2013). https://doi.org/10.1007/s11856-012-0139-4

3. Almeida, J., Costa, A.: A geometric interpretation of the Schützenberger group of a minimal subshift. Ark. Math. **54**(2), 243–275 (2016). https://doi.org/10.1007/s11512-016-0233-7

4. Balková, L., Pelantová, E., Steiner, W.: Sequences with constant number of return words. Monatsh. Math. **155**(3–4), 251–263 (2008). https://doi.org/10.1007/s00605-008-0001-2

5. Berstel, J., Séébold, P.: A remark on morphic Sturmian words. RAIRO - Theor. Inform. Appl. **28**(3–4), 255–263 (1994). https://doi.org/10.1051/ita/1994283-402551

6. Berstel, J., De Felice, C., Perrin, D., Reutenauer, C., Rindone, G.: Bifix codes and Sturmian words. J. Algebra **369**, 146–202 (2012). https://doi.org/10.1016/j.jalgebra.2012.07.013

7. Berstel, J., Perrin, D., Reutenauer, C.: Codes and Automata. Cambridge University Press (2009). https://doi.org/10.1017/cbo9781139195768

8. Berstel, J., Séébold, P.: A characterization of Sturmian morphisms. In: Borzyszkowski, A., Sokołowski, S. (eds.) Lecture Notes in Computer Science, vol. 711, pp. 281–290. Springer, Berlin Heidelberg (1993). https://doi.org/10.1007/3-540-57182-5_20

9. Berthé, V., Dolce, F., Durand, F., Leroy, J., Perrin, D.: Rigidity and substitutive dendric words. Int. J. Found. Comput. **29**(05), 705–720 (2018). https://doi.org/10.1142/S0129054118420017

10. Berthé, V., et al.: Acyclic, connected and tree sets. Monatsh. Math. **176**(4), 521–550 (2015). https://doi.org/10.1007/s00605-014-0721-4

11. Berthé, V., et al.: Maximal bifix decoding. Discrete Math. **338**(5), 725–742 (2015). https://doi.org/10.1016/j.disc.2014.12.010

12. Costa, A.: Conjugacy invariants of subshifts: An approach from profinite semigroup theory. Int. J. Algebra Comput. **16**(4), 629–655 (2006). https://doi.org/10.1142/s0218196706003232

13. Durand, F.: A generalization of Cobham's theorem. Theoret. Comput. Sci. **31**(2), 169–185 (1998). https://doi.org/10.1007/s002240000084

14. Durand, F.: A characterization of substitutive sequences using return words. Discrete Math. **179**(1–3), 89–101 (1998). https://doi.org/10.1016/S0012-365X(97)00029-0

15. Durand, F., Host, B., Skau, C.: Substitutional dynamical systems, Bratteli diagrams and dimension groups. Ergod. Theory Dyn. Syst. **19**(4), 953–993 (1999). https://doi.org/10.1017/S0143385799133947

16. Durand, F., Leroy, J.: The constant of recognizability is computable for primitive morphisms. J. Integer S. **20** (2017)

17. Durand, F., Perrin, D.: Dimension Groups and Dynamical Systems. Cambridge Studies in Advanced Mathematics, Cambridge University Press (2022). https://doi.org/10.1017/9781108976039

18. Goulet-Ouellet, H.: Pronilpotent quotients associated with primitive substitutions. J. Algebra **606**, 341–370 (2022). https://doi.org/10.1016/j.jalgebra.2022.05.021
19. Goulet-Ouellet, H.: Suffix-connected languages. Theoret. Comput. Sci. **923**, 126–143 (2022). https://doi.org/10.1016/j.tcs.2022.05.001
20. Kapovich, I., Myasnikov, A.: Stallings foldings and subgroups of free groups. J. Algebra **248**(2), 608–668 (2002). https://doi.org/10.1006/jabr.2001.9033
21. Klouda, K.: Bispecial factors in circular non-pushy D0L languages. Theoret. Comput. Sci. **445**, 63–74 (2012). https://doi.org/10.1016/j.tcs.2012.05.007
22. Kyriakoglou, R.: Iterated morphisms, combinatorics on words and symbolic dynamical systems. Ph.D. thesis, Université Paris-Est (2019)
23. Lothaire, M.: Algebraic combinatorics on words, Encyclopedia of Mathematics and its Applications, vol. 90. Cambridge University Press, Cambridge (2002). https://doi.org/10.1017/CBO9781107326019
24. de Luca, A., Mione, L.: On bispecial factors of the Thue-Morse word. Inf. Process. Lett. **49**(4), 179–183 (1994). https://doi.org/10.1016/0020-0190(94)90008-6
25. Lyndon, R.C., Schupp, P.E.: Combinatorial Group Theory. Springer, Berlin Heidelberg (2001). https://doi.org/10.1007/978-3-642-61896-3
26. Mignosi, F., Séébold, P.: Morphismes sturmiens et règles de Rauzy. J. Théor. Nr. Bordx. **5**(2), 221–233 (1993). https://doi.org/10.5802/jtnb.91
27. Mossé, B.: Puissance de mots et reconnaissabilité des points fixes d'une substitution. Theoret. Comput. Sci. **99**(2), 327–334 (1992). https://doi.org/10.1016/0304-3975(92)90357-L
28. Queffélec, M.: Substitution dynamical systems–spectral analysis, Lecture Notes in Mathematics, vol. 1294. Springer-Verlag, Berlin, second edn. (2010). https://doi.org/10.1007/978-3-642-11212-6
29. Séébold, P.: On the conjugation of standard morphisms. Theoret. Comput. Sci. **195**(1), 91–109 (1998). https://doi.org/10.1016/s0304-3975(97)00159-x
30. Touikan, N.W.M.: A fast algorithm for Stallings' folding process. Int. J. Algebra Comput. **16**(6), 1031–1045 (2006). https://doi.org/10.1142/S0218196706003396
31. Vuillon, L.: A characterization of Sturmian words by return words. Europ. J. Comb. **22**(2), 263–275 (2001). https://doi.org/10.1006/eujc.2000.0444
32. Wen, Z.X., Wen, Z.Y.: Local isomorphisms of the invertible substitutions. C. R. Acad. Sci. Paris **318**, 299–304 (1994)

Recurrence and Frequencies

Valérie Berthé[1][(✉)] and Ahmed Mimouni[2]

[1] IRIF, Université Paris Cité, 75013 Paris, France
berthe@irif.fr
[2] LACL, Université Paris-Est-Créteil, 94000 Créteil, France
amimouni@lacl.fr

Abstract. Given an infinite word with values in a finite alphabet that admit frequencies for all its factors, the smallest frequency function associates with a given positive integer the smallest frequency for factors of this length, and the recurrence function measures the gaps between successive occurrences of factors. We develop in this paper the relations between the growth orders of the smallest frequency and of the recurrence functions. As proved by M. Boshernitzan, linearly recurrent words are characterized in terms of the smallest frequency function. In this paper, we see how to extend this result beyond the case of linearly recurrent words. We first establish a general relation between the smallest frequency and the recurrence functions. Given a lower bound for the smallest frequency function, this relation provides an upper bound for the recurrence function which involves the primitive of this lower bound. We then see how to improve this relation in the case of Sturmian words where the product of the smallest frequency and the recurrence function is bounded for most of the lengths of factors. We also indicate how to construct Sturmian words having a prescribed behaviour for the smallest frequency function, and for the product of the smallest frequency and the recurrence function.

Keywords: Recurrence function · Frequencies · Smallest frequency function · Factor complexity · Sturmian words

1 Introduction

There exist various combinatorial natural measures of disorder that occur simultaneously in word combinatorics, symbolic dynamics, and in the context of aperiodic order. Aperiodic order refers to the mathematical formalization of quasicrystals, i.e., solids with an aperiodic atom structure. This involves different mathematical objects, such as tilings, cut-and-project sets, Delone point sets, or infinite words over a finite alphabet, and various functions allow the quantification of order. For words, these notions involve factors, and for tilings and point sets, patches or patterns. In the case of infinite words (which is the case on which we focus here), it is natural to count how many different factors exist (factor complexity), whether or not they reappear (recurrence), and how often they occur (frequencies and shift invariant measures).

This work was supported by the Agence Nationale de la Recherche through the project "Codys" (ANR-18-CE40-0007).

© The Author(s), under exclusive license to Springer Nature Switzerland AG 2023
A. Frid and R. Mercaş (Eds.): WORDS 2023, LNCS 13899, pp. 91–103, 2023.
https://doi.org/10.1007/978-3-031-33180-0_7

More precisely, given an infinite word x with values in a finite alphabet, the *factor complexity* p_x counts the number of factors of a given size of an infinite word. The *recurrence function* R_x measures the gaps between successive occurrences of factors, i.e., how often they come back. It is also called repetitivity in the tiling setting, such as introduced in the seminal paper [15], see also [16]. Lastly, the frequency of a factor w in the infinite word x is defined as the limit when n tends towards infinity, if it exists, of the number of occurrences of w in $u_0 u_1 \cdots u_{n-1}$ divided by n. If the infinite word x admits frequencies for all factors, we then consider the *smallest frequency function* e_x for factors of a given length.

The relevance in this context of the quantity $ne_x(n)$ has been highlighted by M. Boshernitzan. He introduced indeed in [6] the so-called ne_n condition; see also [8, Chap. 7] for more details and Theorem 1 below. In a nutshell, $\limsup ne_x(n) > 0$ implies that frequencies are uniform (in dynamical terms, one has unique ergodicity), and $\liminf ne_x(n) > 0$ is equivalent to linear recurrence, which means that there exists a constant $c > 0$ such that factors of length n in x occur in any factor of length cn in x. This is an important property in the context of aperiodc order as described in the survey [1]; see also [11].

In this paper, we see how to extend the strategy developed by M. Boshernitzan for his characterization of linear recurrence, by exploring the relations between these notions. This illustrates the fact that recurrence and frequencies are closely related. We then focus on the Sturmian case which allows to produce examples of infinite words with a prescribed lower bound on e_x.

Basic notions and general statements relating recurrence, frequencies and factor complexity are recalled in Sect. 2. Our main result relating frequencies and recurrence, namely Theorem 2, is stated in Sect. 3. We then focus on the case of Sturmian words in Sect. 4. We provide conditions on the sequence of partial quotients in order to get prescribed behaviour for the smallest frequency function (see Theorem 5 and 6). We also prove that the product $e_x R_x$ takes bounded values for most integer values (see Theorem 7), even in the case where the partial quotients in the continued fraction expansion of the angle of the Sturmian word takes non-bounded values.

2   Basic Notions

Let $\mathcal{A}$ be an alphabet and let $\mathcal{A}^*$ stand for the set of all (finite) words over $\mathcal{A}$. Let x be an infinite word in $\mathcal{A}^{\mathbb{N}}$. Let $L_x(n)$ stand for the set of factors of length n of x, and L_x for the set of all its factors. The word x is said to be *uniformly recurrent* if every factor appears infinitely often and with bounded gaps (or, equivalently, if for every integer n, there exists an integer m such that every factor of x of length m contains every factor of length n). The recurrence functions R_x is then defined as

$$R_x(n) := \min\{m : \forall w \in L_x(m), L_x(n) \text{ is included in the set of factors of } w\}$$

and the infinite word x is then said to be *linearly recurrent* if there exists a positive integer $c > 0$ such that for every positive integer n, one has

$$R_x(n) \le cn.$$

We then define $\mu_x(w)$ as the *frequency* of the word $w \in L_x$ in x, if it exists:

$$\mu_x(w) = \lim_{n \to \infty} \frac{|x_0 x_1 \cdots x_n|_w}{n+1}.$$

If every factor of x has a frequency, we will say that x has frequencies.

If x has frequencies, then x is said to have *uniform frequencies* if for every w, and for every k, $\frac{|x_k x_{k+1} \cdots x_{k+n}|_w}{n+1}$ tends with n to $\mu_x(w)$ uniformly in k.

The *smallest frequency* function $e_x(n)$ is then defined as

$$e_x(n) := \min_{w \in L_x(n)} \mu_x(w).$$

Note that every linearly recurrent word has (uniform) frequencies, by [9]. However there exist uniformly recurrent words that do not admit frequencies (see e.g. [14] for examples of codings of interval exchanges on four letters).

The following proposition is usually stated in terms of shift-invariant measures; see e.g. [8, Chapter 7].

Proposition 1. *Let x be an infinite word that has frequencies. Let $E \subseteq L_x$ be a set of words such that no element of E is prefix of another element of E. We then have*

$$\sum_{w \in E} \mu_x(w) \le 1.$$

Proof. Suppose that E has at least two elements (otherwise the proof comes easily). Suppose that $\sum_{w \in E} \mu_x(w) > 1$. Then there exists a positive integer n such that $\sum_{w \in E} \frac{|x_0 \cdots x_n|_w}{n+1} > 1$, i.e., $\sum_{w \in E} |x_0 \cdots x_n|_w > n + 1$. Then the pigeonhole principle gives us the existence of a positive integer i, and of v, v' in E, such that v and v' appear at position i in x. That would mean that v is a prefix of v', or the other way around, contradicting the hypothesis.

The followng result relates in a natural way recurrence and factor complexity.

Proposition 2 [18]. *Let x be an infinite word. One has*

$$R_x(n) \ge p_x(n) + n - 1 \text{ for all } n.$$

We now state a first bridge between recurrence and frequencies, by splitting words into smaller words in which we are sure to find some fixed factor.

Proposition 3. *Let x be an infinite word with frequencies. One has*

$$e_x(n) \ge \frac{1}{R_x(n)} \text{ for all } n. \tag{1}$$

Proof. Let us show that for every factor w of x of size m, $\mu_x(w) \geq \frac{1}{R_x(|w|)}$. Let n be given. We split the prefix $x_0 \cdots x_n$ of x into smaller factors of size $R_x(|w|)$, in which we must have at least one occurrence of w, by definition; hence

$$|x_0 \cdots x_n|_w \geq \lfloor (n+1)/R_x(|w|) \rfloor,$$

which gives us

$$\mu_x(w) = \lim_{n \to \infty} \frac{|x_0 \cdots x_n|_w}{n+1} \geq \lim_{n \to \infty} \frac{\lfloor (n+1)/R_x(|w|) \rfloor}{n+1} = \frac{1}{R_x(|w|)}.$$

Another easy-to-build bridge is the one giving a lower bound on the complexity function knowing the smallest frequency function e_x. It comes almost directly from Proposition 1.

Proposition 4. *Let x be an infinite word with frequencies, and n a positive integer such that $e_x(n) > 0$. We then have*

$$p_x(n) \leq \frac{1}{e_x(n)}.$$

Proof. By Proposition 1, one gets

$$1 \geq \sum_{v \in L_x(n)} \mu_x(v) \geq \sum_{v \in L_x(n)} e_x(n) \geq p_x(n) \cdot e_x(n).$$

Remark 1. Morse-Hedlund's theorem states that for any infinite word x, if there exists n such that $p_x(n) \leq n$, then, the word x is ultimately periodic. Hence, for any aperiodic infinite word x and for any integer n, one has $p_x(n) > n$. Proposition 4 then implies that $e_x(n) < \frac{1}{n}$ for all n.

Let x be an infinite word that admits frequencies. The condition

$$\limsup n e_x(n) > 0$$

implies that the frequencies are uniform, as shown by M. Boshernitzan [6]. Following the approach developed by M. Boshernitzan, we investigate in the next section the condition $\liminf n e_x(n) > 0$ and see how it allows one to relate recurrence and frequencies.

3   Relating Recurrence and Frequencies

The following result, due to M. Boshernitzan, provides a characterization of linear recurrence in terms of frequencies.

Theorem 1 [8, Ex. 174]. *An infinite word x is linearly recurrent if and only if $\liminf n e_x(n) > 0$.*

We now see how to extend Boshernitzan's result, namely Theorem 1, as a generalisation of [8, Ex. 174].

Theorem 2. *Let* $f : \mathbb{R}_+^* \rightarrow \mathbb{R}_+^*$ *be a continuous non-increasing function and* F *one of its primitives. Let* x *be a uniformly recurrent infinite word that has frequencies. We assume that for* n *large enough, we have*

$$e_x(n) \geq f(n).$$

We then have, for n *large enough,*

$$F(n) + 1 \geq F(R_x(n)).$$

Proof. The statement is true if x is constant. We now assume that x is a non-constant word. Let u be a factor of x. Let

$$r_x(u) := \min\{n : \forall w \in L_x(n), \ u \text{ is a factor of } w\}.$$

Let n be an integer, and $u = u_1 u_2 \cdots u_n$ a factor of x of size n such that $r_x(u) = R_x(n)$ (such u word exists because we work on a finite alphabet). The uniform recurrence of x gives us the existence of a factor $v = v_1 v_2 \cdots v_m$ of x of maximal length such that u appears only once in vu. Let us show that $|v| + |u| = m + n = r_x(u) = R_x(n)$.

Indeed, by definition of $r_x(u)$, and because $r_x(u) > 1$ (x is not constant), there exists a factor $w = w_1 \cdots w_{r_x(u)-1}$ of x such that u does not occur in w, but u occurs in wa, with a being a letter of $\mathcal{A}$, i.e., u appears only once in wa. Then, we can write wa as $w'u$, with $w' = w_1 \cdots w_{r_x(u)-n}$. Hence, since v is of maximal length, $|v| = m \geq r_x(u) - n$, i.e., $m + n \geq r_x(u)$.

Moreover, we suppose by contradiction that $m + n > r_x(u)$. Taking away the last letter of vu gives us the factor $vu_1 \cdots u_{n-1}$, of size $m+n-1 \geq r_x(u)$, which thus contains u. However, it cannot contain u, since u occurred only once in vu, hence the desired contradiction.

Now, suppose that we have $i > j \geq 0$, such that $v_i \cdots v_m u$ is a prefix of $w := v_j \cdots v_m u$. This gives

$$v_j \cdots v_m u = w_1 \cdots w_{n+(m-j)} = v_i \cdots v_m u w_{n+(m-i)+1} \cdots w_{n+(m-j)}.$$

This shows that u appears twice in $w = v_j \cdots v_m u$, i.e., u appears twice in vu, contradicting the hypothesis: there do not exist $i \neq j$ such that $v_i \cdots v_m u$ is a prefix of $v_j \cdots v_m u$. Proposition 1 then gives us

$$1 \geq \mu_x(u) + \sum_{i=1}^{m} \mu_x(v_i \cdots v_m u) \geq \sum_{i=0}^{m} e_x(m - i + n) \geq \sum_{i=0}^{m} f(m-i+n) = \sum_{i=n}^{m+n} f(i),$$

and hence, since f is non-increasing and positive,

$$1 \geq \int_{n}^{m+n+1} f(t) \, dt;$$

consequently, since f is positive and thus F increasing, one gets

$$F(n) + 1 \geq F(R_x(n) + 1) \geq F(R_x(n)). \tag{2}$$

The essence of the proof lies in the sequence of factors of increasing length ranging from n to $R_x(n)$, on which we can use Proposition 1. This creates the inequality in which appear $f(n)$ and $f(R_x(n))$, which then, when integrated, relates $F(n)$ and $F(R_x(n))$.

As a first application we recover Theorem 1, with a function F growing slowly towards infinity, i.e., $f(n) := \frac{c}{n}$ and $F = \int f = c \cdot \log$.

We now consider other lower bounds on $e_x(n)$ with Corollary 1.

Corollary 1. *Let x be a uniformly recurrent infinite word that has frequencies. If there exist $\beta \in (0,1]$ and $c > 0$ such that for n large enough*

$$e_x(n) \geq \frac{c}{n(\log n)^\beta} \; ,$$

then there exists $c' > 1$ such that

$$R_x(n) = O(n^{c'}) \quad and \quad p_x(n) = O(n(\log n)^\beta).$$

Proof. We first consider the case of $\beta = 1$. Suppose that there exists $c > 0$ such that, for n large enough, we have $n \log(n) e_x(n) \geq c$. Then, with $f(n) := c \cdot \frac{1/n}{\log(n)}$, and thus $F(n) := c \cdot \log(\log(n))$, Theorem 2 gives us

$$\log(\log(n)) + \frac{1}{c} \geq \log(\log(R_x(n))),$$

which yields

$$n^{\exp(\frac{1}{c})} \geq R_x(n),$$

and finally

$$R_x(n) = O(n^{\exp(\frac{1}{c})}).$$

The result for the complexity comes from Proposition 4.

We now consider the case of $\beta \in (0,1)$. Suppose that there exists $c > 0$ such that for n large enough, we have $n(\log n)^\beta e_x(n) \geq c$. Theorem 2 with $F(n) := \frac{c}{1-\beta} \cdot (\log n)^{1-\beta}$ gives us

$$\frac{c}{1 - \beta} \cdot (\log(n))^{1-\beta} + 1 \geq \frac{c}{1 - \beta} \cdot \log(R_x(n))^{1-\beta},$$

which yields, for n large enough and for some $\epsilon > 0$,

$$(1 + \epsilon)(\log(n))^{1-\beta} \geq (\log(n))^{1-\beta} + \frac{1 - \beta}{c} \geq \log(R_x(n))^{1-\beta},$$

and finally

$$R_x(n) \leq n^{(1+\epsilon)^{\frac{1}{1-\beta}}},$$

or, with $c' := (1 + \epsilon)^{\frac{1}{1-\beta}} > 1$,

$$R_x(n) = O(n^{c'}).$$

Remark 2. We may want a priori to consider other values for β in the second statement of Corollay 1. Proposition 4 yields that for n large enough, $ne_x(n) \leq 1$ implies $p_x(n) \leq n$, which thus implies that x is ultimately periodic, by Morse-Hedlund's theorem. Hence, if $\beta < 0$, then for n large enough, $e_x(n) \geq \frac{c}{n(\log n)^\beta} \geq \frac{1}{n}$, which would thus imply x to be ultimately periodic. On the other hand, if $\beta > 1$, then F is a negative function that converges towards 0, making us unable to reach any conclusion using (2). This is also why we do not consider functions of the form $\frac{1}{n^\beta}$ for $\beta > 1$.

4 Frequencies and Recurrence for Sturmian Words

We now focus on the family of Sturmian words where the bounds provided by Theorem 1 can be improved, by investigating in more details the case where $\liminf ne_x(n) = 0$.

4.1 First Properties

Sturmian words are defined as the (one-sided) words having exactly $n + 1$ factors of length n, for every positive integer n. They are equivalently defined as symbolic codings (with respect to two-interval partitions) of the irrational translations R_α of the unit circle (that is, the one-dimensional torus $\mathbb{T} = \mathbb{R}/\mathbb{Z}$), where $R_\alpha \colon \mathbb{R}/\mathbb{Z} \to \mathbb{R}/\mathbb{Z}$, $x \mapsto x + \alpha \mod 1$. More precisely, the infinite word $x = (x_n)_{n \in \mathbb{N}} \in \{0,1\}^{\mathbb{N}}$ is a Sturmian word if there exist $\alpha \in (0,1)$, $\alpha \notin \mathbb{Q}$ (called its *angle*), $s \in \mathbb{R}$ such that

$$\forall n \in \mathbb{N}, \ x_n = i \iff R_\alpha^n(s) = n\alpha + s \in I_i \pmod{1},$$

with either $I_0 = [0, 1 - \alpha)$, $I_1 = [1 - \alpha, 1)$, or $I_0 = (0, 1 - \alpha]$, $I_1 = (1 - \alpha, 1]$. For more on Sturmian words, see the corresponding chapters in [17,20] and the references therein.

Let x be a Sturmian word with angle $\alpha \in (0,1) \setminus \mathbb{Q}$. Note that Sturmian words have frequencies. The continued fraction expansion of the angle α allows the expression of the smallest frequency function e_x (see Theorem 4) and of the recurrence function R_x (see Sect. 4.3). Before stating them, let us introduce some notation. Consider the continued fraction expansion of α, with

$$\alpha = \cfrac{1}{a_1 + \cfrac{1}{a_2 + \cfrac{1}{a_3 + \cdots}}}$$

The positive integer digits a_n are called *partial quotients*. The rational numbers p_n/q_n, with p_n, q_n being coprime positive integers being defined as

$$\frac{p_n}{q_n} = \cfrac{1}{a_1 + \cfrac{1}{a_2 + \cfrac{\ddots}{\quad + \cfrac{1}{a_n}}}},$$

are called *convergents*. Let

$$\theta_n = (-1)^n(q_n\alpha - p_n) = |q_n\alpha - p_n|$$

for all nonnegative n. One has $q_{-1} = 0$, $p_{-1} = 1$, $q_0 = 1$, $p_0 = 0$, and for all n

$$q_{n+1} = a_{n+1}q_n + q_{n-1}, \quad p_{n+1} = a_{n+1}p_n + p_{n-1}, \quad \text{and} \quad \theta_{n-1} = a_{n+1}\theta_n + \theta_{n+1}. \tag{3}$$

4.2   On the Frequencies of Sturmian Words

In this section, we indicate how to obtain Sturmian words with a prescribed behaviour for the smallest frequency function (see in particular Remark 4).

Theorem 3 [2].   *Let x be a Sturmian word of angle α, with $\alpha \in (0,1)$, α irrational. Let $m \geq 1$.*
   Assume that $kq_n + q_{n-1} < m < (k+1)q_n + q_{n-1}$, with $n \geq 1$ and $1 \leq k \leq a_{n+1}$. The frequencies of factors of length m belong to the set

$$\{\theta_n, \theta_{n-1} - k\theta_n, \theta_{n-1} - (k-1)\theta_n\} = \{\theta_n, \theta_{n+1} + (a_{n+1} - k)\theta_n, \theta_{n+1} + (a_{n+1} - k + 1)\theta_n\}.$$

Assume $m = kq_n + q_{n-1}$, with $n \geq 1$ and $1 \leq k \leq a_{n+1}$. The frequencies of factors of length m belong to the set

$$\{\theta_n, \theta_{n-1} - k\theta_n\} = \{\theta_n, \theta_{n+1} + (a_{n+1} - k)\theta_n\}.$$

We thus deduce the following, with the notation e_α standing for the smallest frequency function for the Sturmian words of angle α.

Theorem 4.   *Let α be an irrational number, $m > 0$, and n be the unique integer such that $q_n < m \leq q_{n+1}$. Then, $e_\alpha(m) = \theta_n = |q_n\alpha - p_n|$.*

The following estimates hold for the quantity $\theta_n = |q_n\alpha - p_n|$.

Proposition 5.   *For every integer n, we have*

$$\frac{1}{2q_{n+1}} \leq \frac{1}{q_n + q_{n+1}} \leq \theta_n \leq \frac{1}{q_{n+1}} \quad \text{and} \quad \prod_{k=1}^{n} a_k \leq q_n \leq \prod_{k=1}^{n}(a_k + 1).$$

Proof. The first statement is a classical property of continued fractions (see e.g. [10]) based on the fact that, for all n, there exists $\alpha_n \in (0,1)$ such that one has

$$\left| \alpha - \frac{p_n}{q_n} \right| = \left| \frac{p_n + \alpha_n p_{n-1}}{q_n + \alpha_n q_{n-1}} - \frac{p_n}{q_n} \right|.$$

The second statement comes from the recurrence relation $q_{n+1} = a_{n+1} q_n + q_{n-1}$ (see (3) together with the fact that the sequence $(q_n)_n$ is increasing.

Theorem 5. *Let x be a Sturmian word of angle α, with α being an irrational number in $(0,1)$. Let $f : \mathbb{R}_+^* \to \mathbb{R}_+^*$ be a non-increasing function. If*

$$a_{n+1} = O\left(\frac{1}{q_n f(q_n)} \right),$$

then there exists $c > 0$ such that, for all n

$$e_\alpha(n) \geq c \cdot f(n).$$

Proof. The assumptions on f imply that there exists $c > 0$ such that, for n large enough, one has

$$a_{n+1} + 2 \leq \frac{c}{q_n f(q_n)}.$$

Let m be large enough, and let n be the unique integer such that $q_n < m \leq q_{n+1}$. We have that $\frac{q_{n-1}}{q_n} \leq 1$, hence

$$a_{n+1} + 1 + \frac{q_{n-1}}{q_n} \leq \frac{c}{q_n f(q_n)}, \quad \text{i.e., } q_{n+1} + q_n \leq \frac{c}{f(q_n)}.$$

By Theorem 4, $e_\alpha(m) = \theta_n$, and by Proposition 5,

$$e_\alpha(m) = \theta_n \geq \frac{1}{q_{n+1} + q_n} \geq \frac{f(q_n)}{c}.$$

Finally, since f is non-increasing,

$$e_\alpha(m) \geq \frac{f(q_n)}{c} \geq \frac{f(m)}{c}.$$

Remark 3. Theorem 5 (see also Theorem 6 below) relies on assumptions involving relations between both sequences $(a_n)_{n \in \mathbb{N}}$ and $(q_n)_{n \in \mathbb{N}}$. However, the following inequality allows the statement of hypothesis to involve only the sequence $(a_n)_{n \in \mathbb{N}}$. Indeed, let $f : \mathbb{R}^+ \to \mathbb{R}^+$ be a non-increasing function. By Proposition 5, one has, for all n,

$$\frac{1}{(\prod_{k=1}^n a_k) f(\prod_{k=1}^n a_k)} \leq \frac{1}{q_n f(q_n)}.$$

The next statement is the counterpart of Theorem 5 in terms of the existence of an upper bound for infinitely many integers for e_x.

Theorem 6. *Let x be a Sturmian word of angle α, with α being an irrational number in $(0,1)$. Let $f : \mathbb{R}_+^* \to \mathbb{R}_+^*$. If $a_{n+1} \geq \frac{1}{q_n f(q_n+1)}$ for infinitely many n, then $e_\alpha(n) \leq f(n)$, for infinitely many n.*

Proof. Let n be such that $a_{n+1} \geq \frac{1}{q_n f(q_n+1)}$. One has

$$\frac{1}{q_n f(q_n + 1)} \leq a_{n+1} \leq a_{n+1} + \frac{q_{n-1}}{q_n}.$$

Multiplying by q_n together with (3) yields

$$\frac{1}{f(q_n + 1)} \leq q_{n+1}.$$

We take $m = q_n + 1$. By Theorem 4, $e_\alpha(m) = \theta_n$, and by Proposition 5, one has

$$e_\alpha(q_n + 1) = \theta_n \leq \frac{1}{q_{n+1}} \leq f(q_n + 1).$$

Remark 4. Let us illustrate Theorem 5 and 6 with the function $f : x \mapsto \frac{1}{x \log x}$ from Corollary 1. Let x be a Sturmian word of angle α, with α being an irrational number. Theorems 5 and 6 give the following. If $a_{n+1} \leq \log q_n$ for n large enough, then there exists $c > 0$ such that $e_\alpha(n) \geq \frac{c}{n \log n}$, for all n. If $a_{n+1} \geq \frac{q_n+1}{q_n} \log(q_n + 1)$ for infinitely many n, then $e_\alpha(n) \leq \frac{1}{n \log n}$, for infinitely many n.

4.3   On the Recurrence of Sturmian Words

Let x be a Sturmian word with angle α ($\alpha \notin \mathbb{Q}$, $\alpha \in (0,1)$). Let $(q_k)_{k \in \mathbb{N}}$ denote the sequence of denominators of the convergents of the continued fraction expansion of α. One has, by [19], the following:

$$R_x(m) = m - 1 + q_n + q_{n+1}, \text{ for } q_n \leq m < q_{n+1}. \tag{4}$$

In particular, a Sturmian word is linearly recurrent if and only if its angle has bounded partial quotients a_k. For more on the recurrence function of Sturmian words, see e.g. [4,7,21].

Next theorem shows that, beyond the case of linear recurrence (that is, even when $\liminf ne_x(n) = 0$), the smallest frequency and the recurrence functions are still related in the Sturmian case: the functions e_x and $1/R_x$ have the same growth order for most of the integers. The statement about almost every α in the next theorem refers to a set of full Lebesgue measure. We also revisit the condition from Theorems 5 and 6 for the functions $e_x R_x$, R_x, and e_x.

Theorem 7. *Let x be a Sturmian word of angle α, with $\alpha \in (0,1)$, α irrational. Let $(a_n)_{n \in \mathbb{N}}$ be the sequence of partial quotients and let $(q_n)_{n \in \mathbb{N}}$ be the sequence* *of denominators of the convergents in the continued fraction expansion of α. One has, for all positive integers m that are not in the sequence $(q_n)_{n \in \mathbb{N}}$,*

$$1 \leq e_x(m) R_x(m) \leq 3.$$

Moreover, for almost all $\alpha \in (0, 1)$,

$$\limsup_n e_x(q_n) R_x(q_n) = +\infty.$$

More precisely, let $f : \mathbb{R}_+^* \to \mathbb{R}_+^*$ be a non-increasing function. If $a_{n+1} = O\left(\frac{1}{q_n f(q_n)}\right)$, then there exist $c, c' > 0$ such that

$$e_x(q_n) R_x(q_n) \leq \frac{c}{q_n f(q_n)}, \quad R_x(n) \leq \frac{c}{f(n)} \quad \text{and} \quad e_x(n) \geq c' f(n) \text{ for all } n.$$

If $a_{n+1} \geq \frac{1}{q_n f(q_n+1)}$, then the following properties hold for infinitely many n:

$$e_\alpha(q_n) R_x(q_n) \geq \frac{1}{2 q_n f(q_n)}, \quad R_x(n) \geq \frac{1}{f(n)} \quad \text{and} \quad e_x(n) \leq f(n).$$

Proof. Let x be a Sturmian word of angle α and let m be a positive integer. The lower bound for $e_x(m) R_x(m)$ comes from Proposition 3.

Now let n be such that $q_n \leq m < q_{n+1}$. Let us first assume $m \neq q_n$. By Theorem 4, $e_x(m) = \theta_n$ and moreover, $R_x(m) = m - 1 + q_n + q_{n+1}$. Regarding the upper bound, one thus gets, by Proposition 5, that

$$e_x(m) R_x(m) = \theta_n (m - 1 + q_{n+1} + q_n) \leq \frac{1}{q_{n+1}} (m - 1 + q_{n+1} + q_n) \leq 3,$$

since $m < q_{n+1}$. We thus have proved that $1 \leq e_x(m) R_x(m) \leq 3$ for $m \neq q_n$.

We now assume $m = q_n$. Then, by Proposition 5, one has $e_x(m) = \theta_{n-1}$ and

$$e_x(q_n) R_x(q_n) = \theta_{n-1}(q_n - 1 + q_{n+1} + q_n) \leq \frac{1}{q_n}(q_n - 1 + q_{n+1} + q_n),$$

hence

$$1 \leq e_x(q_n) R_x(q_n) \leq a_{n+1} + 3. \tag{5}$$

Moreover, again by by Proposition 5, one has

$$e_x(m) R_x(m) = \theta_{n-1}(q_n - 1 + q_{n+1} + q_n) \geq \frac{1}{q_n + q_{n-1}}(q_n - 1 + q_{n+1} + q_n).$$

Hence

$$e_x(m) R_x(m) \geq \frac{(a_{n+1} + 2)q_n + q_{n-1} - 1}{q_n + q_{n-1}} \geq \frac{a_{n+1}}{2}. \tag{6}$$

We now consider $f : \mathbb{R}^+ \to \mathbb{R}^+$ to be a non-increasing function. We assume $a_{n+1} = O\left(\frac{1}{q_n f(q_n)}\right)$. There exists $c > 0$ such that, for n large enough, $a_{n+1} + 3 \leq \frac{c}{q_n f(q_n)}$. Hence, by (5), one has

$$1 \leq e_x(q_n) R_x(q_n) \leq \frac{c}{q_n f(q_n)}.$$

Moreover, let m be given and let n be such that $q_n \leq m < q_{n+1}$. Then, by (4), there exists $c > 0$ such that

$$R_x(m) \leq (2a_{n+1} + 3)q_n \leq \frac{2c}{f(q_n)} \leq \frac{2c}{f(m)}$$

since f is non-increasing. The lower bound on e_x comes from Proposition 3.

We now assume $a_{n+1} \geq \frac{1}{q_n f(q_n+1)}$ for infinitely many n. Then, by (6), one gets, for infinitely many n, $e_x(q_n) R_x(q_n) \geq \frac{1}{2q_n f(q_n)}$. Moreover, let $m = q_n + 1$. Then, by (4), $R_x(q_n + 1) \geq q_{n+1} \geq a_{n+1} q_n \geq \frac{1}{f(q_n+1)}$. The upper bound on e_x comes from the proof of Theorem 6.

It remains to prove the statement on the generic behaviour of $e_x(q_n) R_x(q_n)$. According to [5], one has $\limsup a_n = +\infty$, for Lebesgue almost all $\alpha \in (0, 1)$. Consequently, by (6), $\limsup_n e_x(q_n) R_x(q_n) = +\infty$ for almost all α.

Remark 5. Let us continue with the example of the function $f(x) = \frac{1}{x \log x}$, such as considered in Remark 4. We assume $a_{n+1} = O(\log q_n)$. Then, there exists $c > 0$ such that $e_x(n) \geq \frac{c}{n \log n}$, for all n, and $R_x = O(n \log n)$, which improves the bound of Corollary 1.

5   Concluding Remarks

We have seen with Theorem 2 that the approach developed by M. Boshernitzan for his characterization of linearly recurrent words in terms of the smallest frequency function (see Theorem 1) can be extended beyond the case of linearly recurrent words. Corollary 1 shows for instance that a lower bound of order $\frac{1}{n \log}$ for the smallest frequency function induces a polynomial upper bound for the recurrence function. A strategy in order to provide concrete examples of words having a prescribed behaviour for the growth of the smallest frequency function is to work with explicit S-adic expansions, such as the expansions provided by continued fractions for Sturmian words. The precise arithmetic descriptions of the smallest frequency and of the recurrence functions then allow to improve in Theorems 5 and 7 the results obtained in Theorem 2, as stressed in Remark 4 and 5, when comparing with Corollary 1.

The study of multidimensional words is natural in the context of aperiodic order. We plan to investigate the case of multidimensional Sturmian words (see [3]) in order to see how to extend the relation from Theorem 7, i.e., the quantity $e_x R_x$. Note that the study of the frequencies for factors is related to the three-length theorem. See [12,13] for higher-dimensional generalizations.

References

1. Aliste-Prieto, J., Coronel, D., Cortez, M.I., Durand, F., Petite, S.: Linearly repetitive Delone sets. In: Kellendonk, J., Lenz, D., Savinien, J. (eds.) Mathematics of Aperiodic Order. PM, vol. 309, pp. 195–222. Springer, Basel (2015). https://doi.org/10.1007/978-3-0348-0903-0_6
2. Berthé, V.: Fréquences des facteurs des suites sturmiennes. Theor. Comput. Sci. **165**(2), 295–309 (1996)
3. Berthé, V., Vuillon, L.: Tilings and rotations on the torus: a two-dimensional generalization of Sturmian sequences. Disc. Math. **223**(2000), 27–53 (2000)
4. Berthé, V., Cesaratto, E., Rotondo, P., Vallée, B., Viola, A.: Recurrence function on Sturmian words: a probabilistic study. In: Italiano, G.F., Pighizzini, G., Sannella, D.T. (eds.) MFCS 2015. LNCS, vol. 9234, pp. 116–128. Springer, Heidelberg (2015). https://doi.org/10.1007/978-3-662-48057-1_9
5. Émile Borel, M.: Les probabilités dénombrables et leurs applications arithmétiques. Rendiconti del Circolo Matematico di Palermo (1884–1940) **27**(1), 247–271 (1909). https://doi.org/10.1007/BF03019651
6. Boshernitzan, M.: A condition for unique ergodicity of minimal symbolic flows. Ergod. Theor. Dyn. Syst. **12**(3), 425–428 (1992)
7. Cassaigne, J.: Limit values of the recurrence quotient of Sturmian sequences. Theor. Comput. Sci. **218**(1), 3–12 (1999)
8. Berthé, V., Rigo, M. (eds.): Combinatorics, Automata and Number Theory. Encyclopedia of Mathematics and its Applications, vol. 135. Cambridge University Press, Cambridge (2010)
9. Durand, F., Host, B., Skau, C.: Substitutions, Bratteli diagrams and dimension groups. Ergod. Theor. Dyn. Syst. **19**, 952–993 (1999)
10. Hardy, G.H., Wright, E.M.: An Introduction to the Theory of Numbers. Oxford Science Publications, Published by Oxford University Press, Oxford (1980)
11. Haynes, A., Koivusalo, H., Walton, J.: A characterization of linearly repetitive cut and project sets. Nonlinearity **31**(2), 515–539 (2018)
12. Haynes, A., Marklof, J.: Higher dimensional Steinhaus and Slater problems via homogeneous dynamics. Ann. Sci. Éc. Norm. Supér. **4**(53), 537–557 (2020)
13. Haynes, A., Marklof, J.: A five distance theorem for Kronecker sequences. Int. Math. Res. Not. IMRN **24**, 19747–19789 (2022)
14. Keane, M.: Non-ergodic interval exchange transformations. Israel J. Math. **26**, 188–196 (1977)
15. Lagarias, J.C., Pleasants, P.A.B.: Local complexity of Delone sets and crystallinity. Can. Math. Bull. **45**(4), 634–652 (2002)
16. Lagarias, J.C., Pleasants, P.A.B.: Repetitive Delone sets and quasicrystals. Ergod. Theor. Dyn. Syst. **23**, 831–867 (2003)
17. Lothaire, M.: Algebraic Combinatorics on Words. Encyclopedia of Mathematics and its Applications, vol. 90. Cambridge University Press, Cambridge (2002)
18. Morse, M., Hedlund, G.A.: Symbolic dynamics. Am. J. Math. **60**, 815–866 (1938)
19. Morse, M., Hedlund, G.A.: Symbolic dynamics II. Sturmian trajectories. Am. J. Math. **62**, 1–42 (1940)
20. Fogg, N.P.: Substitutions in Dynamics, Arithmetics and Combinatorics. Lecture Notes in Mathematics, vol. 1794. Springer, Heidelberg (2002). https://doi.org/10.1007/b13861
21. Rotondo, P., Vallée, B: The recurrence function of a random Sturmian word. In: 2017 Proceedings of the Fourteenth Workshop on Analytic Algorithmics and Combinatorics (ANALCO), pp. 100–114. SIAM, Philadelphia (2017)

Sturmian and Infinitely Desubstitutable Words Accepted by an ω-Automaton

Pierre Béaur[ID] and Benjamin Hellouin de Menibus[(✉)][ID]

Université Paris-Saclay, CNRS,
Laboratoire Interdisciplinaire des Sciences du Numérique, 91400 Orsay, France
pierre.beaur@universite-paris-saclay.fr, hellouin@lisn.fr

Abstract. Given an ω-automaton and a set of substitutions, we look at which accepted words can also be defined through these substitutions, and in particular if there is at least one. We introduce a method using desubstitution of ω-automata to describe the structure of preimages of accepted words under arbitrary sequences of homomorphisms: this takes the form of a meta-ω-automaton.

We decide the existence of an accepted purely substitutive word, as well as the existence of an accepted fixed point. In the case of multiple substitutions (non-erasing homomorphisms), we decide the existence of an accepted infinitely desubstitutable word, with possibly some constraints on the sequence of substitutions (*e.g.* Sturmian words or Arnoux-Rauzy words). As an application, we decide when a set of finite words codes *e.g.* a Sturmian word. As another application, we also show that if an ω-automaton accepts a Sturmian word, it accepts the image of the full shift under some Sturmian morphism.

Keywords: Substitutions · ω-automata · Sturmian words · decidability

1 Introduction

One-dimensional symbolic dynamics is the study of infinite words and their associated dynamical structures, and is linked with combinatorics on words. Two classical methods to generate words are the following: on the one hand, sofic shifts are the set of infinite walks on a labeled graph (which can be considered as an ω-automaton) [9]; on the other hand, the substitutive approach consists in iterating a word homomorphism on an initial letter. The latter method was introduced by Axel Thue as a way to create counterexamples to conjectures in combinatorics on words [3].

These two constructions usually build words and languages which are of a very different nature. On the one hand, substitutive words tend to have a self-similar structure, and are used to generate minimal aperiodic subshifts; on the other hand, sofic shifts always contain ultimately periodic words and cannot be minimal if they contain a non-periodic word. We aim at deciding when a given ω-automaton accepts a word with a given substitutive structure, and study the properties of sets of such accepted words. Carton and Thomas provided a

© The Author(s), under exclusive license to Springer Nature Switzerland AG 2023
A. Frid and R. Mercaş (Eds.): WORDS 2023, LNCS 13899, pp. 104–116, 2023.
https://doi.org/10.1007/978-3-031-33180-0_8

method to decide this question in the case of substitutive or morphic words on Büchi ω-automata, using verification theory and semigroups of congruence [5]. This result was partially reproved by Salo [15], using a more combinatorial point of view. For the last 20 years, the substitutive approach (iterating a single homomorphism) has been generalized to the S-adic approach [6] that lets one alternate between multiple substitutions. This more general framework lets us describe other natural classes, such as the family of Sturmian words.

In this paper, we develop a new method based on desubstitutions of ω-automata. We can express the preimages of an ω-automaton by any sequence of substitutions through a meta-ω-automaton, whose vertices are ω-automata and whose edges are labeled by substitutions. We use this meta-ω-automaton to decide whether an ω-automaton accepts a purely substitutive word (giving an alternative proof of [5]), or a fixed point of a substitution, or a morphic word, or an infinitely desubstitutable word (by a set of substitutions). The method is flexible enough to enforce additional constraints on the directive sequences of substitutions, which is powerful enough for example to decide whether an ω-automaton accepts a Sturmian word. A consequence is the decidability of whether a given set of finite words codes some Sturmian word (or from any family of words with an S-adic characterization). We also describe the set of directive sequences of words accepted by some ω-automaton, which is an ω-regular set.

The meta-ω-automaton also provides a more combinatorial insight on how Sturmian words and ω-regular languages interact: namely, that an ω-automaton accepts a Sturmian word if, and only if, it accepts the image of the full shift under a Sturmian morphism.

2 Definitions

2.1 Words and ω-automata

An alphabet $\mathcal{A}$ is a finite set of symbols. The set of finite words on $\mathcal{A}$ is denoted as $\mathcal{A}^*$, and contains the empty word. A (mono)infinite word is an element of $\mathcal{A}^{\mathbb{N}}$. It is usual to write $x = x_0 x_1 x_2 x_3 \ldots$ where $x_i = x(i) \in \mathcal{A}$. If x is a word, $|x|$ is the length of the word (if x is infinite, then $|x| = \infty$). For a word x and $0 \leqslant j \leqslant k < |x|$, $x_{[\![j,k]\!]}$ is the word $x_j x_{j+1} x_{j+2} \ldots x_{k-1} x_k$. We denote $w \sqsubseteq_p x$ when w is a prefix of x, that is, $w = x_{[\![0,k]\!]}$.

It is possible to endow $\mathcal{A}^{\mathbb{N}}$ with a topology, called *the prodiscrete topology*. The prodiscrete topology is defined by the clopen basis $[w]_n = \{x \in \mathcal{A}^{\mathbb{N}} \mid x_n x_{n+1} \ldots x_{n+|w|-1} = w\}$ for $w \in \mathcal{A}^*$. To this topology, we can adjunct a dynamic with the shift operator S:

$$S : \left(\begin{array}{ccc} \mathcal{A}^{\mathbb{N}} & \to & \mathcal{A}^{\mathbb{N}} \\ x = x_0 x_1 x_2 x_3 \ldots & \mapsto & S(x) = x_1 x_2 x_3 x_4 \ldots \end{array} \right)$$

A set $X \subseteq \mathcal{A}^{\mathbb{N}}$ is called *a shift (space)* if it is stable by S and closed for the prodiscrete topology. In particular, $\mathcal{A}^{\mathbb{N}}$ is a shift space, called *the full shift (space)*.

We now introduce the main computational model of this paper: ω-automata.

Definition 1 (ω-automaton). *An ω-automaton $\mathfrak{A}$ is a tuple $(\mathcal{A}, Q, I, T)$, where $\mathcal{A}$ is an alphabet, Q is a finite set of states, $I \subseteq Q$ is the set of initial states, $T \subseteq Q \times \mathcal{A} \times Q$ is the set of transitions of $\mathfrak{A}$.*

We extend several classical notions from finite automata. We write transitions as $q_s \xrightarrow{a} q_t \in T$.

Definition 2 (Computations and walks). *For $n \geqslant 1$ or $n = \infty$, a sequence $(q_k)_{0 \leqslant k \leqslant n}$ with $q_k \in Q$ is a walk in $\mathfrak{A}$ if there is $(a_k)_{1 \leqslant k \leqslant n} \subseteq \mathcal{A}$ such that for all $0 \leqslant k \leqslant n-1$, $q_k \xrightarrow{a_{k+1}} q_{k+1} \in T$. We then write $q_0 \xrightarrow{a_1} q_1 \xrightarrow{a_2} q_2 \xrightarrow{a_3} \cdots \xrightarrow{a_n} q_n$. The word $w = (a_k)_{1 \leqslant k \leqslant n}$ labels the walk, and we call* computation *a labeled walk. If the computation begins with an initial state, w is* accepted *by $\mathfrak{A}$.*

In the literature, ω-automata usually have an acceptance condition (such as the Büchi condition [17]). In this paper, we will consider ω-automata to have the largest acceptance condition: every walk beginning with an initial state is accepting. This is a weaker model than Büchi ω-automata.

Definition 3 (Language of an ω-automaton). *Let $\mathfrak{A}$ be an ω-automaton. The language of **finite** words of $\mathfrak{A}$ is $\mathcal{L}_F(\mathfrak{A}) = \{w \in \mathcal{A}^* \mid w \text{ is accepted by } \mathfrak{A}\}$. The language of **infinite** words of $\mathfrak{A}$ is $\mathcal{L}_\infty(\mathfrak{A})) \{w \in \mathcal{A}^{\mathbb{N}} \mid w \text{ is accepted by } \mathfrak{A}\}$. Then, the language of $\mathfrak{A}$ is $\mathcal{L}(\mathfrak{A}) = \mathcal{L}_F(\mathfrak{A}) \cup \mathcal{L}_\infty(\mathfrak{A})$.*

If all states of $\mathfrak{A}$ are initial ($I = Q$), its language of infinite words is a shift, called a *sofic shift* [9].

2.2 Substitutions

Definition 4 (Homomorphisms and substitutions). *A homomorphism is a function $\sigma : \mathcal{A}^* \to \mathcal{A}^*$ such that $\sigma(uv) = \sigma(u)\sigma(v)$ (concatenation) for all $u, v \in \mathcal{A}^*$. The homomorphism σ is extended to $\mathcal{A}^{\mathbb{N}} \to \mathcal{A}^{\mathbb{N}}$ by $\sigma(x_0 x_1 x_2 \ldots) = \sigma(x_0)\sigma(x_1)\sigma(x_2)\ldots$ A substitution is a* nonerasing *homomorphism, that is, $\sigma(a) \neq \varepsilon$ for all letters $a \in \mathcal{A}$.*

Definition 5 (Fixed points, purely substitutive, substitutive and morphic words). *Let $\sigma, \tau : \mathcal{A}^{\mathbb{N}} \to \mathcal{A}^{\mathbb{N}}$ be two homomorphisms. An infinite word $x \in \mathcal{A}^{\mathbb{N}}$ is:*

- *a fixed point of σ if $\sigma(x) = x$;*
- *a purely substitutive word generated by σ if there is a letter $a \in \mathcal{A}$ such that $x = \lim_{n \to \infty} \sigma^n(a)$, where the limit is well-defined;*
- *a morphic word generated by σ and τ if $x = \tau(y)$, where y is a purely substitutive word generated by σ;*
- *a substitutive word generated by σ if x is a morphic word generated by σ and a coding τ, i.e. $\tau(\mathcal{A}) \subseteq \mathcal{A}$.*

It is now possible to extend these definitions to the case where we use multiple homomorphisms. However, most of literature revolves around the use of multiple non-erasing homomorphisms (substitutions), and we will stick to this case. Let $(\sigma_n)_{n\in\mathbb{N}}$ be a sequence of substitutions. The equivalent of a fixed-point of one homomorphism is an *infinitely desubstitutable word* by a sequence of substitutions:

Definition 6 (Infinitely desubstitutable words and directive sequences).
Let S be a finite set of substitutions on a single alphabet $\mathcal{A}$, and let $(\sigma_n)_{n\in\mathbb{N}} \subseteq S$. An infinite word x is infinitely desubstitutable by $(\sigma_n)_{n\in\mathbb{N}}$ (called a directive sequence *of x) if, and only if, there exists a sequence of infinite words $(x_n)_{n\in\mathbb{N}}$ such that $x_0 = x$ and $x_n = \sigma_n(x_{n+1})$. An infinite word x is infinitely desubstitutable by S if x is infinitely desubstitutable by some directive sequence $(\sigma_n)_{n\in\mathbb{N}} \subseteq S$.*

Just like for words, we write $\sigma_{[\![i,j]\!]} = \sigma_i \circ \sigma_{i+1} \circ \cdots \circ \sigma_j$. Then, by compactness of $\mathcal{A}^{\mathbb{N}}$, x is infinitely desubstitutable by $(\sigma_n)_{n\in\mathbb{N}}$ if, and only if, there is a sequence of infinite words $(x_n)_{n\in\mathbb{N}}$ such that $x = \sigma_{[\![0,n]\!]}(x_{n+1})$ for all $n \geqslant 0$.

3   Finding Substitutive and Infinitely Desubstitutable Words in ω-Automata

3.1   Desubstituting ω-Automata

In this section, we explain our main technical tool: an effective transformation of ω-automaton, called desubstitution. We define it for the broad case of possibly erasing homomorphisms.

Definition 7 (Desubstitution of an ω-automaton). *Let $\mathfrak{A} = (\mathcal{A}, Q, I, T)$ be an ω-automaton, and σ a homomorphism. We define $\sigma^{-1}(\mathfrak{A})$ as the ω-automaton $(\mathcal{A}, Q, I, T')$ where, for all $q_1, q_2 \in Q$ and $a \in \mathcal{A}$, $q_1 \xrightarrow{a} q_2 \in T'$ iff $q_1 \xrightarrow{\sigma(a)}{}^* q_2$ is a computation in $\mathfrak{A}$.*

In particular, in this case, we consider that $q \xrightarrow{\varepsilon} q$ is a computation. Thus, if $\sigma(a) = \varepsilon$, the desubstituted automaton $\sigma^{-1}(\mathfrak{A})$ has a loop labeled by a on every state.

For example, consider the following ω-automaton $\mathfrak{A}$ and substitution σ (Fig. 1(a, b)).

We build the ω-automaton $\sigma^{-1}(\mathfrak{A})$. Start from an empty automaton on the same set of states. For every computation in $\mathfrak{A}$ labeled by $01 = \sigma(0)$ — say, $q \xrightarrow{\sigma(0)}{}^* r$ — add an edge $q \xrightarrow{0} r$ to the automaton (Fig. 1(c)). To conclude, do this with $\sigma(1) = 0$ (Fig. 1(d)).

Stability by inverse morphism is a classical concept in the theory of finite automata [8], and desubstitution satisfies the following property:

Proposition 1. *An infinite word u is accepted by $\sigma^{-1}(\mathfrak{A})$ if and only if $\sigma(u)$ is accepted by $\mathfrak{A}$. In other words, $\mathcal{L}_\infty(\sigma^{-1}(\mathfrak{A})) = \sigma^{-1}(\mathcal{L}_\infty(\mathfrak{A}))$.*

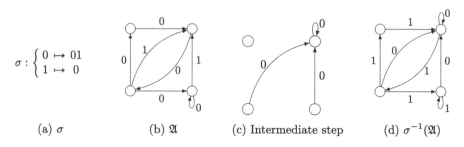

$$\sigma : \begin{cases} 0 \mapsto 01 \\ 1 \mapsto 0 \end{cases}$$

(a) σ (b) $\mathfrak{A}$ (c) Intermediate step   (d) $\sigma^{-1}(\mathfrak{A})$

Fig. 1. Desubstitution of the ω-automaton $\mathfrak{A}$ by σ

Proof. Let u be accepted by $\sigma^{-1}(\mathfrak{A})$. Consider the associated accepting walk $(q_i)_{i \in \mathbb{N}}$. By definition of $\sigma^{-1}(\mathfrak{A})$, for every $i \in \mathbb{N}$, there exists a computation $q_i \xrightarrow{\sigma(u_i)}{}^* q_{i+1}$ in $\mathfrak{A}$. By concatenating these computations, we get an infinite computation $q_0 \xrightarrow{\sigma(u_0)}{}^* q_1 \xrightarrow{\sigma(u_1)}{}^* q_2 \xrightarrow{\sigma(u_2)}{}^* \cdots$ in $\mathfrak{A}$ that accepts $\sigma(u)$ in $\mathfrak{A}$.

Conversely, suppose there is a word of the form $\sigma(u)$ accepted by $\mathfrak{A}$. Consider the states $(q_i)_{i \in \mathbb{N}}$ obtained after reading each $\sigma(a)$ for $a \in \mathcal{A}$. This defines an accepting computation labeled by u in $\sigma^{-1}(\mathfrak{A})$.

This proof actually provides a similar result for finite words:

Proposition 2. *Let w be a finite word, $\mathfrak{A}$ an ω-automaton and σ a homomorphism. Then $q_s \xrightarrow{\sigma(w)}{}^* q_t$ is a computation in $\mathfrak{A}$ iff $q_s \xrightarrow{w}{}^* q_t$ is a computation in $\sigma^{-1}(\mathfrak{A})$.*

An easy but significant property is the composition of desubstitution of ω-automata:

Proposition 3. *Let $\mathfrak{A}$ be an ω-automaton, and σ and τ be two homomorphisms. Then, $(\sigma \circ \tau)^{-1}(\mathfrak{A}) = \tau^{-1}(\sigma^{-1}(\mathfrak{A}))$.*

Proof. These two ω-automata share the same sets of states and of initial states. We prove that they have the same transitions. We have indeed:

$$q_s \xrightarrow{a} q_t \text{ in } (\sigma \circ \tau)^{-1}(\mathfrak{A}) \iff q_s \xrightarrow{\sigma \circ \tau(a)}{}^* q_t \text{ in } \mathfrak{A}$$

$$\iff q_s \xrightarrow{\tau(a)}{}^* q_t \text{ in } \sigma^{-1}(\mathfrak{A}), \text{ by Proposition 2}$$

$$\iff q_s \xrightarrow{a} q_t \text{ in } \tau^{-1}(\sigma^{-1}(\mathfrak{A})), \text{ by Proposition 2 again.}$$

3.2 The Problem of the Purely Substitutive Walk

We underline the following property of desubstitutions of ω-automata:

Fact 1. *Let $\mathfrak{A}$ be an ω-automaton, let $\mathfrak{S}(\mathfrak{A})$ be the set of all ω-automata which have the same alphabet, the same set of states and the same initial states as $\mathfrak{A}$. For any homomorphism σ on $\mathcal{A}$, $\sigma^{-1}(\mathfrak{A})$ is an element of $\mathfrak{S}(\mathfrak{A})$.*

The crucial point is that $\mathfrak{S}(\mathfrak{A})$ is finite: given $\mathfrak{A} = (\mathcal{A}, Q, I, T)$, an element of $\mathfrak{S}(\mathfrak{A})$ is identified by its transitions, which form a subset of $(Q \times \mathcal{A} \times Q)$, so $\mathrm{Card}(\mathfrak{S}(\mathfrak{A})) = 2^{|Q|^2 \times |\mathcal{A}|}$. We could work on a subset of $\mathfrak{S}(\mathfrak{A})$ by identifying ω-automata with the same language [2], but finiteness is sufficient for our results.

Given $\mathfrak{A}$ an ω-automaton and σ a homomorphism, σ^{-1} defines a dynamic on the finite set $\mathfrak{S}(\mathfrak{A})$. By the pigeonhole principle:

Fact 2. *Let $\mathfrak{A}$ be an ω-automaton, and σ be a homomorphism. Then there exist $n < m \leqslant |\mathfrak{S}(\mathfrak{A})| + 1$ such that $\sigma^{-n}(\mathfrak{A}) = \sigma^{-m}(\mathfrak{A})$.*

In the remainder of the section, we prove that, given an ω-automaton $\mathfrak{A}$ and a substitution σ, the problems of finding a fixed point of σ or a purely substitutive word generated by σ accepted by $\mathfrak{A}$ are decidable.

A purely substitutive word generated by an erasing homomorphism σ is also generated by a non-erasing homomorphism τ (that is, a substitution) that can be effectively constructed: remove every erased letter from $\mathcal{A}$ and from the images of σ, and repeat the process. Thus, we assume σ itself is a substitution.

Proposition 4. *Let $\mathfrak{A}$ be an ω-automaton, let σ be a substitution and let $n < m \leqslant |\mathfrak{S}(\mathfrak{A})| + 1$ such that $\sigma^{-n}(\mathfrak{A}) = \sigma^{-m}(\mathfrak{A})$. Then, $\mathfrak{A}$ accepts a fixed point for σ^k for some $k \geqslant 1$ iff $\mathcal{L}_\infty(\sigma^{-n}(\mathfrak{A}))$ is nonempty.*

Proof. If $\mathcal{L}_\infty(\sigma^{-n}(\mathfrak{A}))$ is empty, then, by Propositions 1 and 3, $\mathcal{L}_\infty(\sigma^{-p}(\mathfrak{A}))$ is empty for every $p \geqslant n$. Let $k \geqslant 1$: if there were a fixed point x for σ^k accepted by $\mathfrak{A}$, we would have $x = \sigma^k(x) = \sigma^{kn}(x)$ by iterating. So x would be in $\mathcal{L}_\infty(\sigma^{-kn}(\mathfrak{A}))$ which is empty. By contradiction, there is no fixed point for any σ^k.

If $\mathcal{L}_\infty(\sigma^{-n}(\mathfrak{A}))$ is nonempty, let x be a word accepted by $\sigma^{-n}(\mathfrak{A})$. Again by Propositions 1 and 3, because $\sigma^{-n}(\mathfrak{A}) = \sigma^{-m}(\mathfrak{A}) = \sigma^{-(m-n)}(\sigma^{-n}(\mathfrak{A}))$, x is accepted by $\sigma^{-j(m-n)}(\sigma^{-n}(\mathfrak{A})) = \sigma^{-(n+j(m-n))}(\mathfrak{A})$ for all $j \in \mathbb{N}$. This means that $\sigma^{n+j(m-n)}(x)$ is accepted by $\mathfrak{A}$ for all $j \in \mathbb{N}$. Consider an adherence value $\tilde{x}$ of the sequence $(\sigma^{n+j(m-n)}(x))_{j\in\mathbb{N}}$. By compactness of the language of an ω-automaton, $\tilde{x} \in \mathcal{L}_\infty(\sigma^{-n}(\mathfrak{A}))$.

We define $Q_{\sigma^{m-n}} \subseteq \mathcal{A}$ the set of quiet letters for σ^{m-n}: $a \in Q_{\sigma^{m-n}}$ if $|\sigma^{j(m-n)}(a)| = 1$ for all $j \geqslant 1$. Let $k = \inf\{i \in \mathbb{N} \mid \tilde{x}_i \notin Q_{\sigma^{m-n}}\}$ (k may be infinite). Then, for $i < k$, because σ is a (nonerasing) substitution and every letter in $\tilde{x}_{[\![0,k-1]\!]}$ is quiet, $\sigma^{(m-n)}(\tilde{x})_i = \sigma^{(m-n)}(\tilde{x}_i)$. In addition, because $\tilde{x}$ is an adherence value of $(\sigma^{n+j(m-n)}(x))_{j\in\mathbb{N}}$, there is $r(\tilde{x}_i) \geqslant 1$ such that $\sigma^{r(\tilde{x}_i)\cdot(m-n)}(\tilde{x}_i) = \tilde{x}_i$ for every position $i < k$. Since $\mathcal{A}$ is finite, $(r(\tilde{x}_i))_{0\leqslant i<k}$ contains only finitely many values, so we can define $r = \mathrm{lcm}\{r(\tilde{x}_i)\}$.

When $k < \infty$, there exists $q \geqslant 1$ such that $|\sigma^{q(m-n)}(\tilde{x}_k)| > 1$ and $\tilde{x}_k \sqsubseteq_p \sigma^{q(m-n)}(\tilde{x}_k)$, for the same reason that $\tilde{x}$ is an adherence value of $(\sigma^{n+j(m-n)}(x))_{j\in\mathbb{N}}$. If $k = \infty$, we set $q = 1$.

Then, by concatenation, $\tilde{x}_{[\![0,k]\!]} \sqsubseteq_p \sigma^{rq(m-n)}(\tilde{x}_{[\![0,k]\!]})$. Thus, $(\sigma^{jrq(m-n)}(\tilde{x}))_{j\in\mathbb{N}}$ has a limit, which is a fixed point for $\sigma^{rq(m-n)}$, and by compactness of $\mathcal{L}_\infty(\mathfrak{A})$, is accepted by $\mathfrak{A}$.

Because the emptiness of the language of an ω-automaton is decidable:

Corollary 1. *The following problem is decidable:*

Input: *An ω-automaton $\mathfrak{A}$ and a substitution σ*
Question: *Does $\mathfrak{A}$ accept a fixed point of σ^k for some k?*

As is, this method alone cannot determine, for instance, whether $\mathfrak{A}$ accepts a fixed point for σ itself (without power). This problem is still decidable, as we show later in Proposition 6 with a refinement of this method. In appendix, we provide examples where $\mathfrak{A}$ accepts fixed points for some σ^k where k does not correspond to $m - n$ where $n < m$ are the minimal powers such that $\sigma^{-n}(\mathfrak{A}) = \sigma^{-m}(\mathfrak{A})$.

Now, we come back to purely substitutive words. A purely substitutive word generated by σ is also a fixed point for some σ^k (in fact, it is a fixed point for every σ^j with $j \geqslant 1$).

Proposition 5. *Let $\mathfrak{A}$ be an ω-automaton, σ a substitution and $n < m \leqslant |\mathfrak{S}(\mathfrak{A})| + 1$ such that $\sigma^{-n}(\mathfrak{A}) = \sigma^{-m}(\mathfrak{A})$. Let $RP_\sigma \subseteq \mathcal{A}$ be the set of letters b that are right-prolongable for σ, i.e. $b \sqsubseteq_p \sigma(b)$ and $b \neq \sigma(b)$. Then, $\mathfrak{A}$ accepts a purely substitutive word generated by σ iff $\sigma^{-n}(\mathfrak{A})$ accepts an infinite word beginning with an element of RP_σ.*

Proof. If $\mathfrak{A}$ accepts a purely substitutive word u generated by σ, $u = \lim_{j \to \infty} \sigma^j(b)$ begins by an element of RP_σ. Since $\sigma(u) = u$, $\sigma^n(u)$ is accepted by $\mathfrak{A}$ so u is accepted by $\sigma^{-n}(\mathfrak{A})$.

On the converse, suppose that $\sigma^{-n}(\mathfrak{A})$ accepts an infinite word beginning by $b \in RP_\sigma$. Then, $\sigma^{m-n}(b)$ labels an accepting computation on $\sigma^{-m}(\mathfrak{A}) = \sigma^{-n}(\mathfrak{A})$. By iteration, for every $k \geqslant 1$, we have that $\sigma^{k(m-n)}(b)$ labels an accepting computation on $\sigma^{-n}(\mathfrak{A})$, so $\sigma^{n+k(m-n)}(b)$ always labels an accepting computation on $\mathfrak{A}$. By compactness, $u = \lim_{k \to \infty} \sigma^{n+k(m-n)}(b)$ is accepted by $\mathfrak{A}$. Now, because $b \in RP_\sigma$, the word $\lim_{j \to \infty} \sigma^j(b)$ is defined and equal to u. Therefore u, the purely substitutive word generated by σ on the letter b, is accepted by $\mathfrak{A}$.

The following result already appeared in [15], but an erratum clarified that some cases were not covered [16]. It is a parallel to a result in [5]. Our proof is essentially the same, but writing the proof through the lens of desubstitution makes it easier to extend the result to other decision problems.

Corollary 2. *The problem of the purely substitutive walk is decidable:*

Input: *an ω-automaton $\mathfrak{A}$, a homomorphism σ.*
Question: *Does $\mathfrak{A}$ accept some purely substitutive word generated by σ?*

This result extends to morphic words: to find a morphic word generated by σ and τ accepted by $\mathfrak{A}$, find a purely substitutive word generated by σ accepted by $\tau^{-1}(\mathfrak{A})$.

We now extend the method used to prove Proposition 5 to solve the question of finding a pure fixed point for a substitution σ in an ω-automaton. This improves Proposition 4 where we found a fixed point for some power of σ.

Proposition 6. *The problem of the fixed point walk is decidable:*

Input: *an ω-automaton $\mathfrak{A}$, a substitution σ.*
Question: *Does $\mathfrak{A}$ accepts a fixed point for σ?*

Proof. Let x be a fixed point for σ and define $FP_\sigma = \{b \in \mathcal{A} \mid \sigma(b) = b\}$ be the set of letters which are fixed points under σ. There are two cases:

1. x is an infinite word on the alphabet FP_σ.
2. there is a letter a appearing in x such that $\sigma(a) \neq a$. Suppose that a is the first such letter in x. Then x can be written as $x = pax'$ where p is a word on FP_σ. We have that $x = \sigma(x) = \sigma(p)\sigma(a)\sigma(x') = p\sigma(a)\sigma(x')$. So $a \sqsubseteq_p \sigma(a)$: a is right-prolongable for σ, so $\lim_{n \to \infty} \sigma^n(a)$ exists. Since $x = \sigma^n(x) = p\sigma^n(a)\sigma^n(x')$ for every $n \in \mathbb{N}$, by compactness, $x = p \lim_{n \to \infty} \sigma^n(a)$.

The algorithm works as follows. First (case 1), check whether $\mathfrak{A}$ accepts a word on the alphabet FP_σ. Second (case 2), define a new automata $\mathfrak{A}'$ which is equal to $\mathfrak{A}$ except that the set of initial states is all the states reachable in $\mathfrak{A}$ by words in FP_σ, and check (by the previous algorithm) if $\mathfrak{A}'$ accepts a purely substitutive word generated by σ.

The algorithm outputs "yes" if either case is satisfied, and "no" otherwise.

3.3   The Problem of the Infinitely Desubstitutable Walk

In this section, we suppose that $\mathfrak{A}$ is an ω-automaton and $\mathcal{S}$ is a finite set of substitutions (i.e. nonerasing homomorphisms, as is usual when studying multiple homomorphisms) on a single alphabet $\mathcal{A}$. We prove that the problem of finding an infinitely desubstitutable (infinite) word accepted by $\mathfrak{A}$ is decidable. To study this question, we introduce a meta-ω-automaton: each symbol is a substitution, and each state is an ω-automaton.

Definition 8 (The meta-ω-automaton $\mathcal{S}^{-\infty}(\mathfrak{A})$). *We define the ω-automaton $\mathcal{S}^{-\infty}(\mathfrak{A}) = (\mathcal{S}, D(\mathfrak{A}), \{\mathfrak{A}\}, \mathcal{T})$ with the alphabet $\mathcal{S}$, the set of states $D(\mathfrak{A}) = \{\sigma^{-1}(\mathfrak{A}), \sigma \in \mathcal{S}^*\}$, $\mathfrak{A}$ the only initial state and set of transitions $\mathcal{T} = \{\mathfrak{B} \xrightarrow{\sigma} \sigma^{-1}(\mathfrak{B}) \mid \mathfrak{B} \in D(\mathfrak{A}), \sigma \in \mathcal{S}\}$.*

Because $D(\mathfrak{A}) \subseteq \mathfrak{S}(\mathfrak{A})$ is finite (see Fact 1), $\mathcal{S}^{-\infty}(\mathfrak{A})$ is computable. We prove that directive sequences of words accepted by $\mathfrak{A}$ correspond to *non-nilpotent* walks in $\mathcal{S}^{-\infty}(\mathfrak{A})$, that is, walks $(\mathfrak{B}_n)_{n \in \mathbb{N}}$ such that $\mathcal{L}_\infty(\mathfrak{B}_n) \neq \varnothing$ for all n.

Proposition 7. *There exists x an infinite word infinitely desubstitutable by $(\sigma_n)_{n \in \mathbb{N}}$ accepted by $\mathfrak{A}$ if, and only if, there is a non-nilpotent infinite walk in $\mathcal{S}^{-\infty}(\mathfrak{A})$ labeled by $(\sigma_n)_{n \in \mathbb{N}}$.*

Corollary 3. *The set of directive sequences of infinitely desubstitutable words accepted by $\mathfrak{A}$ is the language of some ω-automaton.*

Proof (of Proposition 7). First, let x be an infinitely desubstitutable word with directive sequence $(\sigma_n)_{n\in\mathbb{N}}$, and let $(x_n)_{n\in\mathbb{N}}$ be the sequence of desubstituted words. Then, by Proposition 1, $x_n \in \mathcal{L}_\infty((\sigma_1 \circ \cdots \circ \sigma_{n-1})^{-1}(\mathfrak{A}))$. So the walk $(\sigma_{[\![0,n]\!]}^{-1}(\mathfrak{A}))_{n\in\mathbb{N}}$ is non-nilpotent and labeled by $(\sigma_n)_{n\in\mathbb{N}}$.

Second, let $(\sigma_n)_{n\in\mathbb{N}}$ label a non-nilpotent infinite walk in $\mathcal{S}^{-\infty}(\mathfrak{A})$. It means that each language $(\sigma_1 \circ \cdots \circ \sigma_k)^{-1}(\mathcal{L}_\infty(\mathfrak{A}))$ is nonempty. Now, consider the sequence $((\sigma_1 \circ \cdots \circ \sigma_n)(\mathcal{L}_\infty((\sigma_1 \circ \cdots \circ \sigma_n)^{-1}(\mathfrak{A}))))_{n\in\mathbb{N}}$. It satisfies the following:

1. each element of the sequence is included in $\mathcal{L}_\infty(\mathfrak{A})$;
2. because $\mathcal{L}_\infty((\sigma_1 \circ \cdots \circ \sigma_n)^{-1}(\mathfrak{A})))$ is compact and nonempty, and $(\sigma_1 \circ \cdots \circ \sigma_n)$ is continuous, every element of the sequence is compact and nonempty;
3. the sequence is decreasing for inclusion.

By Cantor's intersection theorem, there is a point x in the intersection of every element of the sequence. This point x is desubstitutable by any $\sigma_1 \circ \cdots \circ \sigma_k$, thus it is infinitely desubstitutable by the sequence $(\sigma_n)_{n\in\mathbb{N}}$.

With Proposition 7, we can deduce the decidability of the existence of an infinitely desubstitutable word accepted by an ω-automaton $\mathfrak{A}$. First, build $\mathcal{S}^{-\infty}(\mathfrak{A})$; second, remove the states corresponding to ω-automata with an empty language; last, check whether there is an infinite walk.

Proposition 8. *The problem of the infinitely desubstitutable walk is decidable:*

Input: *a finite set of substitutions $\mathcal{S}$, an ω-automaton $\mathfrak{A}$*
Question: *does $\mathcal{L}_\infty(\mathfrak{A})$ contain a word which is infinitely desubstitutable by $\mathcal{S}$?*

3.4   The Problem of the Büchi Infinitely Desubstitutable Walk

Proposition 8 does not apply directly to Sturmian words. Indeed, the classical characterization of Sturmian words restricts the possible directive sequences.

$\mathcal{S}_{St}$ is the set containing the four following substitutions, called (elementary) Sturmian morphisms, as described by [10].

$$L_0 : \begin{cases} 0 \mapsto 0 \\ 1 \mapsto 01 \end{cases}, \quad L_1 : \begin{cases} 0 \mapsto 10 \\ 1 \mapsto 1 \end{cases}, \quad R_0 : \begin{cases} 0 \mapsto 0 \\ 1 \mapsto 10 \end{cases}, \quad R_1 : \begin{cases} 0 \mapsto 01 \\ 1 \mapsto 1 \end{cases}$$

Theorem 1 ([13]). *A word is Sturmian iff it is infinitely desubstitutable by a directive sequence $(\sigma_n)_{n\in\mathbb{N}} \subset \mathcal{S}_{St}$ that alternates infinitely in type, i.e.: $\nexists N \in \mathbb{N}, (\forall n \geqslant N, \sigma_n \in \{L_0, R_0\})$ or $(\forall n \geqslant N, \sigma_n \in \{L_1, R_1\})$.*

This characterization is usually expressed in the S-adic framework, but is equivalent in this context [14]. In this section, we generalize Proposition 8 to Sturmian words and more general restrictions on the directive sequence.

Proposition 9. *The problem of the Sturmian walk is decidable:*

Input: *an ω-automaton $\mathfrak{A}$.*

Question: *is there a Sturmian infinite word accepted by $\mathfrak{A}$?*

Proof. Consider the associated representation automaton $\mathcal{S}_{St}^{-\infty}(\mathfrak{A})$. According to Proposition 7 combined with Theorem 1, there is a Sturmian infinite word accepted by $\mathfrak{A}$ if, and only if, there is an infinite computation accepted by $\mathcal{S}_{St}^{-\infty}(\mathfrak{A})$ labeled by a word $(\sigma_n)_{n\in\mathbb{N}}$ which alternates infinitely in type. This last condition is decidable: compute the strong connected components of $\mathfrak{A}$, and check that there is at least one strongly connected component C which contains two edges labeled by substitutions in $\{L_0, R_0\}$ and $\{L_1, R_1\}$, respectively.

In this case, the condition of alternating infinitely in type is easy to check: it can actually be described using a Büchi ω-automaton on the alphabet $\mathcal{S}$. Proposition 9 generalizes to every such condition.

Definition 9. *Let $\mathcal{S}$ be a set of substitutions, and $\mathfrak{R}$ a Büchi ω-automaton on the alphabet $\mathcal{S}$. Define $X_{\mathfrak{R}}$ as $\{x \in \mathcal{A}^{\mathbb{N}} \mid \exists(\sigma_n)_{n\in\mathbb{N}} \in \mathcal{L}_\infty(\mathfrak{R}), x$ is inf. desub. by $(\sigma_n)\}$.*

Proposition 10. *The following problem is decidable:*

Input: *an ω-automaton $\mathfrak{A}$, a finite set of substitutions $\mathcal{S}$, a Büchi ω-automaton $\mathfrak{R}$ on the alphabet $\mathcal{S}$*
Question: *is there an infinite word of $X_{\mathfrak{R}}$ accepted by $\mathfrak{A}$?*

Proof. The question of the problem is equivalent to: is $\mathcal{L}_\infty(\mathfrak{R}) \cap \mathcal{L}_\infty(\mathcal{S}^{-\infty}(\mathfrak{A})) \neq \varnothing$? The intersection between a Büchi ω-automaton and an ω-automaton is a Büchi ω-automaton that can be effectively constructed [11], and checking the non-emptiness of a Büchi ω-automaton is decidable.

The interest of Proposition 10 is that there exists a zoology of families of words which have a characterization by infinite desubstitution. For instance, Proposition 10 applies to Arnoux-Rauzy words [1] and to minimal dendric ternary words [7]. We also characterize the set of allowed directive sequences akin to Corollary 3: the set of directive sequences on $\mathcal{S}$ accepted by the Büchi ω-automaton $\mathfrak{R}$ that define a word accepted by $\mathfrak{A}$ is itself recognized by a Büchi ω-automaton.

Let us translate Proposition 10 in more dynamical terms:

Proposition 11. *The following problem is decidable:*

Input: *a set of substitutions $\mathcal{S}$, a Büchi ω-automaton $\mathfrak{R}$ on the alphabet $\mathcal{S}$ and a sofic shift $\mathbb{S}$.*
Question: *Is $\mathbb{S} \cap X_{\mathfrak{R}}$ empty?*

3.5 Application to the Coding of Sturmian Words

Here is an example of a natural question from combinatorics on words that we solve on Sturmian words, even though the method generalizes easily. Let W be a finite set of finite words on $\{0, 1\}$. Consider W^ω the set of infinite concatenations of elements of W, i.e. $W^\omega = \{x \in \{0, 1\}^{\mathbb{N}} \mid \exists(w_n)_{n\in\mathbb{N}} \subseteq W, x = \lim_{n\to\infty} w_0 w_1 \dots w_n\}$.

Proposition 12. *The following problem is decidable:*

Input: W *a finite set of words on* $\{0,1\}$
Question: *does* W^ω *contain a Sturmian word?*

Proof. The language W^ω is ω-regular: there is an ω-automaton $\mathfrak{A}_W$ such that $\mathcal{L}_\infty(\mathfrak{A}_W) = W^\omega$. Then, W^ω contains a Sturmian word iff $\mathfrak{A}_W$ accepts a Sturmian word, which is decidable by Proposition 9.

4   About ω-Automata Recognizing Sturmian Words

In this Section, we focus on Sturmian words and show that the language of Sturmian words is as far as possible from being regular, in the sense that an ω-automaton may only accept a Sturmian word if it accepts the image of the full shift under a Sturmian morphism.

Theorem 2. *Let* $\mathcal{S} = \mathcal{S}_{St}$ *be the set of elementary Sturmian morphisms as defined earlier, and let* $\mathfrak{A}$ *be an* ω*-automaton. If* $\mathfrak{A}$ *accepts a Sturmian word, then* $\exists \sigma \in \mathcal{S}_{St}^*, \sigma(\mathcal{A}^{\mathbb{N}}) \subseteq \mathcal{L}_\infty(\mathfrak{A})$.

This is equivalent to the presence of a total automaton in $\mathcal{S}^{-\infty}(\mathfrak{A})$: an ω-automaton $\mathfrak{A}$ is total if $\mathcal{L}_\infty(\mathfrak{A}) = \mathcal{A}^{\mathbb{N}}$. Totality is a stable property under any desubstitution.

To prove Theorem 2, we introduce the following technical tools.

Definition 10. *Let* $\mathfrak{A}$ *be an* ω*-automaton on* $\mathcal{A} = \{0,1\}$. *A state* q *of* $\mathfrak{A}$ *has property* (H) *if* $(\exists q_s, q \xrightarrow{0} q_s \to^\omega \cdots \in \mathfrak{A}) \Leftrightarrow (\exists q_t, q \xrightarrow{1} q_t \to^\omega \cdots \in \mathfrak{A})$, *where* $q_s \to^\omega \cdots$ *means that there is an infinite computation starting from* q_t *in* $\mathfrak{A}$.

If all states of $\mathfrak{A}$ have property (H), there are two possibilities: if there is no infinite computation starting on an initial state, the infinite language of $\mathfrak{A}$ is empty; otherwise, $\mathfrak{A}$ is total.

Lemma 1. *Let* $\mathfrak{C}$ *be an* ω*-automaton, and* $\phi \in \mathcal{S}_{St}^*$ *starting with* L_0 *and ending with* L_1 *such that* $\phi^{-1}(\mathfrak{C}) = \mathfrak{C}$. *Then, every state of* $\mathfrak{C}$ *has property* (H).

Proof (of Lemma 1). Let $\mathfrak{C} = (\{0,1\}, Q_\mathfrak{C}, I_\mathfrak{C}, T_\mathfrak{C})$, and $q \in Q_\mathfrak{C}$. First, suppose that $q \xrightarrow{0} q_t \to^\omega \cdots$ is a computation in $\mathfrak{C}$. Then $q \xrightarrow{0} q_t$ is also a transition of $\phi^{-1}(\mathfrak{C})$. So $q \xrightarrow{\phi(0)}{}^* q_t$ is a computation in $\mathfrak{C}$. Because ϕ ends with L_1, $\phi(1) \sqsubseteq_p \phi(0)$. So $q \xrightarrow{\phi(1)}{}^* q_u \xrightarrow{m}{}^* q_t \to^\omega \cdots$ is a computation in $\mathfrak{C}$, with some $q_u \in Q_\mathfrak{C}$ and $\phi(0) = \phi(1)m$. Now, using $\mathfrak{C} = \phi^{-1}(\mathfrak{C})$, $q \xrightarrow{1} q_u \xrightarrow{m}{}^* q_t \to^\omega \cdots$ is a computation in $\mathfrak{C}$.

Conversely, if $q \xrightarrow{1} q_t \to^\omega \cdots$ is a computation in $\mathfrak{C} = \phi^{-1}(\mathfrak{C})$, there is also $q \xrightarrow{\phi(1)}{}^* q_t \to^\omega \cdots$ Because ϕ begins with L_0, $\phi(1) = 0m$ for some finite m. So the last computation can be written $q \xrightarrow{0} q_u \xrightarrow{m}{}^* q_t \to^\omega \cdots$

Proof (of Theorem 2). Let x be a Sturmian word accepted by $\mathfrak{A}$. Consider the transformation of ω-automata forget : $(\mathcal{A}, Q, I, T) \mapsto (\mathcal{A}, Q, Q, T)$ which makes all states initial. Then, forget($\mathfrak{A}$) also accepts x, and $\mathcal{L}_{\infty}(\text{forget}(\mathfrak{A}))$ is a sofic shift. Then $\bigcup_{n \geq 0} S^n(x)$, which is the orbit of x under the shift S, is contained in $\mathcal{L}_{\infty}(\text{forget}(\mathfrak{A}))$. Let $\chi(x)$ be the Sturmian characteristic word associated with x (see [12]): it belongs to the orbit of x, so it is accepted by forget($\mathfrak{A}$). Then, $\chi(x) = \lim_{n \to \infty} \sigma_0 \circ \cdots \circ \sigma_n(a_n)$ with $(\sigma_n)_{n \in \mathbb{N}} \subseteq \mathcal{S}_{St}$ a sequence that alternates infinitely in type (see Theorem 1). Besides, because $\chi(x)$ is a characteristic word, it represents the orbit of zero from the point of view of circle rotation (see [12]): when combined with Proposition 2.7 of [4], it yields that $(\sigma_n)_{n \in \mathbb{N}} \subseteq \{L_0, L_1\}^{\mathbb{N}}$. By the pigeonhole principle, there is an ω-automaton $\mathfrak{B}$ that appears infinitely often in the sequence $(\sigma_{[\![0,n]\!]}^{-1}(\text{forget}(\mathfrak{A})))_{n \in \mathbb{N}} \subseteq \mathfrak{S}(\text{forget}(\mathfrak{A}))$. Thus, we can find a substitution τ such that $\mathfrak{B} = \tau^{-1}(\mathfrak{B})$ and $\tau \in \{L_0, L_1\}^* \backslash (L_0^* \cup L_1^*)$. Because τ contains both L_0 and L_1, there are two cases:

1. $L_1 L_0 \sqsubseteq_f \tau$: we can write $\tau = p_\tau L_1 L_0 s_\tau$. Let $\mathfrak{B}' = (p_\tau \circ L_1)^{-1}(\mathfrak{B})$ and $\tau' = L_0 \circ s_\tau \circ p_\tau \circ L_1$: we have that $\tau'^{-1}(\mathfrak{B}') = \mathfrak{B}'$.
2. $L_1 L_0 \not\sqsubseteq_f \tau$: then, τ begins with a L_0 and ends with a L_1.

In both cases, we can come back to the case where τ begins with a L_0 and ends with a L_1.

Now, we apply Lemma 1 to show that every state of $\mathfrak{B}$ has property (H). $\mathfrak{B}$ can be written as $\psi^{-1}(\text{forget}(\mathfrak{A}))$ for some Sturmian morphism ψ. Since the transformation forget does not modify the transitions of an ω-automaton, this yields that every state of $\psi^{-1}(\mathfrak{A})$ also has property (H). Since by assumption $\psi^{-1}(\mathfrak{A})$ accepts an infinite word, it follows that it is total.

Let f be the Fibonacci word, i.e. the substitutive word associated with the substitution $\sigma_f(0) = 01, \sigma_f(1) = 0$. Since Lemma 1 holds when $\phi = \sigma_f^n$ $(n \geq 1)$, by adapting the proof of Theorem 2, we obtain an equivalent statement for f:

Corollary 4. *Let $\mathfrak{A}$ be an ω-automaton which accepts f. Then, there exists $n \in \mathbb{N}$ such that $\sigma_f^{-n}(\mathfrak{A})$ is total.*

This combinatorial result can be thought in dynamical terms:

Corollary 5. *A sofic subshift $\mathbb{S}$ contains f iff $\mathbb{S}$ contains some $\sigma_f^n(\mathcal{A}^{\mathbb{N}})$.*

Because the Fibonacci word is aperiodic, containing f means that there is a substitution τ such that $\tau(\mathcal{A}^{\mathbb{N}})$ is contained in $\mathbb{S}$. Because the Fibonacci word is Sturmian, Berstel and Séébold [10] established that τ had to be a Sturmian morphism. This new analysis specifies that τ can be chosen a power of σ_f.

5   Open Questions

– Following Proposition 8, find an algorithm to find an accepted $\mathcal{S}$-adic word. There are technical difficulties to take into account the growth of the directive sequence, which should be solvable using results from [14].

– Can our methods extend to Büchi ω-automata, as in [5]? The difficulty is that the language of Büchi ω-automata is not always compact, so Proposition 4 does not apply. It may be possible to extend methods from [5].
– For which sets of substitutions does Theorem 2 hold?

References

1. Arnoux, P., Rauzy, G.: Représentation géométrique de suites de complexité 2n+1. Bulletin de la Société mathématique de France **119**(2), 199–215 (1991). https://doi.org/10.24033/bsmf.2164
2. Bassino, F., David, J., Sportiello, A.: Asymptotic enumeration of minimal automata. In: Dürr, C., Wilke, T. (eds.) 29th International Symposium on Theoretical Aspects of Computer Science (STACS 2012). Leibniz International Proceedings in Informatics (LIPIcs), vol. 14, pp. 88–99. Schloss Dagstuhl-Leibniz-Zentrum fuer Informatik, Dagstuhl (2012). https://doi.org/10.4230/LIPIcs.STACS.2012.88
3. Berstel, J.: Axel Thue's papers on repetitions in words: a translation. In: Monographies du LaCIM, vol. 11, pp. 65–80. LaCIM (1992). https://hal.science/hal-00620702
4. Berthé, V., Holton, C., Zamboni, L.Q.: Initial powers of sturmian sequences. Acta Arithmetica **122**, 315–347 (2006). https://doi.org/10.4064/aa122-4-1
5. Carton, O., Thomas, W.: The monadic theory of morphic infinite words and generalizations. Inf. Comput. **176**(1), 51–65 (2002). https://doi.org/10.1006/inco.2001.3139
6. Ferenczi, S.: Rank and symbolic complexity. Ergodic Theor. Dyn. Syst. **16**(4), 663–682 (1996). https://doi.org/10.1017/S0143385700009032
7. Gheeraert, F., Lejeune, M., Leroy, J.: S-adic characterization of minimal ternary dendric subshifts. CoRR abs/2102.10092 (2021). https://arxiv.org/abs/2102.10092
8. Hopcroft, J.E., Ullman, J.D.: Introduction to Automata Theory, Languages, and Computation. Addison-Wesley Publishing Company, Boston (1979)
9. Lind, D., Marcus, B.: An Introduction to Symbolic Dynamics and Coding. Cambridge University Press, Cambridge (1995). https://doi.org/10.1017/CBO9780511626302
10. Lothaire, M.: Sturmian words. In: Algebraic Combinatorics on Words. Encyclopedia of Mathematics and its Applications, vol. 2, pp. 45–110. Cambridge University Press, Cambridge (2002). https://doi.org/10.1017/CBO9781107326019.003
11. Perrin, D., Pin, J.E.: Infinite Words, Pure and Applied Mathematics, vol. 141. Elsevier, Amsterdam (2004). https://doi.org/10.1016/S0079-8169(04)80002-3
12. Perrin, D., Restivo, A.: A note on sturmian words. Theor. Comput. Sci. **429**, 265–272 (2012). https://doi.org/10.1016/j.tcs.2011.12.047
13. Pytheas Fogg, N., Berthé, V., Ferenczi, S., Mauduit, C., Siegel, A.: Sturmian sequences. In: Substitutions in Dynamics, Arithmetics and Combinatorics, vol. 6, pp. 143–198. Springer, Heidelberg (2002). https://doi.org/10.1007/3-540-45714-3_6
14. Richomme, G.: On sets of indefinitely desubstitutable words. Theor. Comput. Sci. **857**, 97–113 (2021). https://doi.org/10.1016/j.tcs.2021.01.004
15. Salo, V.: Decidability and universality of quasiminimal subshifts. J. Comput. Syst. Sci. **89**, 288–314 (2017). https://doi.org/10.1016/j.jcss.2017.05.017
16. Salo, V.: Notes and errata on Decidability and Universality of Quasiminimal Subshifts (2022). https://villesalo.com/notes/DaUoQSNotes.html
17. Thomas, W.: Handbook of Formal Languages: Volume 3 Beyond Words. Springer, Heidelberg (1997). https://doi.org/10.1007/978-3-642-59126-6

String Attractors for Factors
of the Thue-Morse Word

Francesco Dolce[(✉)]

FIT, Czech Technical University in Prague, Prague, Czech Republic
dolcefra@fit.cvut.cz

Abstract. In 2020 Kutsukake et al. showed that every for every $n \geq 4$ the prefix of length 2^n of the Thue-Morse word has a string attractor of size 4. In this paper we extend their result by constructing a smallest string attractor for any given factor of the Thue-Morse word. In particular, we show that these string attractors have size at most 5 and that this upper bound is sharp.

Keywords: string attractors · Thue-Morse word · factorial languages

1 Introduction

String attractors were introduced by Kempa and Prezza in [6] in the context of dictionary-based data compression. A string attractor for a word w is a set of positions of the word such that all factors of w have an occurrence containing at least one of the elements of the set. Intuitively, the more repetitive is w the lower is the size of a smallest string attractor for w. Actually, the smallest size of a string attractor for a word is a lower bound for several other repetitiveness measures associated with the most common compression schemes, including the number of phrases in the LZ77 parsing and the number of equal-letter runs produced by the Burrows-Wheeler Transform (see [6, 10, 12]).

While it is trivial to construct a string attractor for a given word (e.g., by taking all possible positions), finding a smallest one is a NP-complete problem.

Mantaci et al. studied in [10] the size of a smallest string attractor of several infinite families of words. In particular they showed that every standard Sturmian word different than a letter has a smallest string attractor of size 2 (see also [5] for a generalization of this results to episturmian words), while the de Brujin word of length n has a smallest string attractor of size $\frac{n}{\log n}$. In the same paper they also studied the well-known Thue-Morse word $\mathbf{t}$, also known as Prouhet-Thue-Morse word, since first studied by Prouhet before being rediscovered by Thue and Morse, between others (see [11, 13, 16]). In a preliminary version of their paper ([9]) Mantaci et al. conjectured that prefixes of size 2^n of $\mathbf{t}$ have a smallest string attractor of size n. This conjecture has been proven to be wrong by Katsukake et al. in [7], who showed that for any such prefix it is possible to find a string attractor of size at most 4.

Schaeffer and Shallit introduced in [15] the notion of string attractor profile function for infinite words by evaluating the size of a smallest attractor for each

© The Author(s), under exclusive license to Springer Nature Switzerland AG 2023
A. Frid and R. Mercaş (Eds.): WORDS 2023, LNCS 13899, pp. 117–129, 2023.
https://doi.org/10.1007/978-3-031-33180-0_9

prefix (see also [14]). If instead of prefixes we consider a generic factor of a (finite or infinite) word the situation get more complicated. Indeed, the measure of a smallest string attractor is not monotone, meaning that a factor w of a word u can have a smallest string attractor bigger than a string attractor of u (see Proposition 2).

In this article we prove and explicitly construct a smallest string attractor for any given factor of the Thue-Morse word. In particular, our main result is the following.

Theorem 1. *Let w be a non-empty finite factor of* **t**. *Then there exists a string attractor for w of size at most 5.*

2   Preliminaries

For all undefined notation we refer to [8]. Let $\mathcal{A}$ be an *alphabet*, that is a finite set of symbols called *letters*. A (finite) *word* over $\mathcal{A}$ of *length* n is a concatenation $u = u_1 \cdots u_n$, where $u_i \in \mathcal{A}$ for all $i \in \{1, \ldots, n\}$. The length of u is denoted by $|u|$. The set of all finite words over $\mathcal{A}$ together with the operation of concatenation form a monoid, denoted by $\mathcal{A}^*$, whose neutral element is the *empty word* ε. We also denote $\mathcal{A}^+ = \mathcal{A}^* \setminus \{\varepsilon\}$. Similarly, given a set of words $S \subset \mathcal{A}^*$, we denote by S^* (resp., S^+) the set of all possible concatenations (resp., non-empty concatenations) of elements of S. When $\mathcal{A} = \{\mathsf{a}, \mathsf{b}\}$ is a binary alphabet we denote by $\overline{w}$ the word obtained from w by changing every a in b and vice versa. Formally $\overline{w}$ is obtained from w by applying the involution $\bar{\ }: a \mapsto b; b \mapsto a$.

Let $u = pfs$ for some $p, f, s \in \mathcal{A}^*$. We call p a *prefix* of w, s a *suffix* of w and f a *factor* of w. The prefix p (resp. suffix s) is called *proper* if it is different than u. If both p and s are non-empty we call f an *internal factor* of u. The set $\mathrm{Pref}(u)$ (resp., $\mathrm{Suf}(u)$) is the set of all non-empty prefixes (resp., suffixes) of u. The *language* of u, denoted by $\mathcal{L}(u)$, is the set of all finite factors of u.

An *infinite word* over $\mathcal{A}$ is a sequence $\mathbf{u} = u_1 u_2 \cdots$, where $u_i \in \mathcal{A}$ for every positive integer i. The notions above (prefix, suffix, etc.) naturally extend to infinite words.

Example 1. The Thue-Morse word is the infinite binary word

$$\mathbf{t} = \lim_{n \to \infty} t_n = \mathsf{abbabaabbaababbabaabbaabbabaabbaabbabaababab} \cdots,$$

where $t_0 = \mathsf{a}$ and $t_{n+1} = t_n \overline{t_n}$ for any $n > 0$. Note that for any $n \in \mathbb{N}$ we have $|t_n| = |\overline{t_n}| = 2^n$.

Given a set $M \subset \mathbb{Z}$ and an integer $q \in \mathbb{Z}$, we denote $M + q = \{m + q \mid m \in M\}$.

3   String Attractors

Let $w, u \in \mathcal{A}^+$, with $w \in \mathcal{L}(u)$, we say that w has an *occurrence starting at position i* in u, if it is possible to write $w = u_i u_{i+1} \cdots u_{i+|w|-1}$, with the convention that the empty word has an occurrence at every position. Clearly a word

w could have multiple occurrences in u. Given a position j with $1 \leq j \leq |u|$, we also say that an occurrence of w in u *contains* the position j if such occurrence starts at position i with $i \leq j < i + |w|$.

Example 2. Let us consider the words t_n as in Example 1. The word $w = $ bba has three occurrences in $t_4 = $ abbabaabbaababba starting respectively at positions 2, 8 and 14. The second occurrence is the only one containing the position 10.

Given a word $u \in \mathcal{A}^+$ a set Γ of positions is a *string attractor* for u if for every factor $w \in \mathcal{L}(u)$ there exists a $\gamma \in \Gamma$ such that at least one occurrence of w is of the form $w = u_i u_{i+1} \cdots u_{i+|w|-1}$ with $i \leq \gamma < i + |w|$.

The set $\{1, 2, \ldots, |u|\}$ is trivially a string attractor for a word u. On the other hand, a trivial lower bound for the size of a string attractor is given by the number of different letters appearing in u. Moreover, if Γ is a string attractor for u, so is Γ' for every superset $\Gamma' \supset \Gamma$. Note that a word can have different string attractors of the same size and, more generally, different string attractors that are not included into each other.

Example 3. Let t_n and $\overline{t_n}$ be defined as in Example 1. The set $\Gamma_0 = \{1\}$ is a string attractor for both words $t_0 = \underline{a}$ and $\overline{t_0} = \underline{b}$ (the positions of the string attractor are underlined). Similarly, the set $\Gamma_1 = \{1, 2\}$ is a string attractor for $t_1 = \underline{ab}$ and for $\overline{t_1} = \underline{ba}$. Such string attractor is the smallest one, since both letters a and b must be covered.

The set $\Gamma_2 = \{1, 2, 4\}$ is a string attractor for the word $t_2 = \underline{ab}b\underline{a}$ (resp., for $\overline{t_2}$). Notice that $\Gamma_2' = \{2, 4\}$ is also a string attractor for $t_2 = a\underline{b}b\underline{a}$ (resp., for $\overline{t_2}$). Since both letters appear in t_2, the minimal size for a string attractor is 2. It is easy to check that $\{2, 5, 7\}$ is a string attractor for the word $t_3 = a\underline{b}ba\underline{b}a\underline{a}b$ (resp., for $\overline{t_3}$), while the same word does not have any string attractor of size 2. A larger string attractor for t_3 is given by $\Gamma_3 = \{2, 3, 4, 6\}$.

It is possible to check that the sets $\Gamma_4 = \{4, 6, 8, 12\}$, $\Gamma_5 = \{8, 12, 16, 24\}$, $\Gamma_6 = \{16, 24, 32, 48\}$ and $\Gamma_7 = \{32, 48, 64, 96\}$ are smallest string attractors respectively for the words $t_4 = abb\underline{a}b\underline{a}a\underline{b}baa\underline{b}abba$ (resp., for $\overline{t_4}$), t_5 (resp., for $\overline{t_5}$), t_6 (resp. $\overline{t_6}$) and t_7 (resp., $\overline{t_7}$).

The following two interesting combinatorial results are proved in [10, Propositions 12 and 14].

Proposition 1 ([10]). *Let $u, v \in \mathcal{A}^+$, Γ_u a string attractor for u and Γ_v a string attractor for v. Then $\Gamma_u \cup \{|u|\} \cup (\Gamma_v + |u|)$ is a string attractor for uv.*

Example 4. Let t_2, $\overline{t_2}$ and Γ_2' as in Example 3. A string attractor for $t_3 = t_2 \overline{t_2}$ is given by $\Gamma_2' \cup \{4\} \cup (\Gamma_2' + 4) = \{2, 4, 6, 8\}$. Note that such a string attractor is not a smallest one.

Proposition 2 ([10]). *The size of a smallest string attractor for a word is not a monotone measure.*

The previous proposition says that if w is a factor of u, then it could be possible for u to have a string attractor of size smaller than the size of a smallest string attractor for w.

Example 5. Let t_n be as in Example 1. As seen in Example 3, the word t_7 has a smallest string attractor of size 4. However, it is possible to check that the word $w = $ abaababbaabbabaabbaababbaabbabaababbabaabbaababbab $= a\overline{t_4}\,\overline{t_5}\,b \in \mathcal{L}(t_7)$ has no string attractor of size 4. Note that $\Gamma = \{9, 13, 25, 33, 41\}$ is a string attractor of size 5 of w.

4   Proof of the Main Result

In [2] Brlek shows several combinatorial results concerning the factors in $\mathcal{L}(\mathbf{t})$. In particular he provides an explicit formula of the factor complexity of $\mathbf{t}$. Part of it is stated in the following.

Proposition 3 ([2]). *Let $n \in \mathbb{N}$ and $w \in \mathcal{L}(t_n)$ with $|w| \geq 2^{n-2} + 1$. The word t_n has exactly one occurrence of w.*

An important ingredient of our proof is [7, Theorem 2].

Theorem 2 ([7]). *Let $n \geq 4$. The set*

$$\Gamma_n = \{2^{n-2},\ 3 \cdot 2^{n-3},\ 2^{n-1},\ 3 \cdot 2^{n-2}\}$$

is a string attractor both for t_n and $\overline{t_n}$.

Note that in their papers Kutsukake et al. only state the result for t_n, but the same argument actually works also for $\overline{t_n}$.

We are now ready to prove Theorem 1.

Proof of Theorem 1. It can easily checked that every factor of t_n (resp., of $\overline{t_n}$) with $n \leq 5$ has a string attractor of size at most 4. Let us suppose the property true for all factors in $\mathcal{L}(t_n) \cup \mathcal{L}(\overline{t_n})$ and let us consider the case of $w \in \mathcal{L}(t_{n+1}) = \mathcal{L}(t_n \overline{t_n})$, with $n \geq 6$ (the case $w \in \mathcal{L}(\overline{t_{n+1}})$ being symmetrical).

If $w \in \mathcal{L}(t_n) \cup \mathcal{L}(\overline{t_n})$, then the result follows by induction. Thus, we can suppose that w has an occurrence in t_{n+1} containing the center of $t_n \overline{t_n}$, i.e., the last letter of the prefix t_n. In the following remaining part of the proof we consider all possible such factors of t_{n+1} by increasing their size. The idea is to write each factor as $w = \lambda t \rho$, with the central factor t of the form t_k or $\overline{t_k}$ for a certain $k \in \mathbb{N}$, and $\lambda = st'$ and $\rho = t''p$, where $t', t'' \in \{t_i, \overline{t_i} \mid i \in \mathbb{N}\}^*$ (more precisely we have $t' = t'_{h_\ell} \cdots t'_{h_1}$ and $t'' = t''_{j_1} \cdots t''_{j_r}$ with $h_\ell < \ldots < h_1$ and $j_1 > \ldots > j_r$), and s (resp., p) is a suffix (resp., a prefix) of a some element in $\{t_i, \overline{t_i} \mid i \in \mathbb{N}\}$.

Since the center of $t_n \overline{t_n}$ is also the center of $t_{n-2} \overline{t_{n-2}}$, if $w \in \mathcal{L}(t_{n-2}\overline{t_{n-2}}) = \mathcal{L}(t_{n-1})$, we can conclude by induction. Let us thus suppose that $w \notin \mathcal{L}(t_{n-1})$. We can write w either as $w = \lambda t_{n-2} \rho$, with $\lambda \in \mathrm{Suf}(t_{n-1} \overline{t_{n-2}})$ and $\rho \in \mathrm{Pref}(\overline{t_n})$, or as $w = \lambda \overline{t_{n-2}} \rho$ with $\lambda \in \mathrm{Suf}(t_n)$ and $\rho \in \mathrm{Pref}(t_{n-2} t_{n-1})$. Let us focus on the former case. Since $|\lambda t_{n-2}| > 2^{n-2}$ then $w \notin \mathcal{L}(t_n)$ according to Proposition 3.

In the following we extend, step by step, ρ to the right, and, for every fixed ρ, we extend λ to the left. While the first step is fully developed, we let the reader check the details of Steps 2 to 7.

Step 1. Let us start considering $\rho = p$ with $p \in \mathrm{Pref}(\overline{t_n})$.

i) Let $w = s \, t_{n-2} \, p$, with $s \in \mathrm{Suf}(\overline{t_{n-4}})$ and $p \in \mathrm{Pref}(\overline{t_{n-4}})$. Then

$$\Gamma = \Gamma_{n-2} + |\lambda|,$$

where $\lambda = s$, is a string attractor for w (see Fig. 1, where we represent only the central factor $\overline{t_{n-1}} \, \overline{t_{n-1}}$ of t_{n+1}). Indeed, let $v \in \mathcal{L}(w)$. If $v \in \mathcal{L}(t_{n-2})$, then, by Theorem 2, one of its occurrences contains at least one of the positions of the string attractor Γ_{n-2} shifted by $|\lambda| = |s|$ (see Fig. 1). If v has an occurrence appearing to the left of the left-most position in Γ, then $v \in \mathcal{L}(\overline{t_{n-4}} \, t_{n-4}) \subset \mathcal{L}(t_{n-2})$, and thus it also has another occurrence containing at least one of the positions of $\Gamma_{n-2} + |\lambda|$ (see Fig. 1). Similarly, if v has an occurrence appearing to the right of the right-most position in Γ, then $v \in \mathcal{L}(t_{n-4} \overline{t_{n-4}}) \subset \mathcal{L}(t_{n-2})$ and thus v has another occurrence containing at least one of the positions of $\Gamma_{n-2} + |\lambda|$.

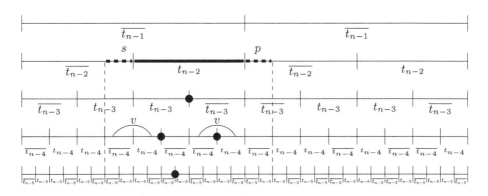

Fig. 1. A smallest string attractor Γ for a factor $w = s \, t_{n-2} \, p$ of t_{n+1}, with $s \in \mathrm{Suf}(\overline{t_{n-4}})$ and $p \in \mathrm{Pref}(\overline{t_{n-4}})$.

ii) Let $w = s \, \overline{t_{n-4}} \, t_{n-2} \, p$, with $s \in \mathrm{Suf}(t_{n-4})$ and $p \in \mathrm{Pref}(\overline{t_{n-4}})$. Then

$$\Gamma = \left((\Gamma_{n-2} \setminus \{3 \cdot 2^{n-4}\}) \cup \{0\} \right) + |\lambda|,$$

where $\lambda = s \, \overline{t_{n-4}}$, is a string attractor for w (see Fig. 2, where we represent only the central factor $\overline{t_{n-1}} \, \overline{t_{n-1}}$ of t_{n+1}). Indeed, let $v \in \mathcal{L}(w)$. If $v \in \mathcal{L}(t_{n-2})$, then, using Theorem 2, we have that v has an occurrence containing at least one of the positions of Γ_{n-2} shifted by $|\lambda|$; if the only position contained by such occurrence is $3 \cdot 2^{n-4} + |\lambda|$, then $v \in \mathcal{L}(\overline{t_{n-4}}\, t_{n-4})$, hence v has another occurrence containing the position $|\lambda|$.

If v has an occurrence appearing to the left of the left-most position in Γ, then $v \in \mathcal{L}(t_{n-4}\, \overline{t_{n-4}}) \subset \mathcal{L}(t_{n-2})$ and we can conclude using again Theorem 2. If v has an occurrence appearing to the right of the right-most position of Γ, i.e., $2^{n-3} + |\lambda|$, then $v \in \mathcal{L}(\overline{t_{n-4}}\, t_{n-4}\, \overline{t_{n-4}})$: either v is fully contained in $\mathcal{L}(t_{n-4}\, \overline{t_{n-4}}) \subset \mathcal{L}(t_{n-2})$ and we can conclude, or v has another occurrence containing the position $|\lambda|$ (see Fig. 2).

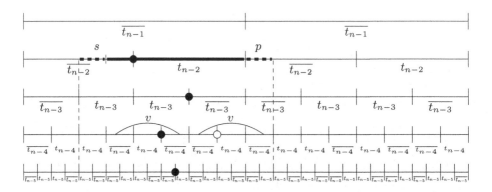

Fig. 2. A smallest string attractor Γ for a factor $w = s\,\overline{t_{n-4}}\, t_{n-2}\, p$ of t_{n+1}, with $s \in \mathrm{Suf}(t_{n-4})$ and $p \in \mathrm{Pref}(\overline{t_{n-4}})$. The position of $\Gamma_{n-2} + |\lambda|$ that is not in Γ is in white.

iii) Let $w = s\,t_{n-3}\,t_{n-2}\,p$, with $s \in \mathrm{Suf}(\overline{t_{n-5}})$ and $p \in \mathrm{Pref}(\overline{t_{n-4}})$. Then

$$\Gamma = \big((\Gamma_{n-2} \setminus \{2^{n-4},\, 3 \cdot 2^{n-4}\}) \cup \{-2^{n-4},\, 0\}\big) + |\lambda|,$$

where $\lambda = s\,t_{n-3}$, is a string attractor for w. Indeed, let $v \in \mathcal{L}(w)$. Similarly to the previous case, if $v \in \mathcal{L}(t_{n-2})$, then, by Theorem 2, v has an occurrence containing at least one of the positions of Γ_{n-2} shifted by $|\lambda|$; if the only position contained in the occurrence of v is $2^{n-4} + |\lambda|$, then $v \in \mathcal{L}(\overline{t_{n-4}}\, t_{n-4})$ and thus there is another occurrence of v containing the position $-2^{n-4} + |\lambda|$; if the only position contained in the occurrence of v is $3 \cdot 2^{n-4} + |\lambda|$, then $v \in \mathcal{L}(\overline{t_{n-4}}\, t_{n-4})$ and thus there is another occurrence of v containing the position $|\lambda|$.

If v appears to the left of the left-most position in Γ, then we can conclude since $v \in \mathcal{L}(\overline{t_{n-5}}\, \overline{t_{n-5}}\, t_{n-5}) \subset \mathcal{L}(t_{n-2})$. If v appears between the positions $-2^{n-4} + |\lambda|$ and $|\lambda|$, then $v \in \mathcal{L}(\overline{t_{n-4}}) \subset \mathcal{L}(t_{n-2})$. If v appears to the right of the right-most position in Γ, i.e., $2^{n-3} + |\lambda|$, then $v \in \mathcal{L}(\overline{t_{n-4}}\, t_{n-4}\, \overline{t_{n-4}})$: either v is fully contained in $\mathcal{L}(t_{n-4}\, \overline{t_{n-4}}) \subset \mathcal{L}(t_{n-2})$, or v has another occurrence containing the position $|\lambda|$.

iv) Let $w = s\,\overline{t_{n-5}}\, t_{n-3}\, t_{n-2}\, p$, with $s \in \mathrm{Suf}(t_{n-3}\, \overline{t_{n-4}}\, t_{n-5})$ and $p \in \mathrm{Pref}(\overline{t_{n-4}})$. Then

$$\Gamma = \big((\Gamma_{n-2} \setminus \{3 \cdot 2^{n-5},\, 3 \cdot 2^{n-4}\}) \cup \{-2^{n-3}, 0\}\big) + |\lambda|,$$

where $\lambda = s\,\overline{t_{n-5}}\,t_{n-3}$, is a string attractor for w. Indeed, let $v \in \mathcal{L}(w)$. As above, if $v \in \mathcal{L}(t_{n-2})$, then, by Theorem 2, v has an occurrence containing at least one of the positions of Γ_{n-2} shifted by $|\lambda|$; if the only position contained in the occurrence is $3 \cdot 2^{n-5} + |\lambda|$ (resp., $3 \cdot 2^{n-4} + |\lambda|$) then v has another occurrence containing the position $-2^{n-3} + |\lambda|$ (resp., $|\lambda|$).
If v appears to the left of the left-most position of Γ, then $v \in \mathcal{L}(t_{n-3}\,\overline{t_{n-3}}) = \mathcal{L}(t_{n-2})$ and we can conclude. If v appears between the positions $-2^{n-3}+|\lambda|$ and $|\lambda|$, then $v \in \mathcal{L}(t_{n-3}) \subset \mathcal{L}(t_{n-2})$ and we can conclude. If v appears to the right of the right-most position of Γ (i.e., $2^{n-3} + |\lambda|$) then it is contained in $\mathcal{L}(\overline{t_{n-4}}\,t_{n-4}\,\overline{t_{n-4}})$: either v is fully contained in $\mathcal{L}(t_{n-4}\,\overline{t_{n-4}}) \subset \mathcal{L}(t_{n-2})$ and we can conclude, or v has another occurrence containing the position $|\lambda|$.

v) Let $w = s\,t_{n-3}\,\overline{t_{n-1}}\,p$, with $s \in \mathrm{Suf}(\overline{t_{n-3}})$ and $p \in \mathrm{Pref}(\overline{t_{n-4}})$. Then

$$\Gamma = \left(\left(\Gamma_{n-1} \setminus \{3 \cdot 2^{n-4}\}\right) \cup \{-2^{n-4}\}\right) + |\lambda|.$$

where $\lambda = s\,t_{n-3}$, is a string attractor for w. Indeed, let $v \in \mathcal{L}(w)$. If $v \in \mathcal{L}(\overline{t_{n-1}})$ then, by Theorem 2, v has an occurrence containing at least one of the positions of Γ_{n-1} shifted by $|\lambda|$; if the only position contained in the occurrence is $3 \cdot 2^{n-4} + |\lambda|$, then $v \in \mathcal{L}(t_{n-4}\,\overline{t_{n-4}})$ and thus there is another occurrence of v containing the position $-2^{n-4} + |\lambda|$.
If v appears to the left of the left-most position of Γ (resp., between $-2^{n-4}+|\lambda|$ and $2^{n-3} + |\lambda|$; resp., to the right of the right-most position of Γ), then it is contained in $\mathcal{L}(\overline{t_{n-4}}\,t_{n-4}\,t_{n-4})$ (resp., $v \in \mathcal{L}(\overline{t_{n-4}}\,\overline{t_{n-4}}\,t_{n-4})$; resp., $v \in \mathcal{L}(\overline{t_{n-4}}\,t_{n-4}\,\overline{t_{n-4}})$) thus it is also contained in $\mathcal{L}(\overline{t_{n-1}})$ and we can conclude.

vi) Let $w = s\,\overline{t_{n-2}}\,\overline{t_{n-1}}\,p$, with $s \in \mathrm{Suf}(t_{n-4})$ and $p \in \mathrm{Pref}(\overline{t_{n-4}})$. This is the first case when it is not enough to "move" some of the positions of a string attractor of the form Γ_k, with $k \in \mathbb{N}$. Indeed, as shown in Example 5, in this case it is not possible to have a string attractor of size 4. On the other hand the set

$$\Gamma = \left(\left(\Gamma_{n-1} \setminus \{3 \cdot 2^{n-4}\}\right) \cup \{-2^{n-3}, \, -2^{n-4}\}\right) + |\lambda|,$$

where $\lambda = s\,\overline{t_{n-2}}$, is a string attractor for w (see Fig. 3). Indeed, let $v \in \mathcal{L}(w)$. If $v \in \mathcal{L}(\overline{t_{n-1}})$ then, by Theorem 2, v has an occurrence containing at least one of the positions of Γ_{n-1}; if the only position contained in the occurrence is $3 \cdot 2^{n-4} + |\lambda|$, then $v \in \mathcal{L}(t_{n-4}\,\overline{t_{n-4}})$ and thus there exists another occurrence of v containing $-2^{n-4} + |\lambda|$.
All the other cases are proved as in the previous cases: namely if v appears to the left of $-2^{n-3} + |\lambda|$, (resp., between $-2^{n-3} + |\lambda|$ and $-2^{n-4} + |\lambda|$; resp., between $-2^{n-4} + |\lambda|$ and $2^{n-3} + |\lambda|$; resp., to the right of $3 \cdot 2^{n-3} + |\lambda|$) then $v \in \mathcal{L}(t_{n-4}\,\overline{t_{n-4}}\,t_{n-4})$ (resp., $v \in \mathcal{L}(t_{n-4})$; resp., $v \in \mathcal{L}(\overline{t_{n-4}}\,\overline{t_{n-4}}\,t_{n-4})$; resp., $v \in \mathcal{L}(\overline{t_{n-4}}\,t_{n-4}\,\overline{t_{n-4}})$), thus it appears in $\mathcal{L}(\overline{t_{n-1}})$ and we can conclude using Theorem 2.

vii) Let $w = s\,t_{n-4}\,\overline{t_{n-2}}\,\overline{t_{n-1}}\,p$, with $s \in \mathrm{Suf}(t_{n-3}\,\overline{t_{n-4}})$ and $p \in \mathrm{Pref}(\overline{t_{n-4}})$. As in the previous case, it is possible to check that there exist no string

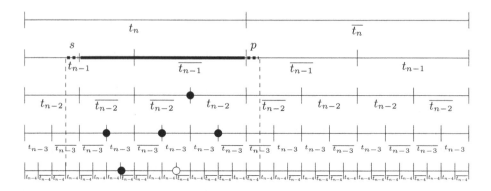

Fig. 3. A smallest string attractor Γ for a factor $w = s\,\overline{t_{n-2}}\,\overline{t_{n-1}}\,p$ of t_{n+1}, with $s \in \mathrm{Suf}(t_{n-4})$, and $p \in \mathrm{Pref}(\overline{t_{n-4}})$. The position of $\Gamma_{n-1} + |\lambda|$ that is not in Γ is in white.

attractor of size 4. However, the set

$$\Gamma = \big((\Gamma_{n-1} \setminus \{3 \cdot 2^{n-4}\}) \cup \{-2^{n-2}, -2^{n-4}\}\big) + |\lambda|,$$

where $\lambda = s\,t_{n-4}\,\overline{t_{n-2}}$, is a string attractor for w. Indeed, let $v \in \mathcal{L}(w)$. If $v \in \mathcal{L}(\overline{t_{n-1}})$, then, by Theorem 2, v has an occurrence containing at least one of the positions of Γ_{n-1}; if the only position contained in the occurrence is $3 \cdot 2^{n-4} + |\lambda|$, then $v \in \mathcal{L}(t_{n-4}\,\overline{t_{n-4}})$ and thus there exists another occurrence of v containing $-2^{n-4} + |\lambda|$.

All the other cases are proved as in the previous cases: namely if v appears to the left of $-2^{n-2}+|\lambda|$, (resp., between $-2^{n-2}+|\lambda|$ and $-2^{n-4}+|\lambda|$; resp., between $-2^{n-4}+|\lambda|$ and $2^{n-3}+|\lambda|$; resp., to the right of $3 \cdot 2^{n-3}+|\lambda|$) then $v \in \mathcal{L}(t_{n-2})$ (resp., $v \in \mathcal{L}(\overline{t_{n-4}}\,t_{n-4}\,t_{n-4})$; resp., $v \in \mathcal{L}(\overline{t_{n-4}}\,\overline{t_{n-4}}\,t_{n-4})$; resp., $v \in \mathcal{L}(\overline{t_{n-4}}\,t_{n-4}\,\overline{t_{n-4}})$), thus it appears in $\mathcal{L}(\overline{t_{n-1}})$ and we can conclude.

The seven cases above are summarized in Table 1. Note that in all previous cases $\overline{t_{n-4}}$ is a prefix of p, and hence a prefix of ρ.

Step 2. We now consider the case of ρ containing $\overline{t_{n-4}}$ as a proper prefix. The seven possible cases of factors are summarized in Table 2. Note that in all these cases t_{n-4} is a prefix of p, and hence $\overline{t_{n-3}} = \overline{t_{n-4}}\,t_{n-4}$ is a prefix of ρ.

Step 3. We now consider the case of ρ containing as a proper prefix $\overline{t_{n-3}}$. The six possible cases of factors are summarized in Table 3. Note that in all these cases t_{n-5} is a prefix of p, and hence $\overline{t_{n-3}}\,t_{n-5}$ is a prefix of ρ.

Step 4. We now consider the case of ρ containing as a proper prefix $\overline{t_{n-3}}\,t_{n-5}$. The six possible cases of factors are summarized in Table 4. Note that in all these cases $\overline{t_{n-5}}\,\overline{t_{n-4}}\,t_{n-4}$ is a prefix of p, and hence $\overline{t_{n-2}}\,t_{n-4}$ is a prefix of ρ.

Step 5. We now consider the case of ρ containing as a proper prefix $\overline{t_{n-2}}\,t_{n-4}$. Since $t_{n-2}\overline{t_{n-2}} = t_{n-1}$ is a factor of w, in the first two cases weconsider as starting point for constructing a string attractor Γ_{n-1} instead of Γ_{n-2} (for the remaining two cases we have as central factor $t = \overline{t_{n-1}}$, so we also use Γ_{n-1}).

Table 1. Summary of **Step 1** of the proof of Theorem 1. For a factor of the form $w = s\,t'\,t\,t\,p$, with $s \in \mathrm{Suf}(\cdot)$, $p \in \mathrm{Pref}(\cdot)$, a smallest string attractor is $\Gamma = ((\Gamma_k \setminus \Gamma') \cup \Gamma'') + |s\,t'|$, with k the integer such that $t = t_k$ or $\overline{t_k}$.

Suf(·)	t'	t	t''	Pref(·)	Γ'	Γ''	$\lvert\Gamma\rvert$
$\overline{t_{n-4}}$	ε	t_{n-2}	ε	$\overline{t_{n-4}}$	$\emptyset$	$\emptyset$	4
t_{n-4}	$\overline{t_{n-4}}$	t_{n-2}	ε	$\overline{t_{n-4}}$	$\{3\cdot 2^{n-4}\}$	$\{0\}$	4
$\overline{t_{n-5}}$	t_{n-3}	t_{n-2}	ε	$\overline{t_{n-4}}$	$\left\{\begin{array}{c}2^{n-4},\\ 3\cdot 2^{n-4}\end{array}\right\}$	$\left\{\begin{array}{c}-2^{n-4},\\ 0\end{array}\right\}$	4
$t_{n-3}\,\overline{t_{n-4}}\,t_{n-5}$	$\overline{t_{n-5}}\,t_{n-3}$	t_{n-2}	ε	$\overline{t_{n-4}}$	$\left\{\begin{array}{c}3\cdot 2^{n-5},\\ 3\cdot 2^{n-4}\end{array}\right\}$	$\left\{\begin{array}{c}-2^{n-3},\\ 0\end{array}\right\}$	4
$\overline{t_{n-3}}$	t_{n-3}	$\overline{t_{n-1}}$	ε	$\overline{t_{n-4}}$	$\{3\cdot 2^{n-4}\}$	$\{-2^{n-4}\}$	4
t_{n-4}	$\overline{t_{n-2}}$	$\overline{t_{n-1}}$	ε	$\overline{t_{n-4}}$	$\{3\cdot 2^{n-4}\}$	$\left\{\begin{array}{c}-2^{n-3},\\ -2^{n-4}\end{array}\right\}$	5
$t_{n-3}\,\overline{t_{n-4}}$	$t_{n-4}\,\overline{t_{n-2}}$	$\overline{t_{n-1}}$	ε	$\overline{t_{n-4}}$	$\{3\cdot 2^{n-4}\}$	$\left\{\begin{array}{c}-2^{n-4},\\ -2^{n-2}\end{array}\right\}$	5

Table 2. Summary of **Step 2** of the proof of Theorem 1. For a factor of the form $w = s\,t'\,t\,t\,p$, with $s \in \mathrm{Suf}(\cdot)$, $p \in \mathrm{Pref}(\cdot)$, a smallest string attractor is $\Gamma = ((\Gamma_k \setminus \Gamma') \cup \Gamma'') + |s\,t'|$, with k the integer such that $t = t_k$ or $\overline{t_k}$.

Suf(·)	t'	t	t''	Pref(·)	Γ'	Γ''	$\lvert\Gamma\rvert$
$\overline{t_{n-4}}$	ε	t_{n-2}	$\overline{t_{n-4}}$	t_{n-4}	$\{3\cdot 2^{n-5}\}$	$\{9\cdot 2^{n-5}\}$	4
t_{n-4}	$\overline{t_{n-4}}$	t_{n-2}	$\overline{t_{n-4}}$	t_{n-4}	$\left\{\begin{array}{c}3\cdot 2^{n-5},\\ 3\cdot 2^{n-4}\end{array}\right\}$	$\left\{\begin{array}{c}0,\\ 9\cdot 2^{n-5}\end{array}\right\}$	4
$\overline{t_{n-5}}$	t_{n-3}	t_{n-2}	$\overline{t_{n-4}}$	t_{n-4}	$\{2^{n-4}\}$	$\{-2^{n-4}\}$	4
$t_{n-3}\,\overline{t_{n-4}}\,t_{n-5}$	$\overline{t_{n-5}}\,t_{n-3}$	t_{n-2}	$\overline{t_{n-4}}$	t_{n-4}	$\left\{\begin{array}{c}2^{n-4},\\ 3\cdot 2^{n-5}\end{array}\right\}$	$\left\{\begin{array}{c}-2^{n-3},\\ -2^{n-4}\end{array}\right\}$	4
$\overline{t_{n-3}}$	t_{n-3}	$\overline{t_{n-1}}$	$\overline{t_{n-4}}$	$t_{n-4}\,t_{n-3}\,t_{n-2}$	$\left\{\begin{array}{c}3\cdot 2^{n-4},\\ 3\cdot 2^{n-3}\end{array}\right\}$	$\left\{\begin{array}{c}0,\\ 2^{n-1}\end{array}\right\}$	4
t_{n-4}	$\overline{t_{n-2}}$	$\overline{t_{n-1}}$	$\overline{t_{n-4}}$	$t_{n-4}\,t_{n-3}\,t_{n-2}$	$\left\{\begin{array}{c}2^{n-3},\\ 3\cdot 2^{n-4},\\ 3\cdot 2^{n-3}\end{array}\right\}$	$\left\{\begin{array}{c}-2^{n-3},\\ 0,\\ 2^{n-1}\end{array}\right\}$	4
$t_{n-3}\,\overline{t_{n-4}}$	$t_{n-4}\,\overline{t_{n-2}}$	$\overline{t_{n-1}}$	$\overline{t_{n-4}}$	$t_{n-4}\,t_{n-3}\,t_{n-2}$	$\left\{\begin{array}{c}3\cdot 2^{n-4},\\ 3\cdot 2^{n-3}\end{array}\right\}$	$\left\{\begin{array}{c}-2^{n-2},\\ 0,\\ 2^{n-1}\end{array}\right\}$	5

The four possible cases of factors are summarized in Table 5. Note that in all these cases $\overline{t_{n-4}}\,\overline{t_{n-3}}$ is a prefix of p, and hence $\overline{t_{n-1}}$ is a prefix of ρ.

Step 6. We now consider the case of ρ containing as a proper prefix $\overline{t_{n-1}}$. As in the previous step, we have $t_{n-2}\,\overline{t_{n-2}} = t_{n-1}$ is a factor of w. For this reason, in

Table 3. Summary of **Step 3** of the proof of Theorem 1. For a factor of the form $w = s\,t'\,t\,t\,p$, with $s \in \mathrm{Suf}(\cdot)$, $p \in \mathrm{Pref}(\cdot)$, a smallest string attractor is $\Gamma = ((\Gamma_k \setminus \Gamma') \cup \Gamma'') + |s\,t'|$, with k the integer such that $t = t_k$ or $\overline{t_k}$.

Suf(·)	t'	t	t''	Pref(·)	Γ'	Γ''	$\|\Gamma\|$
$\overline{t_{n-4}}$	ε	t_{n-2}	$\overline{t_{n-3}}$	t_{n-5}	$\{2^{n-4},\ 3\cdot 2^{n-4}\}$	$\{2^{n-2},\ 5\cdot 2^{n-5}\}$	4
t_{n-4}	$\overline{t_{n-4}}$	t_{n-2}	$\overline{t_{n-3}}$	t_{n-5}	$\{2^{n-4},\ 3\cdot 2^{n-5},\ 3\cdot 2^{n-4}\}$	$\{-2^{n-5},\ 2^{n-2},\ 5\cdot 2^{n-4}\}$	4
$\overline{t_{n-5}}$	t_{n-3}	t_{n-2}	$\overline{t_{n-3}}$	t_{n-5}	$\{2^{n-4},\ 3\cdot 2^{n-4}\}$	$\{-2^{n-4},\ 2^{n-2},\ 5\cdot 2^{n-4}\}$	5
$t_{n-3}\,\overline{t_{n-4}}\,t_{n-5}$	$\overline{t_{n-5}}\,t_{n-3}$	t_{n-2}	$\overline{t_{n-3}}$	t_{n-5}	$\{3\cdot 2^{n-5},\ 3\cdot 2^{n-4}\}$	$\{-2^{n-3},\ 5\cdot 2^{n-4}\}$	4
$\overline{t_{n-3}}$	t_{n-3}	$\overline{t_{n-1}}$	$\overline{t_{n-3}}$	$t_{n-3}\,t_{n-2}$	$\{3\cdot 2^{n-4},\ 3\cdot 2^{n-3}\}$	$\{0,\ 2^{n-1}\}$	4
t_{n-2}	$\overline{t_{n-2}}$	$\overline{t_{n-1}}$	$\overline{t_{n-3}}$	$t_{n-3}\,t_{n-2}$	$\{2^{n-3},\ 3\cdot 2^{n-4}\}$	$\{-2^{n-3},\ 2^{n-1}\}$	4

Table 4. Summary of **Step 4** of the proof of Theorem 1. For a factor of the form $w = s\,t'\,t\,t\,p$, with $s \in \mathrm{Suf}(\cdot)$, $p \in \mathrm{Pref}(\cdot)$, a smallest string attractor is $\Gamma = ((\Gamma_k \setminus \Gamma') \cup \Gamma'') + |s\,t'|$, with k the integer such that $t = t_k$ or $\overline{t_k}$.

Suf(·)	t'	t	t''	Pref(·)	Γ'	Γ''	$\|\Gamma\|$
$\overline{t_{n-4}}$	ε	t_{n-2}	$\overline{t_{n-3}\,t_{n-5}}$	$\overline{t_{n-5}\,t_{n-4}}\,t_{n-4}$	$\{2^{n-4},\ 3\cdot 2^{n-5}\}$	$\{2^{n-2},\ 3\cdot 2^{n-3}\}$	4
t_{n-4}	$\overline{t_{n-4}}$	t_{n-2}	$\overline{t_{n-3}\,t_{n-5}}$	$\overline{t_{n-5}\,t_{n-4}}\,t_{n-4}$	$\{2^{n-4},\ 3\cdot 2^{n-5},\ 3\cdot 2^{n-4}\}$	$\{0,\ 2^{n-2},\ 3\cdot 2^{n-4}\}$	4
$\overline{t_{n-5}}$	t_{n-3}	t_{n-2}	$\overline{t_{n-3}\,t_{n-5}}$	$\overline{t_{n-5}\,t_{n-4}}\,t_{n-2}$	$\{2^{n-4},\ 3\cdot 2^{n-5}\}$	$\{-2^{n-4},\ 3\cdot 2^{n-3}\}$	4
$t_{n-3}\,\overline{t_{n-4}}\,t_{n-5}$	$\overline{t_{n-5}}\,t_{n-3}$	t_{n-2}	$\overline{t_{n-3}}\,t_{n-5}$	$\overline{t_{n-5}\,t_{n-4}}\,t_{n-2}$	$\{3\cdot 2^{n-5},\ 3\cdot 2^{n-4}\}$	$\{-2^{n-3},\ 5\cdot 2^{n-4},\ 3\cdot 2^{n-3}\}$	5
$\overline{t_{n-3}}$	t_{n-3}	$\overline{t_{n-1}}$	$\overline{t_{n-3}\,t_{n-5}}$	$\overline{t_{n-5}\,t_{n-4}}\,t_{n-2}$	$\{3\cdot 2^{n-4},\ 3\cdot 2^{n-3}\}$	$\{0,\ 2^{n-1}\}$	4
t_{n-2}	$\overline{t_{n-2}}$	$\overline{t_{n-1}}$	$\overline{t_{n-3}\,t_{n-5}}$	$\overline{t_{n-5}\,t_{n-4}}\,t_{n-2}$	$\{2^{n-3},\ 3\cdot 2^{n-4},\ 3\cdot 2^{n-3}\}$	$\{-2^{n-3},\ 0,\ 2^{n-1}\}$	4

the first of the three cases considered we construct the string attractor starting by a shift of Γ_{n-1} (for the remaining cases we also use Γ_{n-1} following the same construction of steps above). The three possible cases of factors are summarized in Table 6. Note that in all these cases t_{n-3} is a prefix of p.

Table 5. Summary of **Step 5** of the proof of Theorem 1. For a factor of the form $w = s\,t'\,t\,t\,p$, with $s \in \mathrm{Suf}(\cdot)$, $p \in \mathrm{Pref}(\cdot)$, a smallest string attractor is $\Gamma = ((\Gamma_{n-1} \setminus \Gamma') \cup \Gamma'') + |s\,t'|$.

$\mathrm{Suf}(\cdot)$	t'	t	t''	$\mathrm{Pref}(\cdot)$	Γ'	Γ''	$|\Gamma|$
$\overline{t_{n-4}}$	ε	t_{n-1}	t_{n-4}	$\overline{t_{n-4}}\,\overline{t_{n-3}}$	$\{3\cdot 2^{n-4}\}$	$\{2^{n-1}\}$	4
$\overline{t_{n-3}}\,t_{n-4}$	$\overline{t_{n-4}}$	t_{n-1}	t_{n-4}	$\overline{t_{n-4}}\,\overline{t_{n-3}}$	$\{3\cdot 2^{n-4}\}$	$\left\{\begin{matrix}0,\\2^{n-1}\end{matrix}\right\}$	5
$\overline{t_{n-3}}$	ε	$\overline{t_{n-1}}$	$t_{n-2}\,t_{n-4}$	$\overline{t_{n-4}}\,\overline{t_{n-3}}$	$\{3\cdot 2^{n-4}\}$	$\{2^{n-1}\}$	4
$t_{n-2}\,\overline{t_{n-3}}$	t_{n-3}	$\overline{t_{n-1}}$	$t_{n-2}t_{n-4}$	$\overline{t_{n-4}}\,\overline{t_{n-3}}$	$\left\{\begin{matrix}3\cdot 2^{n-4},\\3\cdot 2^{n-3}\end{matrix}\right\}$	$\left\{\begin{matrix}0,\\2^{n-1}\end{matrix}\right\}$	4

Table 6. Summary of **Step 6** of the proof of Theorem 1. For a factor of the form $w = s\,t'\,t\,t\,p$, with $s \in \mathrm{Suf}(\cdot)$, $p \in \mathrm{Pref}(\cdot)$, a smallest string attractor is $\Gamma = ((\Gamma_{n-1} \setminus \Gamma') \cup \Gamma'') + |s\,t'|$.

$\mathrm{Suf}(\cdot)$	t'	t	t''	$\mathrm{Pref}(\cdot)$	Γ'	Γ''	$|\Gamma|$
$\overline{t_{n-2}}$	ε	t_{n-1}	t_{n-2}	t_{n-3}	$\left\{\begin{matrix}2^{n-3},\\3\cdot 2^{n-4}\end{matrix}\right\}$	$\left\{\begin{matrix}2^{n-1},\\5\cdot 2^{n-3}\end{matrix}\right\}$	4
t_{n-3}	ε	$\overline{t_{n-1}}$	$\overline{t_{n-1}}$	t_{n-3}	$\left\{\begin{matrix}3\cdot 2^{n-4},\\3\cdot 2^{n-3}\end{matrix}\right\}$	$\left\{\begin{matrix}2^{n-1},\\7\cdot 2^{n-3}\end{matrix}\right\}$	4
$t_{n-2}\,\overline{t_{n-3}}$	t_{n-3}	$\overline{t_{n-1}}$	$\overline{t_{n-1}}$	t_{n-3}	$\left\{\begin{matrix}3\cdot 2^{n-4},\\2^{n-2},\\3\cdot 2^{n-3}\end{matrix}\right\}$	$\left\{\begin{matrix}0,\\2^{n-1},\\3\cdot 2^{n-2}\end{matrix}\right\}$	4

Table 7. Summary of **Step 7** of the proof of Theorem 1. For a factor of the form $w = s\,t'\,t\,t\,p$, with $s \in \mathrm{Suf}(\cdot)$, $p \in \mathrm{Pref}(\cdot)$, a smallest string attractor is $\Gamma = ((\Gamma_{n-1} \setminus \Gamma') \cup \Gamma'') + |s\,t'|$.

$\mathrm{Suf}(\cdot)$	t'	t	t''	$\mathrm{Pref}(\cdot)$	Γ'	Γ''	$|\Gamma|$
$\overline{t_{n-2}}$	ε	t_{n-1}	$t_{n-2}\,t_{n-3}$	$\overline{t_{n-3}}\,\overline{t_{n-2}}$	$\left\{\begin{matrix}3\cdot 2^{n-4},\\3\cdot 2^{n-3}\end{matrix}\right\}$	$\left\{\begin{matrix}2^{n-1},\\3\cdot 2^{n-2}\end{matrix}\right\}$	4
t_{n-3}	ε	$\overline{t_{n-1}}$	$\overline{t_{n-1}}\,t_{n-3}$	$\overline{t_{n-3}}\,\overline{t_{n-2}}$	$\left\{\begin{matrix}2^{n-3},\\3\cdot 2^{n-4}\end{matrix}\right\}$	$\left\{\begin{matrix}2^{n-1},\\2^n\end{matrix}\right\}$	4
$t_{n-2}\,\overline{t_{n-3}}$	t_{n-3}	$\overline{t_{n-1}}$	$\overline{t_{n-1}}\,t_{n-3}$	$\overline{t_{n-3}}\,\overline{t_{n-2}}$	$\left\{\begin{matrix}2^{n-3},\\3\cdot 2^{n-4},\\3\cdot 2^{n-3}\end{matrix}\right\}$	$\left\{\begin{matrix}0,\\2^{n-1},\\2^n\end{matrix}\right\}$	4

Step 7. As last step we consider the case of ρ containing as a proper prefix $\overline{t_{n-1}}\,t_{n-3}$. Similarly to the previous step, since $t_{n-2}\,\overline{t_{n-2}} = t_{n-1}$ is a factor of w,

we construct the string attractor starting from a shift of Γ_{n-1} also in the first of the three cases. The three possible cases of factors are summarized in Table 7.

Thus we proved the result for all factors $w \in \mathcal{L}(t_{n+1})$ containing $x\,t_{n-2}\,y$, with x the last letter of $\overline{t_{n-2}}$ and y the first letter of $\overline{t_n}$, as a proper factor. The case $w = \lambda\,\overline{t_{n-2}}\,\rho$ with $\lambda \in \mathrm{Suf}(t_n)$ and $\rho \in \mathrm{Pref}(t_{n-2}\,t_{n-1})$ is proved in a symmetrical way.

5   Future Works and Different Approaches

The Thue-Morse word has been generalized to larger alphabets in several different ways. One possible generalization is the one given in [2], where $\mathbf{t}_m$ is defined over an alphabet $\mathcal{A}_m = \{a_1, a_2, \ldots, a_m\}$ of cardinality m as the fixed point $\mathbf{t}_m = \lim_{n \to \infty} \varphi_m^n(a_1)$, where $\varphi_m(a_k) = a_k \cdots a_m a_1 \cdots a_{k-1}$ for every $1 \le k \le m$. For instance, we have $\mathbf{t}_3 = \texttt{abcbcacabbcacababccababcabc} \cdots$ over the ternary alphabet $\{\mathsf{a}, \mathsf{b}, \mathsf{c}\}$.

Conjecture 1. For every $m \in \mathbb{N}$ there exist an integer K_m such that every non-empty factor of $\mathbf{t}_m$ has a string attractor of size at most K_m.

Recently Dvořáková proved that every factor of an episturmian word has a sting attractor having size the number of distinct letters appearing in the factor (see [5]). In particular, every factor of a Sturmian word different from a letter has a string attractor of size 2. Such result is based on the construction of (standard) episturmian words by iterated palindromic closure (see [4]). We believe that a similar approach could be used also for the Thue-Morse word, using pseudo-palindromic closure instead (see [1,3]).

Acknowledgements. This research received funding from the Ministry of Education, Youth and Sports of the Czech Republic through the project CZ.02.1.01/0.0/0.0/16_019/0000765. To check some of the examples we used a code written by undergraduate student Veronika Hendrychová.

References

1. Massé, A.B., Paquin, G., Tremblay, H., Vuillon, L.: On generalized pseudo standard words over binary alphabets. J. Integer Seq. **16**(2), 28 (2013). Article 13.2.11
2. Brlek, S.: Enumeration of factors in the Thue-Morse word. Discret. Appl. Math. **24**(1–3), 83–96 (1989)
3. de Luca, A., De Luca, A.: Pseudopalindrome closure operators in free monoids. Theoret. Comput. Sci. **362**(1–3), 282–300 (2006)
4. Droubay, X., Justin, J., Pirillo, G.: Episturmian words and some constructions of de Luca and Rauzy. Theoret. Comput. Sci. **255**(1–2), 539–553 (2001)
5. Dvořáková, Ĺ.: String attractors of episturmian sequences (2022). https://arxiv.org/pdf/2211.01660v2.pdf
6. Kempa, D., Prezza, N.: At the roots of dictionary compression: string attractors. In: STOC 2018–Proceedings of the 50th Annual ACM SIGACT Symposium on Theory of Computing, pp. 827–840. ACM, New York (2018)

7. Kutsukake, K., Matsumoto, T., Nakashima, Y., Inenaga, S., Bannai, H., Takeda, M.: On repetitiveness measures of Thue-Morse words. In: Boucher, C., Thankachan, S.V. (eds.) SPIRE 2020. LNCS, vol. 12303, pp. 213–220. Springer, Cham (2020). https://doi.org/10.1007/978-3-030-59212-7_15

8. Lothaire, M.: Algebraic Combinatorics on Words. Encyclopedia of Mathematics and its Applications, vol. 90. Cambridge University Press, Cambridge (2002)

9. Mantaci, S., Restivo, A., Romana, G., Rosone, G., Sciortino, M.: String attractors and combinatorics on words. In: ICTS, CEUR Workshop Proceedins, vol. 2504, pp. 57–71 (2019)

10. Mantaci, S., Restivo, A., Romana, G., Rosone, G., Sciortino, M.: A combinatorial view on string attractors. Theoret. Comput. Sci. **850**, 236–248 (2021)

11. Harold Marston Morse: Recurrent geodesics on a surface of negative curvature. Trans. Amer. Math. Soc. **22**(1), 84–100 (1921)

12. Navarro, G.: Indexing highly repetitive collections. In: Arumugam, S., Smyth, W.F. (eds.) IWOCA 2012. LNCS, vol. 7643, pp. 274–279. Springer, Heidelberg (2012). https://doi.org/10.1007/978-3-642-35926-2_29

13. Prouhet, E.: Mémoire sur quelques relations entre les puissances des nombres. C.R. Acad. Sci. **31**, 225 (1851)

14. Navarro, G.: Indexing highly repetitive collections. In: Arumugam, S., Smyth, W.F. (eds.) IWOCA 2012. LNCS, vol. 7643, pp. 274–279. Springer, Heidelberg (2012). https://doi.org/10.1007/978-3-642-35926-2_29

15. Schaeffer, L., Shallit, J.: String attractors for automatic sequences (2021). https://arxiv.org/pdf/2012.06840.pdf

16. Thue, A.: Über unendliche zeichenreihenal. orske vid. Selsk. Skr. Mat. Nat. Kl. **7**, 1–22 (1906)

Critical Exponents of Regular Arnoux-Rauzy Sequences

L'ubomíra Dvořáková[1(✉)] and Jana Lepšová[1,2]

[1] FNSPE Czech Technical University in Prague, Prague, Czechia
{lubomira.dvorakova,jana.lepsova}@fjfi.cvut.cz
[2] LaBRI, University of Bordeaux, Bordeaux, France

Abstract. We provide a formula for the critical exponent and the asymptotic critical exponent of regular Arnoux-Rauzy sequences. Over the binary alphabet it coincides with a well-known formula for Sturmian sequences based on their S-adic representation. We show that among regular d-ary Arnoux-Rauzy sequences, the minimal (asymptotic) critical exponent is reached by the d-bonacci sequence.

Keywords: Arnoux-Rauzy sequence · Repetitions · Critical exponent · Asymptotic critical exponent · Return word · Bispecial factor

1 Introduction

The critical exponent of sequences is an extensively studied area of combinatorics on words. Krieger and Shallit [6] proved that every real number greater than 1 is a critical exponent of some sequence. The values of the minimal critical exponent of sequences over a fixed size alphabet were conjectured by Dejean [4] in 1972. Gradually the community inched forward by joint effort [4,7–9,11–13] towards proving the conjecture in 2011.

The goal of this paper is to derive a formula for the (asymptotic) critical exponent of regular d-ary Arnoux-Rauzy (AR) sequences based on their S-adic representation. A formula for the critical exponent of Sturmian sequences based on their S-adic representation was provided by Carpi and de Luca [2] (for a special case) and by Damanik and Lenz [15] (in full generality). It coincides with our formula over the binary alphabet. The critical exponent of Arnoux-Rauzy

We would like to thank Edita Pelantová for fruitful discussions and helpful advice. We thank Sébastien Labbé for his advice in programming. Our thanks belong also to the referees for their useful comments. The first author was supported by the Ministry of Education, Youth and Sports of the Czech Republic through the project CZ.02.1.01/0.0/0.0/16_019/0000778. The second author acknowledges financial support by The French Institute in Prague and the Czech Ministry of Education, Youth and Sports through the Barrande fellowship programme, Agence Nationale de la Recherche through the project Codys (ANR-18-CE40-0007), and the support by Grant Agency of Czech Technical University in Prague, through the project SGS23/187/OHK4/3T/14.

© The Author(s), under exclusive license to Springer Nature Switzerland AG 2023
A. Frid and R. Mercaş (Eds.): WORDS 2023, LNCS 13899, pp. 130–142, 2023.
https://doi.org/10.1007/978-3-031-33180-0_10

sequences fixed by a primitive morphism was studied by Justin and Pirillo [16]. Repetitions in regular (see Definition 4) Arnoux-Rauzy sequences were studied by Glen [17].

Our main contribution is not only the formula, but also its relation to bispecial factors and return words that, to our knowledge, was not realized by the above mentioned authors. Moreover, we use the formula to prove that the minimal (asymptotic) critical exponent among regular d-ary Arnoux-Rauzy sequences is reached by the d-bonacci sequence.

2   Preliminaries

An *alphabet* $\mathcal{A}$ is a finite set and its elements are called *letters*. A *word* u over $\mathcal{A}$ of *length* n is a finite string $u = u_0 u_1 \cdots u_{n-1}$ of letters $u_i \in \mathcal{A}$, $i \in \{0, 1, \ldots, n-1\}$, and the length of u is denoted $|u|$. The set of all finite words over $\mathcal{A}$ is denoted $\mathcal{A}^*$. If $u = xyz$ is a concatenation of words $x, y, z \in \mathcal{A}^*$, then x is a *prefix*, z is a *suffix* and y is a *factor* of u. The set $\mathcal{A}^*$ with concatenation as the monoid operation forms a monoid with the *empty word* ε as the neutral element. Let $d \in \mathbb{N}$ be the cardinality of the alphabet $\mathcal{A}$. The *Parikh vector* $\vec{u} \in \mathbb{N}^d$ of a word $u \in \mathcal{A}^*$ is the vector defined as $(\vec{u})_a = |u|_a$ for all $a \in \mathcal{A}$, where $|u|_a$ is the number of occurrences of the letter $a \in \mathcal{A}$ in the word u.

A *sequence*[1] $\mathbf{u}$ over $\mathcal{A}$ is an infinite string $\mathbf{u} = u_0 u_1 u_2 \cdots$ of letters $u_i \in \mathcal{A}$ for all $i \in \mathbb{N}$. The set of all sequences over $\mathcal{A}$ is denoted $\mathcal{A}^{\mathbb{N}}$. A *factor* of a sequence $\mathbf{u} = u_0 u_1 u_2 \cdots$ is a word $y \in \mathcal{A}^*$ such that $y = u_i u_{i+1} u_{i+2} \cdots u_{j-1}$ for some $i, j \in \mathbb{N}$, $i \leq j$. The integer i is called an *occurrence* of the factor y in the sequence $\mathbf{u}$. In particular, the factor y is the empty word if and only if $i = j$, and the set of all occurrences of the empty word is $\mathbb{N}$. A factor $y \in \mathcal{A}^*$ is a *prefix* of $\mathbf{u}$ if $i = 0$ is an occurrence of y in the sequence $\mathbf{u}$. The *language* $\mathcal{L}(\mathbf{u})$ of a sequence $\mathbf{u}$ is the set of factors occurring in $\mathbf{u}$. A factor w of a sequence $\mathbf{u}$ is *right special* if $wa, wb \in \mathcal{L}(\mathbf{u})$ for at least two distinct letters $a, b \in \mathcal{A}$. A *left special* factor is defined analogously. A factor is called *bispecial* if it is both left and right special. The *factor complexity* of a sequence $\mathbf{u}$ is the map $\mathcal{C}_{\mathbf{u}} : \mathbb{N} \to \mathbb{N}$ defined by $\mathcal{C}_{\mathbf{u}}(n) = \#\{w \in \mathcal{L}(\mathbf{u}) : |w| = n\}$.

A sequence $\mathbf{u}$ is *recurrent* if each factor of $\mathbf{u}$ has infinitely many occurrences in $\mathbf{u}$. Moreover, a recurrent sequence $\mathbf{u}$ is *uniformly recurrent* if the distances between the consecutive occurrences of each factor in $\mathbf{u}$ are bounded. A sequence $\mathbf{u}$ is *eventually periodic* if there exist words $w \in \mathcal{A}^*$ and $v \in \mathcal{A}^* \setminus \{\varepsilon\}$ such that $\mathbf{u}$ can be written as $\mathbf{u} = wvvv \cdots = wv^{\omega}$. In particular, $\mathbf{u}$ is *periodic* if $w = \varepsilon$. A sequence $\mathbf{u}$ is *aperiodic* if it is not eventually periodic.

A *morphism* over $\mathcal{A}$ is a map $\psi : \mathcal{A}^* \to \mathcal{A}^*$ such that $\psi(uv) = \psi(u)\psi(v)$ for all words $u, v \in \mathcal{A}^*$. Morphisms can be naturally extended to $\mathcal{A}^{\mathbb{N}}$ by setting $\psi(u_0 u_1 u_2 \cdots) = \psi(u_0)\psi(u_1)\psi(u_2) \cdots$. A *fixed point* of a morphism ψ is a sequence $\mathbf{u}$ such that $\psi(\mathbf{u}) = \mathbf{u}$.

A sequence $\mathbf{u}$ over $\mathcal{A}$ is *Arnoux–Rauzy* (AR for short) if the language $\mathcal{L}(\mathbf{u})$ is closed under reversal and if there exists exactly one right special factor w of each

---

[1] Note that sequences are denoted in bold characters.

length and $wa \in \mathcal{L}(\mathbf{u})$ for each letter $a \in \mathcal{A}$. An AR sequence $\mathbf{u}$ is *standard* if each of its prefixes is a left special factor of $\mathbf{u}$. For each AR sequence there exists a unique standard AR sequence with the same language. Binary AR sequences are called *Sturmian* sequences. The factor complexity of a d-ary AR sequence satisfies $\mathcal{C}(n) = (d-1)n + 1$ (see [14]).

Consider a factor y of a recurrent sequence $\mathbf{u} = u_0 u_1 u_2 \cdots$. Let $i < j$ be two consecutive occurrences of y in $\mathbf{u}$. Then the word $u_i u_{i+1} \cdots u_{j-1}$ is a *return word* to y in $\mathbf{u}$. The set of all return words to y in $\mathbf{u}$ is denoted by $\mathcal{R}_\mathbf{u}(y)$. The set $\mathcal{R}_\mathbf{u}(y)$ is finite if and only if $\mathbf{u}$ is uniformly recurrent. If y is a prefix of $\mathbf{u}$, then $\mathbf{u}$ can be written as a concatenation $\mathbf{u} = r_{d_0} r_{d_1} r_{d_2} \cdots$ of return words to y. The *derived sequence* of $\mathbf{u}$ to y is the sequence $\mathbf{d}_\mathbf{u}(y) = \mathbf{r}_{d_0} \mathbf{r}_{d_1} \mathbf{r}_{d_2} \cdots$ over the alphabet $\{\mathbf{r}_1, \mathbf{r}_2, \ldots\}$ of cardinality $\#\mathcal{R}_\mathbf{u}(y)$. In the sequel we distinguish, using a different text format, between a return word r to y in $\mathbf{u}$ and the corresponding letter $\mathbf{r}$ in the derived sequence $d_\mathbf{u}(y)$. The concept of derived sequences was introduced by Durand [5]. Each factor of a d-ary AR sequence has exactly d return words [18]. Moreover, the derived sequence to a prefix of a d-ary AR sequence is also a d-ary AR sequence [21].

Example 1. The Tribonacci sequence $\mathbf{u}_T = 1213121121312121213121121312\cdots$ obtained as the fixed point of the morphism $\varphi : 1 \mapsto 12, 2 \mapsto 13, 3 \mapsto 1$ is an AR sequence. The reader is invited to check that the return words to the bispecial prefix $b = 121$ are $r_1 = 1213$, $r_2 = 121$, and $r_3 = 12$. The derived sequence is $\mathbf{d}_{\mathbf{u}_T}(b) = \mathbf{r}_1 \mathbf{r}_2 \mathbf{r}_1 \mathbf{r}_3 \mathbf{r}_1 \mathbf{r}_2 \mathbf{r}_1 \cdots$, which is also the Tribonacci sequence over $\{\mathbf{r}_1, \mathbf{r}_2, \mathbf{r}_3\}$.

3   Critical Exponent

A concatenation of n copies of a non-empty word u is called the *n-th power* of u and denoted u^n. Besides integer powers we can also consider fractional powers. If the word u is of length $\ell > 0$ and e is a positive rational number of the form n/ℓ, then u^e denotes the prefix of length n of the infinite periodic sequence u^ω. For instance, the Czech word *kytky* (flowers) may be written as $(kyt)^{5/3}$. The rational exponent e describes the repetition rate of u in the string u^e.

The *critical exponent* $E(\mathbf{u})$ of an infinite sequence $\mathbf{u} = u_0 u_1 u_2 \cdots$ describes the maximal repetition rate of factors occurring in $\mathbf{u}$. Formally,

$$E(\mathbf{u}) = \sup\{e \in \mathbb{Q} : u^e \in \mathcal{L}(\mathbf{u}), \text{ where } u \neq \varepsilon\}.$$

The *asymptotic critical exponent* $E^*(\mathbf{u})$ of $\mathbf{u}$ expresses the maximal repetition rate of factors of length growing to infinity. It is defined as

$$E^*(\mathbf{u}) = \begin{cases} +\infty, & \text{if } E(\mathbf{u}) = +\infty, \\ \limsup_{n \to \infty}\{e \in \mathbb{Q} : u^e \in \mathcal{L}(\mathbf{u}), \text{ where } |u| = n\}, & \text{otherwise.} \end{cases}$$

Obviously, $E^*(\mathbf{u}) \leq E(\mathbf{u})$ and the equality holds whenever $E(\mathbf{u})$ is irrational.

In [19], a handy formula for the computation of the critical exponent and the asymptotic critical exponent of uniformly recurrent aperiodic sequences is deduced. It uses the shortest return words to the bispecial factors.

Theorem 2 ([19]). *Let* **u** *be a uniformly recurrent aperiodic sequence. Let* $(b_n)_{n\in\mathbb{N}}$ *be the sequence of all bispecial factors in* **u** *ordered by length. For every* $n \in \mathbb{N}$, *let* r_n *be the shortest return word to the bispecial factor* b_n *in* **u**. *Then*

$$E(\mathbf{u}) = 1 + \sup_{n\in\mathbb{N}}\left\{\frac{|b_n|}{|r_n|}\right\} \qquad and \qquad E^*(\mathbf{u}) = 1 + \limsup_{n\to\infty}\frac{|b_n|}{|r_n|}.$$

4   Repetitions in Regular Arnoux–Rauzy Sequences

Our aim is to study repetitions in d-ary AR sequences. We use in the sequel the S-adic representation of d-ary AR sequences based on the following morphisms. For every $\mathbf{i} \in \{1, 2\ldots, \mathbf{d}\}$, let $\varphi_\mathbf{i}$ denote the morphism $\varphi_\mathbf{i}: \{1, 2\ldots, \mathbf{d}\}^* \to \{1, 2\ldots, \mathbf{d}\}^*$ such that $\varphi_\mathbf{i}: \mathbf{i} \mapsto \mathbf{i}$ and $\varphi_\mathbf{i}: \mathbf{j} \mapsto \mathbf{i}\mathbf{j}$ for every $\mathbf{j} \neq \mathbf{i}$.

Theorem 3 ([14]). *For each standard AR sequence over* $\{1, 2, \ldots, \mathbf{d}\}$ *there exists a unique sequence of morphisms* $\Delta = (\psi_n)_{n=1}^{+\infty}$, *where* $\psi_n \in \{\varphi_1, \varphi_2, \ldots, \varphi_\mathbf{d}\}$, *and a unique sequence of standard AR sequences* $(\mathbf{u}^{(n)})_{n=1}^{+\infty}$ *such that*

$$\mathbf{u} = \psi_1\psi_2\psi_3\cdots\psi_n(\mathbf{u}^{(n)}), \text{ for every } n \geq 1.$$

The sequence of morphisms Δ is called the *directive sequence* of **u**. We associate the same directive sequence to each AR sequence having the same language. Hereafter, we consider directive sequences whose generating morphisms alternate regularly.

Definition 4 ([20]). *Let* **u** *be a standard AR sequence over* $\{1, 2, \ldots, \mathbf{d}\}$. *Write* Δ *in the form* $\psi_1^{a_1}\psi_2^{a_2}\cdots$ *with* $\psi_N \neq \psi_{N+1}$ *and* $a_N > 0$ *for all* $N \geq 1$. *If the sequence* $(\psi_i)_{i=1}^{+\infty}$ *equals the periodic sequence with period* $\varphi_1\varphi_2\cdots\varphi_\mathbf{d}$, *then we say that* **u** *is* regular.

Let us emphasize that regularity requires neither periodicity nor eventual periodicity of Δ, hence regular sequences do not coincide with those studied by Justin and Pirillo [16].

Let us associate with a standard regular AR sequence **u**, using Definition 4, the sequence $\theta := (a_N)_{N\geq 0}$, where $a_0 := 0$, which we call the *slope* of **u**. The slope θ of **u** is associated with any AR sequence having the same language as **u** and such sequences are also called *regular*.[2] Regular AR sequences were first studied by Glen [17], the name regular was introduced by Peltomäki [20].

Remark 5. AR sequences have well defined letter frequencies [16]. Moreover, if an AR sequence **u** has the directive sequence $\Delta = \varphi_1^{a_1}\varphi_2^{a_2}\cdots\varphi_\mathbf{d}^{a_d}\cdots$, then we see that the letter $\mathbf{i}$ is more frequent then the letter $\mathbf{j}$ in **u** (we denote this fact $\mathbf{i} \triangleright \mathbf{j}$ in **u**) if and only if $\mathbf{i} < \mathbf{j}$.

---

[2] In the case of a Sturmian sequence **u**, we can proceed similarly and associate to the directive sequence the number θ with the continued fraction $[0; a_1, a_2, a_3, \ldots]$. It holds that θ is equal to the ratio of the less frequent letter to the more frequent letter. The geometrical interpretation of θ is the slope of the straight line producing **u** as a cutting sequence [24].

Example 6. The Tribonacci sequence $\mathbf{u}_T$ is the simplest regular ternary AR sequence in the sense that its directive sequence is $\Delta = (\varphi_1\varphi_2\varphi_3)^\omega$. Therefore its slope is $\theta = (0, \overline{1})$ and it is the fixed point of the morphism $\varphi_1\varphi_2\varphi_3$: $1 \to 1213121$, $2 \to 121312$, $3 \to 1213$. The reader is invited to check that $\mathbf{u}_T$, defined independently in Example 1, is indeed the fixed point of the morphism $\varphi_1\varphi_2\varphi_3$.

4.1   Bispecial Factors and Return Words in AR Sequences

In the sequel, bispecial factors and their return words in AR sequences play an essential role. Bispecial factors in standard AR sequences correspond to the palindromic prefixes [14]. For every standard regular AR sequence with the slope $\theta = (a_N)_{N \geq 0}$ and every $n \in \mathbb{N}$, we associate with the n-th bispecial factor the unique pair of non-negative integers (N, m) such that $n = a_0 + a_1 + a_2 + \cdots + a_N + m$ and $m < a_{N+1}$. A bispecial factor associated with $(N, 0)$ is called *primary*, otherwise it is called *secondary*. We denote $r_1^{(N)}, r_2^{(N)}, \ldots, r_d^{(N)}$ the return words to the primary bispecial factor associated with $(N, 0)$ so that the return words satisfy $\mathbf{r}_1^{(N)} \triangleright \mathbf{r}_2^{(N)} \triangleright \cdots \triangleright \mathbf{r}_d^{(N)}$ as letters in the corresponding derived sequence (see Remark 5). The following lemma enables to determine return words to bispecial factors in standard AR sequences.

Lemma 7 ([14,17]). *Let $\alpha \in \{1, 2, \ldots, d\}$ and let $\mathbf{u}, \hat{\mathbf{u}}$ be standard d-ary AR sequences. Let $r_i \in \mathcal{L}(\hat{\mathbf{u}})$ for every $i \in \{1, 2, \ldots, d\}$. If $\mathbf{u} = \varphi_\alpha(\hat{\mathbf{u}})$, then*

1. *if b is a bispecial factor in $\hat{\mathbf{u}}$, then $\varphi_\alpha(b)\alpha$ is a bispecial factor in $\mathbf{u}$;*
2. *if v is a non-empty bispecial factor in $\mathbf{u}$, then $v = \varphi_\alpha(b)\alpha$ for some bispecial factor b in $\hat{\mathbf{u}}$;*
3. *for every $i \in \{1, 2, \ldots, d\}$, $r_i \in \mathcal{R}_{\hat{\mathbf{u}}}(b)$ if and only if $\varphi_\alpha(r_i) \in \mathcal{R}_\mathbf{u}(\varphi_\alpha(b)\alpha)$;*
4. *for every $i, j \in \{1, 2, \ldots, d\}$ we have that*
   $r_i \triangleright r_j$ *in $\mathbf{d}_{\hat{\mathbf{u}}}(b)$ if and only if $\varphi_\alpha(\mathbf{r}_i) \triangleright \varphi_\alpha(\mathbf{r}_j)$ in $\mathbf{d}_\mathbf{u}(\varphi_\alpha(b)\alpha)$.*

Corollary 8. *Let $n \in \mathbb{N}$. Let $\mathbf{u}$ be a standard AR sequence with $\Delta = (\psi_n)_{n=1}^{+\infty}$ and let b be the n-th bispecial factor of $\mathbf{u}$, when ordered with respect to the length. Let $(\alpha_i)_{i=1}^d$ be the letters of $\mathbf{u}^{(n)}$ (as given in Theorem 3) such that $\alpha_1 \triangleright \alpha_2 \cdots \triangleright \alpha_d$. Let $\psi = \psi_1 \cdots \psi_n$. Then the words $r_i = \psi(\alpha_i)$ are return words to b in $\mathbf{u}$ fulfilling $\mathbf{r}_1 \triangleright \mathbf{r}_2 \cdots \triangleright \mathbf{r}_d$ in the corresponding derived sequence. Moreover, if $\psi_1 = \varphi_j$, then $b = \psi_1(b')j$, where b' is the $(n-1)$-st bispecial factor in $\mathbf{u}^{(1)}$.*

We illustrate on a ternary example how to use Corollary 8 recursively to obtain return words to bispecial factors as morphic images of letters $\{1, 2, 3\}$.

Example 9. Let $\mathbf{u}$ be the standard regular ternary AR sequence with the slope $\theta = (0, \overline{2, 1, 1}) = (0, 2, 1, 1, 2, 1, 1, \ldots)$, hence the prefix of $\mathbf{u}$ is

$$\varphi_1^2\varphi_2\varphi_3(1^2 2) = 1^2 21^2 31^2 21^3 21^2 31^2 21^3 21^2 31^2 2.$$

We determine the first 4 bispecial factors, i.e., the first 4 palindromic prefixes. Then we find their return words applying Corollary 8, obtaining the following table. The return words r_1, r_2, r_3 satisfy $r_1 \rhd r_2 \rhd r_3$ as letters in the derived sequence to the corresponding bispecial factor b.

n	(N, m)	b	ψ	r_1	r_2	r_3
0	$(0,0)$	ε	id	$\psi(1) = 1$	$\psi(2) = 2$	$\psi(3) = 3$
1	$(0,1)$	1	φ_1	$\psi(1) = 1$	$\psi(2) = 12$	$\psi(3) = 13$
2	$(1,0)$	1^2	φ_1^2	$\psi(2) = 1^2 2$	$\psi(3) = 1^2 3$	$\psi(1) = 1$
3	$(2,0)$	$1^2 2 1^2$	$\varphi_1^2 \varphi_2$	$\psi(3) = 1^2 2 1^2 3$	$\psi(1) = 1^2 21$	$\psi(2) = 1^2 2$

We explain the table in more detail. The case $n = 0$ is trivial.

In the case $n = 1 = 0 + 1 = a_0 + m$, we have $\mathbf{u} = \varphi_1(\mathbf{u}^{(1)})$. The directive sequence of $\mathbf{u}^{(1)}$ starts with $\varphi_1 \varphi_2 \varphi_3$, which implies that $1 \rhd 2 \rhd 3$ in $\mathbf{u}^{(1)}$. Moreover, $\mathbf{u} = \varphi_1(\mathbf{u}^{(1)})$ implies, using Corollary 8, the form of return words to the first bispecial factor 1: $r_1 = \varphi_1(1)$, $r_2 = \varphi_1(2)$, $r_3 = \varphi_1(3)$, i.e., in simple terms, the most (least) frequent return word in $\mathbf{u}$ is obtained from the most (least) frequent letter in $\mathbf{u}^{(1)}$.

In the case $n = 2 = 0 + 2 + 0 = a_0 + a_1 + m$, we have $\mathbf{u} = \varphi_1^2(\mathbf{u}^{(2)})$. The directive sequence of $\mathbf{u}^{(2)}$ starts with $\varphi_2 \varphi_3 \varphi_1$, thus $2 \rhd 3 \rhd 1$ in $\mathbf{u}^{(2)}$. Simultaneously, $\mathbf{u} = \varphi_1^2(\mathbf{u}^{(2)})$ implies that we use Corollary 8 with the morphism $\psi = \varphi_1^2$ to obtain the return words $r_1^{(1)} = \varphi_1^2(2)$, $r_2^{(1)} = \varphi_1^2(3)$, $r_3^{(1)} = \varphi_1^2(1)$ to the second bispecial factor 1^2 in $\mathbf{u}$. As 1^2 is a primary bispecial factor associated with $(1,0)$, the return words are denoted $r_1^{(1)}, r_2^{(1)}, r_3^{(1)}$.

Therefore, for every $n \in \mathbb{N}$ and its associated pair (N, m) we deduce from the directive sequence Δ the morphism $\psi = \psi_{(N,m)}$ and the permutation of letters $x, y, z \in \{1, 2, 3\}$ fulfilling $x \rhd y \rhd z$ in the sequence $\mathbf{u}^{(n)}$ to obtain the return words to the nth bispecial factor fulfilling $r_1 \rhd r_2 \rhd r_3$, as illustrated below

$$\Delta = \underbrace{\underbrace{\varphi_1 \ \varphi_1 \ \varphi_2 \varphi_3 \varphi_1}_{\psi_{(1,0)}} \ \varphi_1 \varphi_2 \varphi_3 \varphi_1}_{\psi_{(6,1)}} \ \overbrace{\varphi_1 \varphi_2 \varphi_3}^{1 \rhd 2 \rhd 3} \ \varphi_1 \cdots$$

with $\psi_{(6,1)}$ and $1\rhd2\rhd3$ labeled over and $\psi_{(1,0)}$, $2\rhd3\rhd1$ labeled under.

We deduce recurrence relations for the lengths of return words to bispecial factors in standard AR sequences under the assumption of regularity.

Proposition 10. *Let $\mathbf{u}$ be a regular d-ary AR sequence over the alphabet $\{1, 2, \ldots, d\}$ with $\theta = (a_N)_{N \geq 0}$. Let $(r_1^{(N)})_{N \geq 0}, (r_2^{(N)})_{N \geq 0}, \ldots, (r_d^{(N)})_{N \geq 0}$ denote the sequences of return words to the primary bispecial factors associated with $(N, 0)$, with the convention that $r_1^{(N)} \rhd r_2^{(N)} \rhd \cdots \rhd r_d^{(N)}$ in the corresponding derived sequence. Let $r_1^{(-1)} = d, r_1^{(-2)} = d - 1, \ldots, r_1^{(-d+1)} = 2$ and $a_i = 0$ for* *every* $i \in \{-d+1, \ldots, -1\}$. *Then for every* $N \geq 0$,

$$|r_1^{(N+1)}| = a_{N+1}|r_1^{(N)}| + a_N|r_1^{(N-1)}| + \cdots + a_{N-d+3}|r_1^{(N-d+2)}| + |r_1^{(N-d+1)}|,$$
$$|r_2^{(N+1)}| = a_{N+1}|r_1^{(N)}| + \cdots + a_{N-d+4}|r_1^{(N-d+3)}| + |r_1^{(N-d+2)}|,$$
$$\vdots$$
$$|r_{d-1}^{(N+1)}| = a_{N+1}|r_1^{(N)}| + |r_1^{(N-1)}|,$$
$$|r_d^{(N+1)}| = |r_1^{(N)}|.$$

Let N, m *be integers such that* $0 < m < a_{N+1}$. *If* $b \in \mathcal{L}(\mathbf{u})$ *is a secondary bispecial factor associated with* (N, m) *and* $r_1, r_2, \ldots, r_d \in \mathcal{R}_{\mathbf{u}}(b)$ *are its return words fulfilling* $\mathbf{r}_1 \rhd \mathbf{r}_2 \rhd \cdots \rhd \mathbf{r}_d$ *in* $\mathbf{d}_{\mathbf{u}}(b)$, *then*

$$|r_1| = |r_1^{(N)}|,$$
$$|r_2| = m|r_1^{(N)}| + a_N|r_1^{(N-1)}| + \cdots + a_{N-d+3}|r_1^{(N-d+2)}| + |r_1^{(N-d+1)}|,$$
$$\vdots$$
$$|r_{d-1}| = m|r_1^{(N)}| + a_N|r_1^{(N-1)}| + |r_1^{(N-2)}|,$$
$$|r_d| = m|r_1^{(N)}| + |r_1^{(N-1)}|.$$

Proof. Let $\mathbf{u}$ be a standard regular d-ary AR sequence associated with $\theta = (a_N)_{N \geq 0}$ and let $N \geq 0$. Let ψ be the morphism obtained as in Corollary 8, i.e., a composition of morphisms $\varphi_1, \varphi_2, \ldots, \varphi_d$ (or identity for $N = 0$), such that $r_1^{(N)} = \psi(\alpha_1), r_2^{(N)} = \psi(\alpha_2), \ldots, r_d^{(N)} = \psi(\alpha_d)$ for mutually distinct letters $\alpha_i \in \{1, 2, \ldots, \mathbf{d}\}$. Using the regularity of $\mathbf{u}$, we have

$$r_1^{(N+1)} = \psi\varphi_{\alpha_1}^{a_{N+1}}(\alpha_2) = \psi(\alpha_1^{a_{N+1}}\alpha_2) = \psi(\alpha_1)^{a_{N+1}}\psi(\alpha_2) = \left(r_1^{(N)}\right)^{a_{N+1}} r_2^{(N)},$$
$$r_2^{(N+1)} = \psi\varphi_{\alpha_1}^{a_{N+1}}(\alpha_3) = \psi(\alpha_1^{a_{N+1}}\alpha_3) = \psi(\alpha_1)^{a_{N+1}}\psi(\alpha_2) = \left(r_1^{(N)}\right)^{a_{N+1}} r_3^{(N)},$$
$$\vdots$$
$$r_{d-1}^{(N+1)} = \psi\varphi_{\alpha_1}^{a_{N+1}}(\alpha_d) = \psi(\alpha_1^{a_{N+1}}\alpha_d) = \psi(\alpha_1)^{a_{N+1}}\psi(\alpha_d) = \left(r_1^{(N)}\right)^{a_{N+1}} r_d^{(N)},$$
$$r_d^{(N+1)} = \psi\varphi_{\alpha_1}^{a_{N+1}}(\alpha_1) = \psi(\alpha_1) = r_1^{(N)}.$$

Let $b \in \mathcal{L}(\mathbf{u})$ be the secondary bispecial factor associated with (N, m) and $r_1, r_2, \ldots, r_d \in \mathcal{R}_{\mathbf{u}}(b)$ be its return words fulfilling $\mathbf{r}_1 \rhd \mathbf{r}_2 \rhd \cdots \rhd \mathbf{r}_d$ in $\mathbf{d}_{\mathbf{u}}(b)$. From the regularity of $\mathbf{u}$ we have

$$r_1 = \psi\varphi_{\alpha_1}^m(\alpha_1) = \psi(\alpha_1) = r_1^{(N)},$$
$$r_2 = \psi\varphi_{\alpha_1}^m(\alpha_2) = \psi(\alpha_1^m\alpha_2) = \psi(\alpha_1)^m\psi(\alpha_2) = \left(r_1^{(N)}\right)^m r_2^{(N)},$$
$$\vdots$$
$$r_{d-1} = \psi\varphi_{\alpha_1}^m(\alpha_{d-1}) = \psi(\alpha_1^m\alpha_{d-1}) = \psi(\alpha_1)^m\psi(\alpha_{d-1}) = \left(r_1^{(N)}\right)^m r_{d-1}^{(N)},$$
$$r_d = \psi\varphi_{\alpha_1}^m(\alpha_d) = \psi(\alpha_1^m\alpha_d) = \psi(\alpha_1)^m\psi(\alpha_d) = \left(r_1^{(N)}\right)^m r_d^{(N)}.$$

The desired statements follow directly, using the initial conditions. □

Example 11. Consider Example 9. We have $r_1^{(-1)} = 3$, $r_1^{(0)} = 1$, $r_1^{(1)} = 1^2 2$, $r_1^{(2)} = 1^2 21^2 3$ and $r_2^{(2)} = 1^2 21$ and $r_3^{(2)} = 1^2 2$. Indeed,

$$|r_1^{(2)}| = a_2|r_1^{(1)}| + a_1|r_1^{(0)}| + |r_1^{(-1)}| = 1 \cdot 3 + 2 \cdot 1 + 1 = 6;$$
$$|r_2^{(2)}| = a_2|r_1^{(1)}| + |r_1^{(0)}| = 1 \cdot 3 + 1 = 4;$$
$$|r_3^{(2)}| = |r_1^{(1)}| = 3.$$

Moreover, recurrence relations for the lengths Q_N of the most frequent return words $r_1^{(N)}$ associated with primary bispecial factors can be deduced from Proposition 10.

Corollary 12. *Let* **u** *be a regular d-ary AR sequence with the slope* $\theta = (a_N)_{N \geq 0}$. *Let* $(Q_N)_{N \geq 0}$ *denote the sequence of lengths of the most frequent return words* $r_1^{(N)}$ *to the primary bispecial factors associated with* $(N, 0)$.[3] *Then*

$$Q_N = \sum_{i=1}^{d-1} a_{N-i+1} Q_{N-i} + Q_{N-d}, \text{ for every } N \geq 1, \tag{1}$$

where we set $a_i = 0$ *and* $Q_i = 1$ *for every* $i \in \{-d+1, \ldots, -1\}$.

We show that the length of bispecial factors in AR sequences may be determined from the length of their return words.

Proposition 13. *Let* b *be a bispecial factor in a d-ary AR sequence and let* $r_1, r_2, \ldots, r_d$ *be its return words. Then* $|b| = \frac{1}{d-1}(\sum_{i=1}^d |r_i| - d)$.

Proof. We prove $\vec{b} = \frac{1}{d-1}\left(\sum_{i=1}^d \vec{r_i} - (1,1,\ldots,1)^T\right)$, which implies the desired statement. For every standard d-ary AR sequence **u** we denote by $(b_n^{\mathbf{u}})_{n \in \mathbb{N}}$ the sequence of bispecial factors in **u** ordered by length. We proceed by induction on $n \in \mathbb{N}$. Let **û** be a standard AR sequence and let $b = b_0^{\hat{\mathbf{u}}} = \varepsilon$ be its 0-th bispecial factor. Then the return words to b are 1, 2, ..., d and

$$\vec{b} = \begin{pmatrix} 0 \\ 0 \\ \vdots \\ 0 \end{pmatrix} = \frac{1}{d-1}\left(\begin{pmatrix} 1 \\ 0 \\ \vdots \\ 0 \end{pmatrix} + \begin{pmatrix} 0 \\ 1 \\ \vdots \\ 0 \end{pmatrix} + \begin{pmatrix} 0 \\ 0 \\ \vdots \\ 1 \end{pmatrix} - \begin{pmatrix} 1 \\ 1 \\ \vdots \\ 1 \end{pmatrix}\right) = \frac{1}{d-1}\left(\sum_{i=1}^d \vec{r_i} - \begin{pmatrix} 1 \\ 1 \\ \vdots \\ 1 \end{pmatrix}\right).$$

Induction hypothesis: for some $n \in \mathbb{N}$ it holds that for every standard d-ary AR sequence **û**, the bispecial factor $b' = b_n^{\hat{\mathbf{u}}}$ and its return words $(r_i')_{i=1}^d$ fulfil

$$\vec{b'} = \frac{1}{d-1}\left(\sum_{i=1}^d \vec{r_i'} - \begin{pmatrix} 1 \\ \vdots \\ 1 \end{pmatrix}\right).$$

Let $b = b_{n+1}^{\mathbf{u}}$ be the $(n+1)$-st bispecial factor in a standard d-ary AR sequence **u** and let $(r_i)_{i=1}^d$ be its return words in **u**. Let $\alpha \in \{1, 2, \ldots, d\}$ and let **û** be the

---

[3] The most frequent return words to primary bispecial factors coincide with the generalized standard words defined in [17].

AR sequence such that $\mathbf{u} = \varphi_\alpha(\hat{\mathbf{u}})$. Then by Corollary 8 we have $b = \varphi_\alpha(b')\alpha$ and $r_i = \varphi_\alpha(r'_i)$ for every $i \in \{1, \ldots, d\}$ for some bispecial factor $b' = b_n^{\hat{u}}$ and its return words $(r'_i)_{i=1}^d$ in $\hat{\mathbf{u}}$. Let $\alpha = 1$ (the proof for the other cases is similar). Denote M_1 the incidence matrix of the morphism φ_1. From the relations between b, $(r_i)_{i=1}^d$ and b', $(r'_i)_{i=1}^d$ and the induction hypothesis we have

$$
\vec{b} = M_1\vec{b'} + \begin{pmatrix} 1 \\ 0 \\ \vdots \\ 0 \end{pmatrix} = M_1\left(\frac{1}{d-1}\left(\sum_{i=1}^d \vec{r'_i} - \begin{pmatrix} 1 \\ \vdots \\ 1 \end{pmatrix}\right)\right) + \begin{pmatrix} 1 \\ 0 \\ \vdots \\ 0 \end{pmatrix}
$$

$$
= \frac{1}{d-1}\left(\sum_{i=1}^d M_1\vec{r'_i} - \begin{pmatrix} 1 \\ \vdots \\ 1 \end{pmatrix}\right) + \begin{pmatrix} 1 \\ 0 \\ \vdots \\ 0 \end{pmatrix} = \frac{1}{d-1}\left(\sum_{i=1}^d \vec{r_i} - \begin{pmatrix} 1 \\ \vdots \\ 1 \end{pmatrix}\right). \quad \square
$$

We obtain a formula for the lengths of primary and secondary bispecial factors in regular AR sequences.

Corollary 14. *Let* $\mathbf{u}$ *be a regular d-ary AR sequence with the slope $\theta = (a_N)_{N\geq0}$. Let $(Q_N)_{N\geq0}$ be the sequence from Corollary 12. Let N, m be non-negative integers such that $0 \leq m < a_{N+1}$ and let b be the bispecial factor associated with (N, m). Then*

$$
|b| = \frac{1}{d-1}\left(\left(1 + (d-1)m\right)Q_N + \sum_{i=1}^{d-1}\left((d-i-1)a_{N-i+1} + 1\right)Q_{N-i} - d\right).
$$

Proof. If $N = m = 0$, then $|b| = 0 = \frac{1}{d-1}(Q_0 + \sum_{i=1}^{d-1} Q_{-i} - d)$. Otherwise, it suffices to use Proposition 13, Proposition 10 and Equation (1). $\quad\square$

Example 15. In Example 9, the first values of Q_N are $Q_{-2} = Q_{-1} = Q_0 = 1$, $Q_1 = 3$ and $Q_2 = 6$. Using the formula from Corollary 14, we get the following lengths for bispecial factors associated with (N, m)

$$
(N, m) = (0, 1) : |b_{0,1}| = \tfrac{1}{2}(3Q_0 + Q_{-1} + Q_{-2} - 3) = 1\,;
$$

$$
(N, m) = (1, 0) : |b_{1,0}| = \tfrac{1}{2}(Q_1 + 3Q_0 + Q_{-1} - 3) = 2\,;
$$

$$
(N, m) = (2, 0) : |b_{2,0}| = \tfrac{1}{2}(Q_2 + 2Q_1 + Q_0 - 3) = 5\,.
$$

The obtained lengths are thus in correspondance with the table from Example 9.

5   (Asymptotic) Critical Exponent of Regular AR Sequences

We derive a formula for the (asymptotic) critical exponent of regular AR sequences using Theorem 2 and Sect. 4.1.

Theorem 16. *The critical exponent and the asymptotic critical exponent of a regular d-ary AR sequence* $\mathbf{u}$ *with the slope* $\theta = (a_N)_{N \geq 0}$ *satisfy*

$$E(\mathbf{u}) = \frac{d}{d-1} + \sup_{N \geq 1} \left\{ a_N + \frac{1}{d-1} \frac{\sum_{i=2}^{d} ((d-i)a_{N-i+1} + 1)\, Q_{N-i} - d}{Q_{N-1}} \right\} ;$$

$$E^*(\mathbf{u}) = \frac{d}{d-1} + \limsup_{N \to +\infty} \left(a_N + \frac{1}{d-1} \frac{\sum_{i=2}^{d} ((d-i)a_{N-i+1} + 1)\, Q_{N-i}}{Q_{N-1}} \right).$$

Proof. Recall the formulae in Theorem 2. Let $b_{N,m} \in \mathcal{L}(\mathbf{u})$ denote the bispecial factor associated with (N, m) and let $r_{N,m} \in \mathcal{R}_{\mathbf{u}}(b)$ denote its shortest return word. We show that it suffices to consider primary bispecial factors associated with $N \geq 1$ to compute the supremum (limes superior). From Equation (1), the following relation may be deduced

$$0 \leq \sum_{i=2}^{d} ((d-i)a_{N-i+1} + 1)\, Q_{N-i} - d < (d-1)Q_{N-1}, \quad \text{for every } N \geq 2. \quad (2)$$

If $b_{N,0}$ is primary, then by Proposition 10 we have $r_{N,0} = r_d^{(N)}$ and $|r_{N,0}| = Q_{N-1}$. Applying Corollary 14 and Equation (1), we have for $N \geq 1$

$$\begin{aligned}
\frac{|b_{N,0}|}{|r_{N,0}|} &= \frac{1}{d-1} \frac{Q_N + \sum_{i=1}^{d-1}((d-i-1)a_{N-i+1}+1)Q_{N-i} - d}{Q_{N-1}} \\
&= \frac{1}{d-1} \frac{\sum_{i=1}^{d}((d-i)a_{N-i+1}+1)Q_{N-i} - d}{Q_{N-1}} \\
&= \frac{1}{d-1} + a_N + \frac{1}{d-1} \frac{\sum_{i=2}^{d}((d-i)a_{N-i+1}+1)Q_{N-i} - d}{Q_{N-1}} .
\end{aligned} \quad (3)$$

Applying Equation (2), we obtain for every $N \geq 2$

$$\frac{1}{d-1} + a_N \leq \frac{|b_{N,0}|}{|r_{N,0}|} < \frac{d}{d-1} + a_N . \quad (4)$$

The case $N = 0$ may be omitted when computing supremum as $b_{0,0} = \varepsilon$.

If $b_{N,m}$ is secondary, then by Proposition 10 the shortest return word to $b_{N,m}$ is $r_{N,m} = r_1^{(N)}$ of length Q_N. We show that $\frac{|b_{N,m}|}{|r_{N,m}|} < \frac{|b_{N+1,0}|}{|r_{N+1,0}|}$, hence secondary bispecial factors may be omitted. Applying Corollary 14 and Equations (2) and (4), we have for $N \geq 1$

$$\begin{aligned}
\frac{|b_{N,m}|}{|r_{N,m}|} &= \frac{1}{d-1} \frac{(1+(d-1)m)Q_N + \sum_{i=1}^{d-1}((d-i-1)a_{N-i+1}+1)Q_{N-i} - d}{Q_N} \\
&= \frac{1}{d-1} + m + \frac{1}{d-1} \frac{\sum_{i=1}^{d-1}((d-i-1)a_{N-i+1}+1)Q_{N-i} - d}{Q_N} \\
&\leq \frac{1}{d-1} + a_{N+1} - 1 + \frac{1}{d-1} \frac{\sum_{i=1}^{d-1}((d-i-1)a_{N-i+1}+1)Q_{N-i} - d}{Q_N} \\
&< \frac{1}{d-1} + a_{N+1} \leq \frac{|b_{N+1,0}|}{|r_{N+1,0}|} .
\end{aligned}$$

If $N = 0$, then $b_{0,m} = \alpha^m$, $r_{0,m} = \alpha$ for some $\alpha \in \{1, 2, \ldots, d\}$ and therefore $\frac{|b_{0,m}|}{|r_{0,m}|} = m < a_1 = \frac{|b_{1,0}|}{|r_{1,0}|}$. Thus we obtain from Theorem 2

$$E(\mathbf{u}) = \frac{d}{d-1} + \sup_{N \geq 1} \left\{ a_N + \frac{1}{d-1} \frac{\sum_{i=2}^{d} ((d-i)a_{N-i+1} + 1)\, Q_{N-i} - d}{Q_{N-1}} \right\} ;$$

$$E^*(\mathbf{u}) = \frac{d}{d-1} + \limsup_{N\to+\infty}\left(a_N + \frac{1}{d-1}\frac{\sum_{i=2}^{d}((d-i)a_{N-i+1}+1)Q_{N-i}}{Q_{N-1}}\right). \quad \square$$

From Theorem 16 and Equation (4), we obtain the following upper bounds.

Corollary 17. *The critical exponent and the asymptotic critical exponent of a regular d-ary AR sequence $\mathbf{u}$ with $\theta = (a_N)_{N\geq 0}$ satisfy*

$$E(\mathbf{u}) < 1 + \frac{d}{d-1} + \sup_{N\geq 1}\{a_N\} \quad and \quad E^*(\mathbf{u}) \leq 1 + \frac{d}{d-1} + \limsup_{N\to+\infty} a_N.$$

Using the formulae from Theorem 16, it is straightforward to deduce that the d-bonacci sequence has the minimal (asymptotic) critical exponent among regular d-ary AR sequences.

Theorem 18. *The minimal (asymptotic) critical exponent for regular d-ary AR sequences is attained for the d-bonacci sequence $\mathbf{u}_d$ with $\theta = (a_N) = (0,\overline{1})$. Moreover, the asymptotic critical exponent equals $E^*(\mathbf{u}_d) = 2 + \frac{1}{t-1}$, where t is the dominant real root of the polynomial $x^d - \sum_{i=1}^{d} x^{d-i}$.*

Proof. First, we explain that the critical exponent $E(\mathbf{u}_d)$ is minimal among d-ary regular AR sequences. By Theorem 16, if the slope θ of $\mathbf{u}$ contains at least one element $a_N \geq 2$, then $E(\mathbf{u}) \geq 2+\frac{d}{d-1}$. On the other hand, the d-bonacci sequence $\mathbf{u}_d$ with $\theta = (a_N) = (0,\overline{1})$ satisfies by Corollary 17 that $E(\mathbf{u}_d) < 2 + \frac{d}{d-1}$.

Second, we determine $E^*(\mathbf{u}_d)$. The sequence $\mathbf{u}_d$ satisfies $Q_N = \sum_{i=1}^{d} Q_{N-i}$ for all $N \geq d$ and $Q_0 = 1, Q_1 = 2, Q_2 = 4, \ldots, Q_{d-1} = 2^{d-1}$. Hence, $\lim_{N\to\infty}\frac{Q_N}{Q_{N-1}} = t$, where $t = t(d)$ is the unique real root, larger than the absolute value of all other roots, of the polynomial $x^d - \sum_{i=1}^{d} x^{d-i}$. Observe that $2 - \frac{1}{d} < t < 2$. We obtain using Theorem 16

$$\begin{aligned}
E^*(\mathbf{u}_d) &= 1 + \frac{d}{d-1} + \frac{1}{d-1}\lim_{N\to\infty}\frac{\sum_{i=2}^{d}((d-i)+1)Q_{N-i}}{Q_{N-1}}\\
&= \frac{2d-1}{d-1} + \frac{1}{d-1}\frac{\sum_{i=2}^{d}((d-i)+1)t^{d-i}}{t^{d-1}}\\
&= \frac{2d-2}{d-1} + \frac{t^{d-1}}{(d-1)t^{d-1}} + \frac{1}{d-1}\frac{\sum_{i=2}^{d}((d-i)+1)t^{d-i}}{t^{d-1}}\\
&= 2 + \frac{1}{t-1},
\end{aligned}$$

where the relation $t^d = \sum_{i=1}^{d} t^{d-i}$ was used in the last step.

Finally, we explain that the asymptotic critical exponent $E^*(\mathbf{u}_d)$ is minimal among regular d-ary AR sequences $\mathbf{u}$. If the slope θ of $\mathbf{u}$ contains infinitely many $a_N \geq 2$, then by Theorem 16, $E^*(\mathbf{u}) \geq 3 + \frac{1}{d-1} > 2 + \frac{1}{t-1}$. If $a_N = 1$ for all sufficiently large N, then $E^*(\mathbf{u}) = 2 + \frac{1}{t-1}$. $\quad\square$

Example 19. For illustration, we computed the approximate values of $E^*(\mathbf{u}_d)$ for $d \in \{2,3,\ldots,7\}$ in the following table.

d	2	3	4	5	6	7
$t = t(d)$	1.618	1.839	1.928	1.966	1.984	1.992
$E^*(\mathbf{u}_d)$	3.618	3.191	3.078	3.035	3.017	3.008

The values of the asymptotic critical exponent and the critical exponent coincide in the case of the Tribonacci sequence $\mathbf{u}_T$, see [23]. We believe that it is possible to generalize the equality of the asymptotic critical exponent and the critical exponent for all d-bonacci sequences.

Conjecture 20. The d-bonacci sequence $\mathbf{u}_d$ satisfies $E(\mathbf{u}_d) = E^*(\mathbf{u}_d)$.

We provide a starting point of an incomplete proof. We need to show that $E(\mathbf{u}_d) \leq E^*(\mathbf{u}_d)$ and thus $E(\mathbf{u}_d) = E^*(\mathbf{u}_d)$. We compare the terms from the formula for $E(\mathbf{u}_d)$ for all $N \geq 1$ with the value $E^*(\mathbf{u}_d)$. If $N < d - 1$ then using the fact that $tQ_N \leq Q_{N+1}$ for all $0 \leq N < d - 1$, we can see that $\frac{d}{d-1} + a_N + \frac{1}{d-1} \frac{\sum_{i=2}^{d}((d-i)a_{N-i+1}+1)Q_{N-i}-d}{Q_{N-1}} < 2 + \frac{1}{t-1}$. Let $N \geq d - 1$. The desired inequality $\frac{1}{d-1} + \frac{1}{d-1} \frac{\sum_{i=2}^{d}((d-i)+1)Q_{N-i}-d}{Q_{N-1}} \leq \frac{1}{t-1}$ simplifies to $(t-1)(Q_N + \sum_{i=2}^{d}(d-i)Q_{N-i} - d) \leq (d-1)Q_{N-1}$. We verified this inequality by computer experiments, therefore we stated the above conjecture.

6   Comments and Open Problems

In this paper we derived a formula for the (asymptotic) critical exponent of regular AR sequences. Moreover, we observed using the formula that the minimal (asymptotic) critical exponent among regular d-ary Arnoux-Rauzy sequences is reached by the d-bonacci sequence. We determined the value of the asymptotic critical exponent of the d-bonacci sequence and we conjectured that its critical exponent has the same value.

The following questions remain open.

- Is it possible to derive a formula for the (asymptotic) critical exponent for other classes of AR sequences?
- What is the minimal (asymptotic) critical exponent of d-ary AR sequences? The result is known for Sturmian sequences, where the minimal value $2 + \frac{1+\sqrt{5}}{2}$ is attained for the Fibonacci sequence [15].
- How to compute the critical exponent of colourings of AR sequences by constant gap sequences? Such sequences represent generalization of balanced sequences, which themselves are colourings of Sturmian sequences [10]. We implemented a program for computation of the (asymptotic) critical exponent of regular substitutive (with eventually periodic slopes θ) AR sequences in SageMath [22] which we plan to extend to the colourings of AR sequences.

References

1. Berthé, V., Cassaigne, J., Steiner, W.: Balance properties of Arnoux-Rauzy sequences. Int. J. Algebra Comput. **23**(04), 689–703 (2013)
2. Carpi, A., de Luca, A.: Special factors, periodicity, and an application to Sturmian words. Acta Inf. **36**(12), 983–1006 (2000)

3. Coven, E.M., Hedlund, G.A.: Sequences with minimal block growth. Math. Syst. Theor. **7**, 138–153 (1973)
4. Dejean, F.: Sur un théorème de Thue. J. Combin. Theory. Ser. A **13**, 90–99 (1972)
5. Durand, F.: A characterization of substitutive sequences using return words. Discrete Math. **179**, 89–101 (1998)
6. Krieger, D., Shallit, J.O.: Every real number greater than 1 is a critical exponent. Theoret. Comput. Sci. **381**, 177–182 (2007)
7. Pansiot, J.-J.: A propos d'une conjecture de F. Dejean sur les répétitions dans les mots. Discr. Appl. Math. **7**(1), 297–311 (1984)
8. Moulin-Ollagnier, J.: Proof of Dejean's conjecture for alphabets with $5, 6, 7, 8, 9, 10$ and 11 letters. Theoret. Comput. Sci. **95**, 187–205 (1992)
9. Mohammad-Noori, M., Currie, J.D.: Dejean's conjecture and Sturmian words. European J. Comb. **28**, 876–890 (2007)
10. Hubert, P.: Suites équilibrées. Theoret. Comput. Sci. **242**(1–2), 91–108 (2000)
11. Carpi, A.: On Dejean's conjecture over large alphabets. Theoret. Comput. Sci. **385**, 137–151 (2007)
12. Currie, J.D., Rampersad, N.: A proof of Dejean's conjecture. Math. Comp. **80**, 1063–1070 (2011)
13. Rao, M.: Last cases of Dejean's conjecture. Theoret. Comput. Sci. **412**, 3010–3018 (2011)
14. Droubay, X., Justin, J., Pirillo, G.: Episturmian words and some constructions of de Luca and Rauzy. Theoret. Comput. Sci. **255**(1), 539–553 (2001)
15. Damanik, D., Lenz, D.: The index of Sturmian sequences. J. Eur. J. Comb. **23**, 23–29 (2002)
16. Justin, J., Pirillo, G.: Episturmian words and episturmian morphisms. Theoret. Comput. Sci. **276**, 281–313 (2002)
17. Glen, A.: Powers in a class of A-strict standard episturmian words. Theoret. Comput. Sci. **380**, 330–354 (2007)
18. Justin, J., Vuillon, L.: Return words in Sturmian and episturmian words. RAIRO-Theoret. Inf. Appl. **34**, 343–356 (2000)
19. Dolce, F., Dvořáková, L., Pelantová, E.: On balanced sequences and their critical exponent. Theoret. Comput. Sci. **939**, 18–47 (2023)
20. Peltomäki, J.: Initial nonrepetitive complexity of regular episturmian words and their Diophantine exponents. arXiv (2021) https://arxiv.org/abs/2103.08351
21. Medková, K.: Derived sequences of Arnoux–Rauzy sequences. In: Mercaş, R., Reidenbach, D. (eds.) WORDS 2019. LNCS, vol. 11682, pp. 251–263. Springer, Cham (2019). https://doi.org/10.1007/978-3-030-28796-2_20
22. SageMath package. https://pypi.org/project/colored-arnoux-rauzy-sequences/ Last Accessed 14 Feb 2023
23. Tan, B., Wen, Z.-Y.: Some properties of the Tribonacci sequence. Eur. J. Comb. **28**(6), 1703–1719 (2007)
24. Lothaire, M.: Algebraic Combinatorics on Words. Cambridge University Press, Cambridge (2002)

On a Class of 2-Balanced Sequences

Ľubomíra Dvořáková[(⊠)], Martin Mašek, and Edita Pelantová

FNSPE Czech Technical University in Prague, Prague, Czechia
{lubomira.dvorakova,edita.pelantova}@fjfi.cvut.cz

Abstract. We define a new class of ternary and quaternary sequences that are 2-balanced. These sequences are obtained by colouring of Sturmian sequences. We provide an upper bound on factor and abelian complexity of these sequences. In case of ternary sequences, the factor complexity is at most quadratic, in case of quaternary sequences, at most cubic. We state a conjecture on conditions guaranteeing the maximal complexity and support our conjecture by computer experiments. Furthermore, we focus on a particular ternary 2-balanced sequence obtained by colouring of the Fibonacci sequence. To describe its factor complexity we determine the number of factors of the Fibonacci sequence sharing the same Parikh vector. We deduce a formula based on expansion of integers in a particular non-standard numeration system.

Keywords: Sturmian sequence · Frequency · Balancedness · Numeration system · Factor complexity · Abelian complexity

1 Introduction

A sequence is *C-balanced* for some integer $C \geq 1$ if, for any two factors of the same length and each letter, the number of occurrences of that letter in the two factors differs at most by C. Binary 1-*balanced* sequences were introduced by Hedlund and Morse [11]. Binary 1-balanced aperiodic sequences are called *Sturmian*. Hubert showed for larger alphabets that each recurrent aperiodic balanced sequence can be obtained from a Sturmian sequence by the so called colouring by constant gap sequences [9]. For 2-balanced sequences, such a useful characterization is missing.

One of the well-known generalizations of Sturmian sequences to larger alphabets are *Arnoux-Rauzy (AR) sequences*. They share a lot of properties with Sturmian sequences. However, there exist AR sequences that are not *C*-balanced for any *C*. Berthé, Cassaigne, and Steiner [3] found a sufficient condition for 2-balancedness of AR sequences. It is known that the letter frequencies of any Arnoux-Rauzy sequence belong to the Rauzy gasket [2], a fractal set of Lebesgue measure zero.

A new class of ternary sequences associated with a multidimensional continued fraction algorithm was introduced and studied by Cassaigne, Labbé, and

The work was supported by projects CZ.02.1.01/0.0/0.0/16_019/0000778 and SGS23/187/OHK4/3T/14.

© The Author(s), under exclusive license to Springer Nature Switzerland AG 2023
A. Frid and R. Mercaş (Eds.): WORDS 2023, LNCS 13899, pp. 143–154, 2023.
https://doi.org/10.1007/978-3-031-33180-0_11

Leroy [4]. Almost every sequence in the new class is C-balanced for some C. Moreover, the class contains sequences of any given letter frequencies. A generalization of Sturmian sequences associated with N-continued fraction algorithms was introduced by Langeveld, Rossi, and Thuswaldner [10]. All these sequences are binary and C-balanced for some C. It is possible to characterize those ones among them that are 2-balanced. The class contains sequences of any given rationally independent letter frequencies. All sequences mentioned above have sublinear factor complexity.

In our contribution, we define ternary and quaternary sequences that are 2-balanced. We obtain these sequences by colouring of Sturmian sequences by Sturmian sequences. Our class contains sequences of any given letter frequencies. We provide an upper bound on factor and abelian complexity of these sequences. In case of ternary sequences, the factor complexity is at most quadratic, in case of quaternary sequences, at most cubic. In Sect. 4 we state a conjecture on conditions guaranteeing the maximal complexity. We support our conjecture by computer experiments. In Sect. 5 we study a particular ternary 2-balanced sequence obtained by colouring of the Fibonacci sequence. To describe its factor complexity it is necessary to determine the number of factors of the Fibonacci sequence having the same Parikh vector. We deduce a formula based on expansion of integers in a particular non-standard numeration system.

2   Preliminaries

An *alphabet* $\mathcal{A}$ is a finite set of symbols, called *letters*. A *word* w over $\mathcal{A}$ is a finite string of letters from $\mathcal{A}$. Its *length* $|w|$ equals the number of letters it contains. To denote the number of occurrences of a letter a in w, we use $|w|_a$. The *Parikh vector* of w, denoted $\Psi(w)$, is a vector with coordinates defined by $\Psi(w)_a = |w|_a$. The set of all finite words over $\mathcal{A}$ together with the operation of concatenation forms a monoid, denoted $\mathcal{A}^*$. Its neutral element is the *empty word* ε. A *sequence* $\mathbf{u} = u_0 u_1 u_2 \cdots$ over $\mathcal{A}$ is an infinite string of letters from $\mathcal{A}$, i.e., $u_i \in \mathcal{A}$ for each $i \in \mathbb{N}$. A word w is a *factor* of $\mathbf{u} = u_0 u_1 u_2 \cdots$ if there exists $i \in \mathbb{N}$ such that $w = u_i u_{i+1} \cdots u_{i+|w|-1}$. The *language* $\mathcal{L}(\mathbf{u})$ of $\mathbf{u}$ is the set of all factors of $\mathbf{u}$. A sequence $\mathbf{u}$ is called *eventually periodic* if it can be written in the form $\mathbf{u} = xy^\omega$, where x, y are words, y is non-empty, and y^ω denotes an infinite repetition of y. The sequence $\mathbf{u}$ is *aperiodic* if $\mathbf{u}$ is not eventually periodic. The *factor complexity* is a mapping $\mathcal{C}_\mathbf{u} : \mathbb{N} \to \mathbb{N}$ defined for each $n \in \mathbb{N}$ as

$$\mathcal{C}_\mathbf{u}(n) = \#\{w \ : \ |w| = n, \ w \in \mathcal{L}(\mathbf{u})\},$$

where $\#$ stands for the cardinality. For each aperiodic sequence $\mathbf{u}$, the factor complexity satisfies $\mathcal{C}_\mathbf{u}(n) \geq n + 1$ for all $n \in \mathbb{N}$ [11]. The *abelian complexity* is a mapping $\mathcal{C}_\mathbf{u}^{ab} : \mathbb{N} \to \mathbb{N}$ defined for each $n \in \mathbb{N}$ as

$$\mathcal{C}_\mathbf{u}^{ab}(n) = \#\{\Psi(w) \ : \ |w| = n, \ w \in \mathcal{L}(\mathbf{u})\}.$$

Fici and Puzynina have recently published a comprehensive survey on abelian combinatorics on words [7].

A *morphism* over $\mathcal{A}$ is a mapping $\psi : \mathcal{A}^* \to \mathcal{A}^*$ such that $\psi(uv) = \psi(u)\psi(v)$ for all $u, v \in \mathcal{A}^*$. The morphism ψ can be naturally extended to sequences by setting $\psi(u_0 u_1 u_2 \cdots) = \psi(u_0)\psi(u_1)\psi(u_2)\cdots$.

A sequence $\mathbf{u}$ over $\mathcal{A}$ is *C-balanced* for some integer $C \geq 1$ if, for any two factors u, v of $\mathbf{u}$ of the same length and each letter $a \in \mathcal{A}$, we have $\left| |u|_a - |v|_a \right| \leq C$. Binary 1-*balanced* (also called *balanced*) sequences were introduced by Hedlund and Morse [11]. Binary balanced aperiodic sequences are called *Sturmian*. Sturmian sequences may be equivalently defined as aperiodic sequences having the least possible factor complexity, i.e., $\mathcal{C}_\mathbf{u}(n) = n + 1$ for all $n \in \mathbb{N}$. The *frequency* $f(a)$ of a letter $a \in \mathcal{A}$ in a sequence $\mathbf{u} = u_0 u_1 u_2 \cdots$ is defined as the limit, if it exists,

$$f(a) = \lim_{n \to \infty} \frac{|u_0 u_1 \cdots u_{n-1}|_a}{n}.$$

If $\mathbf{u}$ is a C-balanced sequence, then the frequency $f(a)$ of each letter a in $\mathbf{u}$ is well defined and moreover each factor u of $\mathbf{u}$ satisfies $\left| |u|_a - f(a)|u| \right| \leq C$ [1]. The abelian complexity of $\mathbf{u}$ is bounded if and only if $\mathbf{u}$ is C-balanced for some positive integer C [13].

Sturmian sequences may be equivalently defined as codings of two interval exchange. For a given irrational parameter $\alpha \in (0,1)$, we consider two intervals $I_a = [0, \alpha)$ and $I_b = [\alpha, 1)$. The *two interval exchange transformation* (2iet) $T : [0,1) \to [0,1)$ is defined by

$$T(x) = x - \alpha \mod 1 = \begin{cases} x + 1 - \alpha & \text{if } x \in I_a, \\ x - \alpha & \text{if } x \in I_b. \end{cases} \tag{1}$$

If we take an initial point $\rho \in I_a \cup I_b$, the sequence $\mathbf{u} = u_0 u_1 u_2 \cdots \in \{a, b\}^{\mathbb{N}}$ defined by

$$u_n = \begin{cases} a & \text{if } T^n(\rho) \in I_a, \\ b & \text{if } T^n(\rho) \in I_b, \end{cases}$$

i.e., a coding of the trajectory of the point ρ, is a *2iet sequence* with the *parameter* α. The sequence $\left(T^n(\rho) \right)_{n \in \mathbb{N}}$ is uniformly distributed, in particular, it is dense in $[0, 1]$. It is well known that the language of $\mathbf{u}$ depends only on α, but does not depend on ρ, and the frequency of the letter a in $\mathbf{u}$ is α.

If the parameter $\alpha \in (0, 1)$ in (1) is a rational number, then $\left(T^n(\rho) \right)_{n \in \mathbb{N}}$ is a periodic sequence, and consequently, $\mathbf{u}$ is a periodic 1-balanced sequence.

Remark 1. The abelian complexity of a Sturmian sequence $\mathbf{u}$ satisfies $\mathcal{C}_\mathbf{u}^{ab}(n) = 2$ for all $n \geq 1$ [5]. In other words, only two vectors occur as Parikh vectors of factors of length n in $\mathbf{u}$. Since the frequency of the letter a in a Sturmian sequence $\mathbf{u}$ coding the transformation T is an irrational number, say α, balancedness of $\mathbf{u}$ implies that the Parikh vector of a factor of lenght n is

$$\Psi(u) = \begin{pmatrix} \lceil n\alpha \rceil \\ \lfloor (1-\alpha)n \rfloor \end{pmatrix} \quad \text{or} \quad \Psi(u) = \begin{pmatrix} \lfloor n\alpha \rfloor \\ \lceil (1-\alpha)n \rceil \end{pmatrix}.$$

3   Colouring of Sequences and Balancedness

Definition 1. *Let* **u** *be a sequence over the alphabet* $\{a, b\}$. *Let* **a**, **b** *be two sequences over mutually disjoint alphabets* $\mathcal{A}$ *and* $\mathcal{B}$. *The colouring of* **u** *by* **a**, **b** *is a sequence* **v** $=$ colour$(\mathbf{u}, \mathbf{a}, \mathbf{b})$ *over* $\mathcal{A} \cup \mathcal{B}$ *obtained from* **u** *by replacing the subsequence of all a's with* **a** *and all b's with* **b**.

For **v** $=$ colour$(\mathbf{u}, \mathbf{a}, \mathbf{b})$ we use the notation $\pi(\mathbf{v}) = \mathbf{u}$ and $\pi(v) = u$ for any $v \in \mathcal{L}(\mathbf{v})$ and the corresponding $u \in \mathcal{L}(\mathbf{u})$. We say that **u** (resp. u) is a *projection* of **v** (resp. v). The map $\pi : \mathcal{L}(\mathbf{v}) \to \mathcal{L}(\mathbf{u})$ is clearly a morphism.

Example 1. Let **u** be the Fibonacci sequence over $\{a, b\}$, i.e., $\mathbf{u} = \varphi(\mathbf{u})$ for the morphism φ given by $\varphi(a) = ab$, $\varphi(b) = a$. Furthermore, let $\mathbf{a} = (12)^\omega$ and $\mathbf{b} = 3^\omega$, then

$$\mathbf{u} = a\ b\ a\ a\ b\ a\ b\ a\ a\ b\ a\ a\ b\ a\ b\ a\ a\ b\ a\ b\ a\ a\ b\ a\ a\ b\ a\ b\ a\ a\ b\ a\ a\ b\ a\ b \cdots$$
$$\mathbf{v} = 1\ 3\ 2\ 1\ 3\ 2\ 3\ 1\ 2\ 3\ 1\ 2\ 3\ 1\ 3\ 2\ 1\ 3\ 2\ 3\ 1\ 2\ 3\ 1\ 2\ 3\ 1\ 3\ 2\ 1\ 3\ 2\ 1\ 3\ 2\ 3 \cdots$$

We have $\pi(132132312) = abaababaa$.

In the sequel, we intend to colour 2iet sequences by one 2iet sequence and one constant sequence or by two 2iet sequences. Let us observe what happens with balancedness after colouring.

Lemma 1. *Let* **u** *be a 1-balanced sequence over* $\{a, b\}$, *and* $\mathbf{a} = a_0 a_1 a_2 \ldots$ *and* $\mathbf{b} = b_0 b_1 b_2 \ldots$ *be two 1-balanced sequences over two disjoint alphabets* $\mathcal{A}$ *and* $\mathcal{B}$, *respectively. Then* **v** $=$ colour$(\mathbf{u}, \mathbf{a}, \mathbf{b})$ *is 2-balanced.*

Proof. Let U, V be factors of **v** of the same length. We want to prove that for each letter $c \in \mathcal{A} \cup \mathcal{B}$
$$\big||U|_c - |V|_c\big| \leq 2.$$
WLOG let $c \in \mathcal{A}$. Denote $u = \pi(U)$ and $v = \pi(V)$. Clearly $|u| = |v|$. Thanks to balancedness of **u**, we have $\big||u|_a - |v|_a\big| \leq 1$. Let $\Pi : (\mathcal{A} \cup \mathcal{B})^* \to \mathcal{A}^*$ be a morphism such that $\Pi(x) = x$ if $x \in \mathcal{A}$ and $\Pi(x) = \varepsilon$ if $x \in \mathcal{B}$. It holds that for each factor W of **v** the word $\Pi(W)$ is a factor of **a**. By definition of Π, we have $|U|_c = |\Pi(U)|_c$, $|V|_c = |\Pi(V)|_c$ and by definition of colouring $|\Pi(U)| = |u|_a$, $|\Pi(V)| = |v|_a$.

Since $|u|_a$ and $|v|_a$ differ at most by one, the words $\Pi(U)$ and $\Pi(V)$ are factors of **a** whose lengths differ at most by one.

- Either $|\Pi(U)| = |\Pi(V)|$. Then since **a** is balanced, it follows $\big||\Pi(U)|_c - |\Pi(V)|_c\big| \leq 1$.
- Or (without loss of generality) $|\Pi(U)| = |\Pi(V)| + 1$. Then $\Pi(U) = a_i \ldots a_{i+\ell+1}$ and $\Pi(V) = a_j \ldots a_{j+\ell}$ for some $i, j, \ell \in \mathbb{N}$. Then

$$\big||\Pi(U)|_c - |\Pi(V)|_c\big| \leq \big||a_i \ldots a_{i+\ell}|_c - |a_j \ldots a_{j+\ell}|_c\big| + |a_{i+\ell+1}|_c \leq 2.$$

Since $|U|_c = |\Pi(U)|_c$ and $|V|_c = |\Pi(V)|_c$, we have proved $\big||U|_c - |V|_c\big| \leq 2$. $\square$

Let us show that by colouring a 2iet sequence by one 2iet sequence and one constant sequence, resp. by two 2iet sequences, we are able to construct 2-balanced sequences with prescribed letter frequencies.

Recall that a vector $\boldsymbol{f} = \big(f(1), f(2), \ldots, f(d)\big) \in \mathbb{R}^d$ is called a *frequency vector* if $f(i) > 0$ for each $i = 1, 2, \ldots, d$ and $f(1) + f(2) + \cdots + f(d) = 1$.

Theorem 1. *Let $d \in \{3, 4\}$ and $\boldsymbol{f} \in \mathbb{R}^d$ be a frequency vector. Then there exist infinite 1-balanced sequences $\mathbf{u}, \mathbf{a}, \mathbf{b}$ such that the sequence $\mathbf{v} = \mathrm{colour}(\mathbf{u}, \mathbf{a}, \mathbf{b})$ over a d-letter alphabet $\mathcal{D}$ satisfies*

1. *the frequency of the letter i in $\mathbf{v}$ is $f(i)$ for each $i \in \mathcal{D}$;*
2. *$\mathbf{v}$ is 2-balanced.*

Proof. We will make use of a fact that letter frequencies in $\mathbf{v} = \mathrm{colour}(\mathbf{u}, \mathbf{a}, \mathbf{b})$ can be easily computed from letter frequencies in $\mathbf{u}$, $\mathbf{a}$, and $\mathbf{b}$. More precisely, the letter $i \in \mathcal{A}$ has in $\mathbf{v}$ the frequency $\alpha \gamma_i$, where α is the frequency of the letter a in $\mathbf{u}$ and γ_i is the frequency of the letter i in $\mathbf{a}$.

For $d = 3$ and given positive numbers $f(1), f(2), f(3)$ such that $f(1) + f(2) + f(3) = 1$, we set $\mathbf{v} = \mathrm{colour}(\mathbf{u}, \mathbf{a}, \mathbf{b})$, where $\mathbf{u}$ is a 2iet sequence over $\{a, b\}$ with the frequency of a being $\alpha = f(1) + f(2)$, $\mathbf{a}$ is a 2iet sequence over $\{1, 2\}$, where the frequency of 1 equals $\frac{f(1)}{\alpha}$ and the frequency of 2 equals $\frac{f(2)}{\alpha}$, and $\mathbf{b} = 3^\omega$.

For $d = 4$ and given positive numbers $f(1), f(2), f(3), f(4)$ such that $f(1) + f(2) + f(3) + f(4) = 1$, we set $\mathbf{v} = \mathrm{colour}(\mathbf{u}, \mathbf{a}, \mathbf{b})$, where $\mathbf{u}$ is a 2iet sequence over $\{a, b\}$ with the frequency of a being $\alpha = f(1) + f(2)$, $\mathbf{a}$ is a 2iet sequence over $\{1, 2\}$ with the frequency of 1 equal to $\frac{f(1)}{\alpha}$ and the frequency of 2 equal to $\frac{f(2)}{\alpha}$ and $\mathbf{b}$ is a 2iet sequence over $\{3, 4\}$, where the frequency of 3 equals $\frac{f(3)}{1-\alpha}$ and the frequency of 4 equals $\frac{f(4)}{1-\alpha}$.

2-balancedness of $\mathbf{v}$ is a consequence of Lemma 1. $\qquad\square$

4   On Factor and Abelian Complexity

Using Theorem 1 we can construct for every frequency vector $\boldsymbol{f} \in \mathbb{R}^d, d \in \{3, 4\}$, a 2-balanced sequence $\mathbf{v} = \mathrm{colour}(\mathbf{u}, \mathbf{a}, \mathbf{b})$ having the required frequencies of letters. If the frequency vector is rational, then our construction gives a periodic infinite sequence, in particular, a sequence with bounded factor complexity.

In this section we find an upper bound on factor complexity and on abelian complexity of $\mathbf{v}$ in the case where the binary sequences we used in the construction are Sturmian sequences. For this purpose, we need to determine how many factors of length n in a Sturmian sequence have the same Parikh vector. The form of the Parikh vectors is described in Remark 1.

Definition 2. *Let $\alpha \in (0, 1)$ be an irrational number and let $\mathbf{u}$ be a Sturmian sequence over $\{a, b\}$ with frequency of the letter a equal to α. For each $n \in \mathbb{N}, n \geq 1$, we denote*

$$P_\alpha(n) = \#\left\{ u \in \mathcal{L}(\mathbf{u}) : \Psi(u) = \begin{pmatrix} \lceil n\alpha \rceil \\ \lfloor (1-\alpha)n \rfloor \end{pmatrix} \right\}.$$

The following formula may be derived for the function P_α.

Lemma 2. *Let $\alpha \in (0,1)$ ba an irrational number. Then for each $n \in \mathbb{N}, n \geq 1$,*

$$P_\alpha(n) = \#\Big\{k \in \{1,\ldots,n\} : \ k\alpha - \lfloor k\alpha \rfloor \leq n\alpha - \lfloor n\alpha \rfloor\Big\}.$$

Example 2. Using Lemma 2 we computed the values of $P_\alpha(n)$ for $\alpha = \frac{1}{2+\sqrt{3}}$ and $n \in \mathbb{N}, n \leq 10$. They are provided in Table 1.

Table 1. $\alpha = \frac{1}{2+\sqrt{3}} \sim 0,267$

n	1	2	3	4	5	6	7	8	9	10
$n\alpha - \lfloor n\alpha \rfloor$	0.267	0.535	0.803	0.071	0.339	0.607	0.875	0.143	0.411	0.679
$P_\alpha(n)$	1	2	3	1	3	5	7	2	5	8

4.1   Ternary 2-Balanced Sequences

Theorem 2. *Let $\mathbf{u}$ be a Sturmian sequence over $\{a,b\}$ with the frequency of letter a equal to an irrational number $\alpha \in (0,1)$. Let $\mathbf{a}$ be a Sturmian sequence over the alphabet $\{1,2\}$ and $\mathbf{b} = 3^\omega$. Then the sequence $\mathbf{v} = \mathrm{colour}(\mathbf{u},\mathbf{a},\mathbf{b})$ has the following properties:*

- *$\mathbf{v}$ is 2-balanced; moreover, $\mathbf{v}$ is not 1-balanced if $\alpha \notin \{\frac{1}{1+\gamma}, \frac{1}{2-\gamma}\}$, where $\gamma \in (0,1)$ is the frequency of letter 1 in the sequence $\mathbf{a}$;*
- *the factor and abelian complexity satisfy*

$$\mathcal{C}_\mathbf{v}(n) \leq P_\alpha(n) + (n+1)\lceil n\alpha \rceil \qquad and \qquad \mathcal{C}_\mathbf{v}^{ab}(n) \leq 4, \tag{2}$$

for each $n \in \mathbb{N}, n \geq 1$.

Proof. The sequence $\mathbf{v}$ is 2-balanced by Lemma 1. On one hand, using a theorem by Hubert [9], each ternary 1-balanced aperiodic recurrent sequence over $\{\hat{1},\hat{2},\hat{3}\}$ is a colouring of a Sturmian sequence by sequences $(\hat{1}\hat{2})^\omega$ and $(\hat{3})^\omega$. The frequency vector of the colouring has then evidently two equal coordinates. On the other hand, the frequency vector of the sequence $\mathbf{v}$ is equal to $(\alpha\gamma, \alpha(1-\gamma), 1-\alpha)$. The first two coordinates are not equal by irrationality of γ. The reader easily verifies that if there are two equal coordinates, then $\alpha = \frac{1}{1+\gamma}$ or $\alpha = \frac{1}{2-\gamma}$. Consequently, $\mathbf{v}$ is not 1-balanced for α distinct from those two values.

Fix a factor $w \in \mathcal{L}(\mathbf{u})$ of length n. We colour all letters a in w by a factor of $\mathbf{a}$ of length $|w|_a$. As $\mathbf{a}$ is Sturmian, we can use $|w|_a + 1$ factors of $\mathbf{a}$ for colouring w. Thus, colouring of the fixed factor w gives us at most $|w|_a + 1$ distinct factors of $\mathbf{v}$. Since in $\mathbf{u}$ we have $P_\alpha(n)$ factors with $|w|_a = \lceil n\alpha \rceil$ and $n + 1 - P_\alpha(n)$ factors with $|w|_a = \lfloor n\alpha \rfloor$, the number of factors of length n of the coloured sequence $\mathbf{v}$ is at most

$$\big(1 + \lceil n\alpha \rceil\big)P_\alpha(n) + \big(1 + \lfloor n\alpha \rfloor\big)\big(n + 1 - P_\alpha(n)\big) =$$
$$= P_\alpha(n)\big(\lceil n\alpha \rceil - \lfloor n\alpha \rfloor\big) + (n+1)(1 + \lfloor n\alpha \rfloor).$$

There are two distinct Parikh vectors for factors of length n in $\mathbf{u}$. After colouring, each of them gives rise to at most two distinct Parikh vectors in $\mathbf{v}$. Consequently, $\mathcal{C}_{\mathbf{v}}^{ab}(n) \leq 4$. □

Since $P_{\alpha}(n) \leq n$ for each $n \in \mathbb{N}$, we have $\mathcal{C}_{\mathbf{v}}(n) \leq \alpha n^2\big(1 + o(1)\big)$.

Remark 2. Notice that the fractions $\frac{1}{1+\gamma}$ and $\frac{1}{2-\gamma}$ in the first part of Theorem 2 are larger than $\frac{1}{2}$. Consequently, as soon as $\alpha < \frac{1}{2}$, the sequence $\mathbf{v}$ is not 1-balanced. Let us consider the case α equals $\frac{1}{1+\gamma}$ or $\frac{1}{2-\gamma}$. If $\mathbf{a}$ and $\mathbf{u}$ are standard Sturmian sequences, then the relation between the continued fraction of α and γ implies that $\mathbf{u}$ over $\{a, b\}$ is a morphic image of $\mathbf{a}$ over $\{1, 2\}$ under the morphism $1 \mapsto ab$, $2 \mapsto a$, or by the morphism $1 \mapsto a$, $2 \mapsto ab$.

Example 3. We compared the factor and abelian complexity with the upper bounds from Theorem 2 for $\mathbf{v} = \text{colour}(\mathbf{u}, \mathbf{a}, \mathbf{b})$, where $\mathbf{u}$ is a Sturmian sequence over $\{a, b\}$ and α is the frequency of a, $\mathbf{a}$ is a Sturmian sequence over $\{1, 2\}$ and γ is the frequency of letter 1 in $\mathbf{a}$ and $\mathbf{b} = 3^{\omega}$, for several cases. Two of them are presented in Table 2 and 3. Recall that τ denotes the golden mean.

Table 2. $\alpha = \frac{1}{\tau^2}$ and $\gamma = 2 - \sqrt{3}$

n	1	2	3	4	5	6	7	8	9	10	11	12	13	14	15	16
$P_{\alpha}(n) + (n+1)\lceil n\alpha \rceil$	3	5	9	13	17	23	29	37	45	53	63	73	83	95	107	121
$\mathcal{C}_{\mathbf{v}}(n)$	3	5	9	13	17	23	29	37	45	53	63	73	83	95	107	121
$\mathcal{C}_{\mathbf{v}}^{ab}(n)$	3	3	4	4	4	4	4	4	4	4	4	4	4	4	4	4

Table 3. $\alpha = \frac{1}{\tau^2}$ and $\gamma = 3 - \sqrt{5}$

n	1	2	3	4	5	6	7	8	9	10	11	12	13	14	15	16
$P_{\alpha}(n) + (n+1)\lceil n\alpha \rceil$	3	5	9	13	17	23	29	37	45	53	63	73	83	95	107	121
$\mathcal{C}_{\mathbf{v}}(n)$	3	5	8	11	14	17	20	22	24	26	28	30	32	34	36	38
$\mathcal{C}_{\mathbf{v}}^{ab}(n)$	3	3	4	4	4	4	4	4	4	4	4	4	3	4	4	4

Based on our computer experiments, we formulate the following conjecture.

Conjecture 1. Let $\mathbf{u}, \mathbf{a}, \mathbf{b}$ be as in Theorem 2, and let $\gamma \in (0, 1)$ denote the frequency of letter 1 in the Sturmian sequence $\mathbf{a}$. If α and γ are rationally independent, then in (2) the upper bound on $\mathcal{C}_{\mathbf{v}}(n)$ is attained for every $n \in \mathbb{N}$ and the upper bound on $\mathcal{C}_{\mathbf{v}}^{ab}(n)$ is attained for every $n \in \mathbb{N}, n \geq 4$.

4.2 Quaternary 2-Balanced Sequences

Quaternary 2-balanced sequences obtained when colouring a Sturmian sequence by two Sturmian sequences have at most cubic factor complexity. A more precise upper bound may be derived by analogous arguments as used in the proof of Theorem 2.

Theorem 3. *Let* **u** *be a Sturmian sequence over* $\{a, b\}$ *with the frequency of letter* a *equal to an irrational number* $\alpha \in (0, 1)$. *Let* **a** *and* **b** *be Sturmian sequences over the alphabet* $\{1, 2\}$ *and* $\{3, 4\}$, *respectively. Then the sequence* **v** = colour(**u**, **a**, **b**) *is 2-balanced, but not 1-balanced, and its abelian and factor complexity satisfy for each* $n \in \mathbb{N}, n \geq 1$, *the inequalities:* $\mathcal{C}_{\mathbf{v}}^{ab}(n) \leq 8$ *and*

$$\mathcal{C}_{\mathbf{v}}(n) \leq (n+1)\lceil \alpha n \rceil \left(1 + \lceil (1-\alpha)n \rceil\right) + P_\alpha(n) \left(\lceil (1-\alpha)n \rceil - \lceil \alpha n \rceil\right).$$

Proof. Each quaternary 1-balanced aperiodic recurrent sequence over $\{\hat{1}, \hat{2}, \hat{3}, \hat{4}\}$ is a colouring of a Sturmian sequence $\hat{\mathbf{u}}$ by sequences $(\hat{1}\hat{2})^\omega$ and $(\hat{3}\hat{4})^\omega$, resp. $(\hat{1}\hat{2}\hat{3})^\omega$ and $(\hat{4})^\omega$, resp. $(\hat{1}\hat{2}\hat{1}\hat{3})^\omega$ and $(\hat{4})^\omega$ [9]. Denote the frequency of one letter in the Sturmian sequence $\hat{\mathbf{u}}$ by β, then the frequency vector of the 1-balanced sequence has the form $(\frac{\beta}{2}, \frac{\beta}{2}, \frac{1-\beta}{2}, \frac{1-\beta}{2})$, resp. $(\frac{\beta}{3}, \frac{\beta}{3}, \frac{\beta}{3}, 1 - \beta)$, resp. $(\frac{\beta}{2}, \frac{\beta}{4}, \frac{\beta}{4}, 1 - \beta)$. Denote γ, resp. δ, the frequency of letter 1, resp. 3 in the sequence **a**, resp. **b**. Then the frequency vector of **v** is equal to $(\alpha\gamma, \alpha(1-\gamma), (1-\alpha)\delta, (1-\alpha)(1-\delta))$. Thanks to irrationality of α, γ, and δ, this vector can neither have two pairs of equal coordinates, nor three equal coordinates, nor two equal coordinates and another one being double of them. Hence, the sequence **v** is not 1-balanced, while it is 2-balanced by Lemma 1.

The proof for the factor and abelian complexity is analogous to the proof of Theorem 2. □

Consequently, $\mathcal{C}_{\mathbf{v}}(n) \leq \alpha(1-\alpha)\,n^3\,(1 + o(1))$.

Example 4. We compare again the factor and abelian complexities with the upper bounds from Theorem 3 for **v** = colour(**u**, **a**, **b**), where **u** is a Sturmian sequence over $\{a, b\}$ and α is the frequency of a, **a** is a Sturmian sequence over $\{1, 2\}$ and γ is the frequency of letter 1 in **a** and δ is the frequency of letter 3 in a Sturmian sequence **b** over the alphabet $\{3, 4\}$, see Tables 4 and 5.

Table 4. $\alpha = \frac{1}{3+\sqrt{2}}, \gamma = \frac{1}{3+\sqrt{5}}, \delta = \frac{1}{\sqrt{2}}$

n	1	2	3	4	5	6	7	8	9	10	11	12	13	14	15	16
$\mathcal{C}_{\mathbf{v}}(n)$ bound	4	11	22	41	66	99	140	195	260	337	426	535	658	797	966	1147
$\mathcal{C}_{\mathbf{v}}(n)$	4	11	22	41	66	99	140	195	260	337	426	535	658	797	966	1147
$\mathcal{C}_{\mathbf{v}}^{ab}(n)$	4	6	6	8	8	8	8	8	8	8	8	8	8	8	8	8

Table 5. $\alpha = \frac{1}{3+\sqrt{2}}, \gamma = \frac{1}{3+\sqrt{2}}, \delta = \frac{1}{\sqrt{2}}$

n	1	2	3	4	5	6	7	8	9	10	11	12	13	14	15	16
$\mathcal{C}_{\mathbf{v}}(n)$ bound	4	11	22	41	66	99	140	195	260	337	426	535	658	797	966	1147
$\mathcal{C}_{\mathbf{v}}(n)$	4	11	22	31	41	53	67	82	98	116	136	156	178	202	227	253
$\mathcal{C}_{\mathbf{v}}^{ab}(n)$	4	6	6	6	8	8	6	6	8	8	6	8	8	8	6	8

5 On Computation of the Function P_α

The value of the function P_α may be calculated using the formula from Lemma 2. A disadvantage of the method is that one has to determine the fractional part of $k\alpha$ for all $k \le n$ with sufficient precision.

In the sequel, for $\alpha = \frac{1}{\tau^2}$, where $\tau = \frac{1+\sqrt{5}}{2}$, we will show how to compute the values of $P_\alpha(n)$ in a symbolic way using the greedy representation of integers in a non-standard numeration system associated with α. The number $\beta = \frac{1}{\alpha} = \tau^2$ is the larger root of the equation $x^2 = 3x - 1$. We associate with β the sequence $\mathbf{U} = (U_k)_{k \in \mathbb{N}}$ defined by recurrence relation

$$U_{-1} = 0, \quad U_0 = 1, \quad \text{and} \quad U_{k+1} = 3U_k - U_{k-1} \quad \text{for every } k \in \mathbb{N}. \tag{3}$$

Now for $\beta = \tau^2$, we define the β-expansion of real numbers from $[0, 1)$, and for the sequence $\mathbf{U}$, we define the $\mathbf{U}$-expansion of non-negative integers. For more details on both types of expansion see Sect. 7.3 by Ch. Frougny in the book [8].

A β-representation of a number $x \in [0, 1)$ is an infinite sequence $(x_i)_{i \ge 1}$ of non-negative integers such that $x = \sum_{i=1}^{+\infty} x_i \beta^{-i}$. A particular β-representation – called the β-expansion – can be computed by the greedy algorithm which uses a transformation T_β defined by $T_\beta(x) = \beta x - \lfloor \beta x \rfloor$ for every $x \in [0, 1)$.

If we put $a_i = \lfloor \beta T^{i-1}(x) \rfloor$ for every $i \in \mathbb{N}, i \ge 1$, then $x = \sum_{i=1}^{+\infty} a_i \beta^{-i}$ and $a_i < \beta$ for every $i \ge 1$. The string $(a_i)_{i \ge 1}$ is denoted by $d_\beta(x)$. The function $d_\beta(x)$ is increasing with respect to the lexicographical order. In the space of infinite sequences over $\{a \in \mathbb{N} : a < \beta\}$ equipped with the product topology, we can define the quasi-greedy expansion of 1 as $d_\beta^*(1) = \lim_{x \to 1-} d_\beta(x)$.

As proven by Parry [12], a sequence of non-negative integers $(x_i)_{i \ge 1}$ is the β-expansion of a number $x \in [0, 1)$ if and only if

$$x_i x_{i+i} x_{i+2} \cdots \prec_{lex} d_\beta^*(1) \quad \text{for every } i \in \mathbb{N}, i \ge 1. \tag{4}$$

It is easy to verify that for $\beta = \tau^2$, the quasi-greedy expansion of 1 is $d_\beta^*(1) = 21^\omega$.

A $\mathbf{U}$-representation of a number $n \in \mathbb{N}$ is a finite sequence of non-negative integers $a_N a_{N-1} \cdots a_0$ such that $n = \sum_{k=0}^{N} a_k U_k$. If moreover, $a_N \ne 0$ and $\sum_{k=0}^{i} a_k U_k < U_{i+1}$ for every $i \le N$, then the representation is called the $\mathbf{U}$-expansion of n and will be denoted

$$(n)_{\mathbf{U}} = a_N a_{N-1} \cdots a_1 a_0.$$

The $\mathbf{U}$-expansion of n can be computed by the greedy algorithm. Since our sequence $\mathbf{U}$ satisfies for every $\ell, k \in \mathbb{N}, k \ge 1$, the inequalities

$$3U_\ell \ge U_{\ell+1}$$
$$2U_{\ell+k} + U_{\ell+k-1} + U_{\ell+k-2} + \ldots + U_{\ell+1} + 2U_\ell = U_{\ell+k+1} + U_{\ell-1} \ge U_{\ell+k+1},$$

the finite sequence $(n)_{\mathbf{U}}$ is

a word over $\{0, 1, 2\}$ which does not contain $21^{k-1}2$ as its factor. $\tag{5}$

In fact, the above property under the assumption that the leading coefficient is non-zero characterizes the $\mathbf{U}$-expansion of integers.

Proposition 1. *Let* $\mathbf{U} = (U_k)_{k \in \mathbb{N}}$ *be the sequence defined in* (3), *and let* $\beta = \tau^2 = \left(\frac{1+\sqrt{5}}{2}\right)^2$. *Let* $n \in \mathbb{N}, n \geq 3$, *be an integer with* $(n)_{\mathbf{U}} = a_N a_{N-1} \cdots a_1 a_0$. *Then the* $\mathbf{U}$-*expansion of the integer part of* $\frac{1}{\tau^2} n$ *and the* β-*expansion of the fractional part of* $\frac{1}{\tau^2} n$ *are*

$$\left(\left\lfloor \frac{1}{\tau^2} n \right\rfloor \right)_{\mathbf{U}} = a_N a_{N-1} \cdots a_1 \quad and \quad d_\beta \left(\left\{ \frac{1}{\tau^2} n \right\} \right) = a_0 a_1 \cdots a_{N-1} a_N 0^\omega.$$

Proof. Using (3) we can easily prove by induction that $\frac{1}{\tau^2} U_k = U_{k-1} + \tau^{-2(k+1)}$ for all $k \in \mathbb{N}$. Hence

$$\begin{aligned}
\tfrac{1}{\tau^2} n &= \tfrac{1}{\tau^2} \sum_{k=0}^{N} a_k U_k \\
&= \sum_{k=0}^{N} a_k U_{k-1} + \sum_{k=0}^{N} a_k \tau^{-2(k+1)} \\
&= \sum_{k=1}^{N} a_k U_{k-1} + \sum_{k=1}^{N+1} a_{k-1} \tau^{-2k}.
\end{aligned}$$

As the finite string $a_N a_{N-1} \cdots a_1 a_0$ satisfies (5), the infinite string $a_0 a_1 \cdots a_N 0^\omega$ fulfills (4) with $d_\beta^*(1) = 21^\omega$. It implies that $a_0 a_1 \cdots a_N 0^\omega$ is the β-expansion of a number $x \in [0, 1)$. Together with the fact that $\sum_{k=1}^{N} a_k U_{k-1}$ is an integer and the finite string $a_N a_{N-1} \cdots a_1$ satisfies (5) as well, it gives both parts of the statement. $\square$

Theorem 4. *Let* $\mathbf{U} = (U_k)_{k \in \mathbb{N}}$ *be the sequence defined in* (3) *and* $\alpha = \frac{1}{\tau^2}$. *If the* $\mathbf{U}$-*expansion of a number* $n \in \mathbb{N}$ *is* $(n)_{\mathbf{U}} = a_N a_{N-1} \cdots a_1 a_0$, *then*

$$P_\alpha(n) = a_N (r_N + 1) + a_{N-1}(r_{N-1} + 1) + \cdots + a_1 (r_1 + 1) + a_0,$$

where r_i *is an integer having* $\mathbf{U}$-*representation* $a_0 a_1 \cdots a_{i-1}$ *for every* $i \in \mathbb{N}, 1 \leq i \leq N$.

Proof. We say that a finite string $b_j b_{j-1} \cdots b_0$ is admissible if it satisfies (5). In other words, if we erase the leading zeros (if any) in $b_j b_{j-1} \cdots b_0$, we get the $\mathbf{U}$-expansion of some integer. As the forbidden strings $21^{k-1}2$ listed in (5) are palindromes, a string $B := b_j b_{j-1} \cdots b_0$ is admissible if and only if its reverse $\overline{B} := b_0 b_1 \cdots b_{j-1} b_j$ is admissible.

Let us fix an integer k with $1 \leq k \leq n$. Such k has exactly one admissible $\mathbf{U}$-representation of length $N + 1$, say $b_N b_{N-1} \cdots b_0$. The greediness of the $\mathbf{U}$-expansion guarantees that

$$b_N b_{N-1} \cdots b_1 b_0 \preceq_{lex} a_N a_{N-1} \cdots a_1 a_0. \tag{6}$$

By Proposition 1, the β-expansion of the fractional part of $\frac{1}{\tau^2} k$ is $b_0 b_1 \ldots b_N 0^\omega$. By Lemma 2, such k contributes to $P_\alpha(n)$ if and only if the fractional parts satisfy $\{\frac{1}{\tau^2} k\} \leq \{\frac{1}{\tau^2} n\}$. The greediness of the β-expansions and Proposition 1 allow us to reformulate the inequality into

$$b_0 b_1 \cdots b_{N-1} b_N \preceq_{lex} a_0 a_1 \cdots a_{N-1} a_N. \tag{7}$$

Denote

$$M = \{ b_N b_{N-1} \cdots b_0 \; : \; b_N b_{N-1} \cdots b_0 \text{ admissible and satisfies (6) and (7)} \},$$

then $P_\alpha(n) = \#M - 1$. We subtract one because the zero sequence represents $k = 0$.

Let us fix $i \in \mathbb{N}$, $1 \leq i \leq N$, and consider an admissible sequence $b_0 b_1 \cdots b_{i-1}$ such that

$$b_0 b_1 \cdots b_{i-1} \preceq_{lex} a_0 a_1 \ldots a_{i-1}. \tag{8}$$

Then the sequence $B := a_N a_{N-1} \cdots a_{i+1} x b_{i-1} \cdots b_1 b_0$ with $x \in \mathbb{N}$, $x < a_i$, does not contain any forbidden string $21^{k-1}2$, therefore B is admissible as well. Obviously, $B \prec_{lex} a_N a_{N-1} \cdots a_1 a_0$ and $\overline{B} \prec_{lex} a_0 a_1 \cdots a_{N-1} a_N$. In the statement of the theorem, r_i denotes the non-negative number with a $\mathbf{U}$-representation $a_0 a_1 \ldots a_{i-1}$. Thus there are $r_i + 1$ non-negative integers smaller than or equal to r_i and consequently $r_i + 1$ admissible strings $b_0 b_1 \cdots b_{i-1}$ satisfying (8). Moreover, we have a_i possibilities to choose a non-negative integer $x < a_i$.

For every fixed $i \in \mathbb{N}$, $1 \leq i \leq N$, we have found $a_i(r_i + 1)$ sequences of the form $B := a_N a_{N-1} \cdots a_{i+1} x b_{i-1} \cdots b_1 b_0$ which belong to M. Moreover, M contains also $a_0 + 1$ sequences of the form $B := a_N a_{N-1} \cdots a_1 x$ with $x \leq a_0$. It is easy to see that any element of M is of the form B we considered.   □

Example 5. Let us illustrate the previous theorem. Let $n = 38$. The $\mathbf{U}$-expansion of n in the numeration system defined by (3) is $(n)_\mathbf{U} = 1201$ as $38 = 1 \cdot U_3 + 2 \cdot U_2 + 0 \cdot U_1 + 1 \cdot U_0$.

For $(n)_\mathbf{U} = a_3 a_2 a_1 a_0 = 1201$, we first determine $(r_3)_\mathbf{U} = 102$, $(r_2)_\mathbf{U} = 10$, $(r_1)_\mathbf{U} = 1$, i.e., $r_3 = 10$, $r_2 = 3$, $r_1 = 1$. By the formula from Theorem 4, $P_\alpha(38) = 1 \cdot 11 + 2 \cdot 4 + 0 \cdot 2 + 1 = 20$.

In Table 6, there are the values of $P_\alpha(n)$ for $17 \leq n \leq 29$. Observe that $P_\alpha(38) = P_\alpha(28) = 20$. The $\mathbf{U}$-expansion of $m = 28$ is $(m)_\mathbf{U} = 1021$, i.e., it is the mirror image of the $\mathbf{U}$-expansion of $n = 38$. The equality of $P_\alpha(m)$ and $P_\alpha(n)$ is a consequence of the symmetry of inequalities (6) and (7).

Table 6. The $\mathbf{U}$-expansion of $n \in \mathbb{N}$, $17 \leq n \leq 29$, and the value $P_\alpha(n)$ for $\alpha = \frac{1}{\tau^2}$.

n	17	18	19	20	21	22	23	24	25	26	27	28	29
$(n)_\mathbf{U}$	201	202	210	211	1000	1001	1002	1010	1011	1012	1020	1021	1100
$P_\alpha(n)$	9	16	5	13	1	10	19	5	15	25	9	20	3

6   Comments

- In computer experiments, we counted in a prefix of $\mathbf{v}$ of length $500\,000$ the number of different factors of length n for $n = 1, 2, \ldots, 300$. For ternary sequences, under the assumption in Conjecture 1, the values of factor and abelian complexity we found out coincide with the upper bound. Similarly for quaternary sequences with at least two rationally independent frequencies among α, γ, δ. The Sturmian sequences we used in our experiments are fixed points of Sturmian morphisms, in particular, we limited our consideration to frequencies being quadratic numbers.

- Regarding abelian complexity, Richomme, Saari, and Zamboni [13] showed that ternary aperiodic 1-balanced sequences have constant abelian complexity equal to 3. Currie and Rampersad [6] showed that there are no recurrent sequences of constant abelian complexity equal to 4. Saarela [14] proved for every integer $c \geq 2$ that there exists a recurrent sequence with abelian complexity $\mathcal{C}^{ab}(n) = c$ for all $n \geq c - 1$. We expect that infinitely many ternary 2-balanced sequences we defined are of this type, i.e., with $\mathcal{C}_v^{ab}(n) = 4$ for each $n \geq 3$.
- We believe that the method derived for computing $P_\alpha(n)$ in case of $\alpha = \frac{1}{\tau^2}$ may be generalized to all cases, where α is a quadratic unit. For other numbers α, the Ostrowski numeration system or its modification could be helpful.
- Our future goal is to study combinatorial properties of 2-balanced sequences from the new class more deeply, in particular, besides studying for which sequences the upper bound for the factor and abelian complexity is reached (see Conjecture 1), we aim to describe return words and bispecial factors in order to determine the critical exponent.

References

1. Adamczewski, B.: Balances for fixed points of primitive substitutions. Theoret. Comput. Sci. **307**, 47–75 (2003)
2. Arnoux, P., Starosta, Š: Further developments in fractals and related fields (The Rauzy gasket). Birkhäuser/Springer, Boston (2013)
3. Berthé, V., Cassaigne, J., Steiner, W.: Balance properties of Arnoux-Rauzy sequences. Int. J. Algebra Comput. **23**(04), 689–703 (2013)
4. Cassaigne, J., Labbé, S., Leroy, J.: Almost everywhere balanced sequences of complexity $2n + 1$. Mosc. J. Comb. Number Theory **11**(4), 287–333 (2022)
5. Coven, E.M., Hedlund, G.A.: Sequences with minimal block growth. Math. Syst. Theory **7**, 138–153 (1973)
6. Currie, J., Rampersad, N.: Recurrent words with constant abelian complexity. Adv. Appl. Math. **47**(1), 116–124 (2011)
7. Fici, G., Puzynina, S.: Abelian combinatorics on words: A survey. Comput. Sci. Rev. **47**, 100532 (2023)
8. Lothaire, M.: Algebraic Combinatorics on Words. Cambridge University Press, Cambridge (2002)
9. Hubert, P.: Suites équilibrées. Theoret. Comput. Sci. **242**(1–2), 91–108 (2000)
10. Langeveld, N., Rossi, L., Thuswaldner, J.M.: Generalizations of Sturmian sequences associated with N-continued fraction algorithms. J. Number Theor. **250**, 49–83 (2023)
11. Morse, M., Hedlund, G.A.: Amer. J. Math. Symbolic Dynamics II. Sturmian trajectories **62**, 1–42 (1940)
12. Parry, W.: On the β expansion of real numbers. Acta Mathematica Academiae Scientiarum Hungaricae **11**, 401–416 (1960)
13. Richomme, G., Saari, K., Zamboni, L.Q.: Abelian complexity of minimal subshifts. J. London Math. Soc. **83**(1), 79–95 (2010)
14. Saarela, A.: Ultimately constant abelian complexity of infinite words. J. Autom. Lang. Comb. **14**(3/4), 255–258 (2009)

Order Conditions for Languages

Sébastien Ferenczi[1]([✉]), Pascal Hubert[1], and Luca Q. Zamboni[2]

[1] Aix Marseille Université, CNRS, Centrale Marseille,
Institut de Mathématiques de Marseille, I2M - UMR 7373, 13453 Marseille, France
{sebastien-simon.ferenczi,pascal.hubert}@univ-amu.fr
[2] Institut Camille Jordan, Université Claude Bernard Lyon 1,
43 boulevard du 11 novembre 1918, F69622 Villeurbanne Cedex, France
zamboni@math.univ-lyon1.fr

Abstract. We define a condition on the resolution of bispecials in a language. A language satisfies this order condition if and only if it is the natural coding of a generalized interval exchange transformation, while the order condition plus some additional ones characterize the codings of various more classical interval exchange transformations. Also, a finite word clusters for the Burrows-Wheeler transform if and only if the language generated by its powers satisfies an order condition.

1 Languages Satisfying an Order Condition

1.1 Usual Definitions

Let $\mathcal{A}$ be a finite set called the *alphabet*, its elements being *letters*. A *word* w of *length* $n = |w|$ is $a_1 a_2 \cdots a_n$, with $a_i \in \mathcal{A}$. The *concatenation* of two words w and w' is denoted by ww'.

By a language L over $\mathcal{A}$ we mean a *factorial extendable language*: a collection of sets $(L_n)_{n \geq 0}$ where the only element of L_0 is the *empty word*, and where each L_n for $n \geq 1$ consists of words of length n, such that for each $v \in L_n$ there exist $a, b \in \mathcal{A}$ with $av, vb \in L_{n+1}$, and each $v \in L_{n+1}$ can be written in the form $v = au = u'b$ with $a, b \in \mathcal{A}$ and $u, u' \in L_n$.

The *complexity function* $p : \mathbb{N} \to \mathbb{N}$ is defined by $p(n) = \#L_n$.

A word $v = v_1...v_r$ is a *factor* of a word $w = w_1...w_s$ or an infinite sequence $w = w_1 w_2...$ if for some $i \geq 1$ $v_1 = w_i, ...v_r = w_{i+r-1}$.

The *reverse* of the word $v = v_1...v_r$ is the word $v_r...v_1$.

A word w is *primitive* if it is not equal to v^n for any word v and integer $n > 1$.

Let W be a family of words or (one- or two-sided) infinite sequences. Whenever the set L of all the factors of the words or sequences in W is a language (namely, is factorial and extendable), we say that L is *the language generated by* W and denote it by $L(W)$.

© The Author(s), under exclusive license to Springer Nature Switzerland AG 2023
A. Frid and R. Mercaş (Eds.): WORDS 2023, LNCS 13899, pp. 155–167, 2023.
https://doi.org/10.1007/978-3-031-33180-0_12

A language L is *recurrent* if for each $v \in L$ there exists a nonempty w such that vw is in L and ends with v.

A language L is *uniformly recurrent* if for each $v \in L$ there exists n such that v is a factor of each word $w \in L_n$.

A language L is *aperiodic* if for each nonempty word w in L, there exists n such that w^n is not in L.

For a word w in L, we call *arrival set of w*, denoted by $A(w)$, the set of all letters x such that xw is in L, and call *departure set of w*, denoted by $D(w)$, the set of all letters x such that wx is in L.

A word w in L is called *right special*, resp. *left special* if $\#D(w) > 1$, resp. $\#A(w) > 1$. If $w \in L$ is both right special and left special, then w is called *bispecial*. If $\#L_1 > 1$, the empty word ε is bispecial, with $A(\varepsilon) = D(\varepsilon) = L_1$.

A bispecial word w in L is a *weak bispecial* if $\#\{awb \in L, a \in A(w), b \in D(w)\} < \#A(w) + \#D(w) - 1$.

A bispecial word w in L is a *strong bispecial* if $\#\{awb \in L, a \in A(w), b \in D(w)\} > \#A(w) + \#D(w) - 1$.

To *resolve* a bispecial word w is to find all words in L of the form awb for letters a and b.

The *symbolic dynamical system* associated to a language L is the two-sided shift S acting on the subset X_L of $\mathcal{A}^{\mathbb{Z}}$ consisting of all bi-infinite sequences x such that $x_r \cdots x_{r+s-1} \in L_s$ for each r and s, defined by $(Sx)_n = x_{n+1}$ for all $n \in \mathbb{Z}$.

Thus in the present paper we use two-sided sequences $x \in X_L$, but also their (infinite) *suffixes* $(x_n, n \geq k)$ and (infinite) *prefixes* $(x_n, n \leq k)$.

1.2   Order Conditions: First Properties

The order condition can be traced to [15], but its first explicit mention appears in [11] in the particular case of one order and its reverse.

Definition 1. A language L on an alphabet $\mathcal{A}$ satisfies a *local order condition* if, for each bispecial word w, there exist two total orders on $\mathcal{A}$, denoted by $<_{A,w}$ and $<_{D,w}$, such that whenever awc and bwd are in L with letters $a \neq b$ and $c \neq d$, then $a <_{A,w} b$ if and only if $c <_{D,w} d$.

A language L on an alphabet $\mathcal{A}$ satisfies an *order condition* if it satisfies a local order condition where the orders $<_{A,w}$ and $<_{D,w}$ are the same for all bispecial words.

The first notion in the following definition seems to be new, and its links with the order conditions will be studied below.

Definition 2. In a language L, a *locally strong bispecial word* is a bispecial word w such that there exist nonempty subsets $A' \subset A(w)$, $D' \subset D(w)$ such that $\#\{awb \in L, a \in A', b \in D'\} > \#A' + \#D' - 1$.

If a language L on an alphabet $\mathcal{A}$ satisfies a local order condition, a bispecial word w has a *connection* if there are letters $a <_{A,w} a'$, consecutive in the order $<_{A,w}$, letters $b <_{D,w} b'$, consecutive in the order $<_{D,w}$, such that awb and $a'wb'$ are in L, and neither awb' nor $a'wb$ is in L.

We recall a well-known result which can be deduced from [9] or [10].

Lemma 1. (Lemmas 2 and 5 of [12]) *A language L which has no strong bispecial word has a finite number of weak bispecial words, and the left (resp. right) special words are the prefixes (resp. suffixes) af a finite number of infinite suffixes (resp. prefixes) of sequences of X_L.*

We turn now to combinatorial consequences of order conditions.

Lemma 2. (Lemma 1 of [12]) *A language L which satisfies a local order condition contains no locally strong bispecial word, and thus no strong bispecial word.*

For languages where each word has at most two right (resp. left) extensions, the absence of strong bispecial words, the absence of locally strong bispecial words, and a local order condition are all equivalent. In the general case, it is easy to find bispecials which are locally strong but not strong (suppose for example that the possible xwy are awa, awb, bwa, bwb, cwc), and a local order condition is stricter than the absence of locally strong bispecials.

Example 1. Suppose L is a language whose words of length 2 are ac, ad, ba, bc, cb, cc, da. Then the empty word is not a locally strong bispecial, yet L does not satisfy any local order condition. We can then choose L_3 to be made with acc, ada, bac, bad, bcb, bcc, cba, cbc, ccb, dac, where each word has at most two left (resp. right) extensions, and continue by resolving the bispecials so that they are all neutral. We get a language without locally strong bispecials but not satisfying any local order condition.

A related notion is dendricity ([4] under the name of *tree sets*). This can be interpreted as the following.

Definition 3. A language is *dendric* if it has neither locally strong bispecial words nor weak bispecial words.

Thus, by Lemmas 1 and 2, a language satisfying a local order condition is *ultimately dendric*. But, because of Example 1 and the possibility of weak bispecials, there is no inclusion relation between dendric languages and languages satisfying an order condition, local or not.

Lemma 3. (Lemma 3 and Corollary 4 of [12]) *A language satisfying a local order condition has complexity $p(n) = kn + l$ for all n large enough and with $0 \leq k \leq \#\mathcal{A} - 1$. Moreover, $k = \#\mathcal{A} - 1$ if and only if L has no connection, and in that case $l = 1$.*

1.3 Recurrence

We look at consequences of an order condition, or weaker properties, on the trajectories in X_L.

Definition 4. Let x be a bi-infinite sequence in $\mathcal{A}^{\mathbb{Z}}$, or a suffix of such a sequence. x is *right recurrent* if any factor of x is a factor of each suffix of x.

Let x be a bi-infinite sequence in $\mathcal{A}^{\mathbb{Z}}$, or a prefix of such a sequence. x is *left recurrent* if any factor of x is a factor of each prefix of x.

A bi-infinite sequence x in $\mathcal{A}^{\mathbb{Z}}$ is *recurrent* if it is both left and right recurrent.

Lemma 4. (Lemma 9 of [12]) *Suppose L has no strong bispecial word.*

If a suffix of a sequence in X_L is right recurrent, it generates a uniformly recurrent language.

If a prefix of a sequence in X_L is left recurrent, it generates a uniformly recurrent language.

If a sequence in X_L is left or right recurrent, it is recurrent and generates a uniformly recurrent language.

Proposition 1. (Proposition 17 of [12]) *Suppose L has no strong bispecial word. Then there are at most a finite number of orbits $\{S^n x, n \in \mathbb{Z}\}$, $x \in X_L$, with x not recurrent.*

Proposition 2. (Proposition 12 of [12]) *Let L be a recurrent language satisfying a local order condition. Then L is a finite union of uniformly recurrent languages.*

Lemma 5. (Lemma 18 of [12]) *Given a language L, we denote by L' the sublanguage of L generated by all recurrent sequences in X_L. Then L' is a recurrent language over an alphabet $\mathcal{A}' \subset \mathcal{A}$. Moreover, if L satisfies an order condition, so does L'.*

2 Interval Exchange Transformations

2.1 Definitions

Generalized interval exchange transformations are defined in [1, 26] and do generalize the well-known classical, or standard, interval exchange transformations of [22, 29].

All intervals are open on the right, closed on the left.

Definition 5. A *generalized interval exchange transformation* is a map T defined on $[0, 1)$ partitioned by intervals I_e, $e \in \mathcal{A}$, continuous and (strictly) increasing on each I_e, and such that the $T I_e$, $e \in \mathcal{A}$, are intervals partitioning $[0, 1)$.

The I_e, indexed in $\mathcal{A}$, are called the *defining intervals* of T.

If the restriction of T to each I_e is an affine map, T is an *affine* interval exchange transformation.

If the restriction of T to each I_e is an affine map of slope 1, T is a *standard* interval exchange transformation.

The endpoints of the I_e, resp. TI_e, excluding 0 and 1, will be denoted by γ_i, resp. β_j, for i, resp. j, taking $\#\mathcal{A} - 1$ values.

Definition 6. T is *minimal* if every orbit is dense in $[0, 1)$.

T satisfies the *i.d.o.c. condition* if there is at least one point γ_i (of Definition 5), and there is no i, j, $k \geq 0$, such that $T^k \beta_i = \gamma_j$.

A *wandering interval* is an interval J for which $T^n J$ is disjoint from J for all $n > 0$.

Note that our i.d.o.c. condition is a modified version of the one introduced by Keane [22]; our condition depends on the defining intervals, and is not intrinsic to T as the original Keane's condition.

Definition 7. For a generalized interval exchange transformation T, its *natural coding* is the language $L(T)$ generated by all the *trajectories*, namely the sequences $(x_n, n \in \mathbb{Z}) \in \mathcal{A}^{\mathbb{Z}}$ where $x_n = e$ if $T^n x$ falls into I_e, $e \in \mathcal{A}$.

Thus we can look at the symbolic system associated to $L(T)$. Note that the set $X_{L(T)}$ is the closure in $\mathcal{A}^{\mathbb{Z}}$ of the set of trajectories, for the product topology defined by the discrete topology on $\mathcal{A}$.

Example 2. A *Sturmian language* [27] is the natural coding of the standard interval exchange transformation T sending $[0, 1 - \alpha)$ to $[\alpha, 1)$ and $[1 - \alpha, 1)$ to $[0, \alpha)$ for α irrational; T is conjugate to a rotation of angle α on the 1-torus.

We shall also consider slightly more general codings, by merging into intervals $\tilde{I}_e$ some adjacent intervals I_e whose images by T are also adjacent. This is equivalent to taking the natural coding of another interval exchange transformation $\tilde{T}$, but when T is affine, if we define $\tilde{T}$ by the intervals $\tilde{I}_e$ it will not necessarily be affine by our definition, as the slope is not constant on its defining intervals, see Example 4 below. Thus we define

Definition 8. A language L is a *grouped coding* of an affine interval exchange transformation T if there exist intervals $\tilde{I}_e$, $e \in \tilde{\mathcal{A}}$ such that

- each $\tilde{I}_e$ is an interval, and a disjoint union of defining intervals of T,
- T is a continuous monotone map on each $\tilde{I}_e$,
- L is the coding of T by the $\tilde{I}_e$, that is the language generated by the trajectories $(x_n, n \in \mathbb{Z}) \in \tilde{\mathcal{A}}^{\mathbb{Z}}$ where $x_n = e$ if $T^n x$ falls into $\tilde{I}_e$, $e \in \tilde{\mathcal{A}}$.

2.2 Interval Exchanges Satisfy Order Conditions

Definition 9. A generalized interval exchange transformation defines two orders on $\mathcal{A}$:

- $e <_D f$ whenever the interval I_e is strictly to the left of the interval I_f,

– $e <_A f$ whenever the interval TI_e is strictly to the left of the interval TI_f.

These orders correspond to the two permutations used by Kerckhoff [23] to define standard interval exchange transformations: the unit interval is partitioned into semi-open intervals which are numbered from 1 to k, ordered according to a permutation π_0 and then rearranged according to another permutation π_1; in more classical definitions, there is only one permutation π, which corresponds to π_1 while $\pi_0 = Id$; note that in some papers the orderings are by π_0^{-1} and π_1^{-1}.

Proposition 3. (Proposition 8 of [12]) *Let T be a generalized interval exchange transformation. Then its natural coding $L(T)$ satisfies an order condition.*

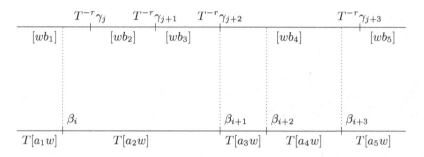

Fig. 1. A bispecial interval

Remark 1. The notion of interval exchange has been extended to *interval exchanges with flips* [28] in the standard case, allowing the slope of T on some defining intervals to be -1. It can be further extended to generalized interval exchanges with flips, allowing T to be decreasing on some defining intervals. The natural codings of such transformations satisfy a *flipped* order condition, which is a local order condition where the order $<_{D,w}$ is always the same order $<_D$, while $<_{A,w}$ is allowed to be either an order $<_A$ or its reverse, according to the number of letters in w corresponding to flipped intervals. Indeed, all the results in Sect. 2.3 hold for (generalized) interval exchanges with flips, mutatis mutandis (Fig. 1).

Remark 2. The language of an *interval translation mapping* [5] does not necessarily satisfy a local order condition: it is possible that TI_a intersects TI_b for $a \neq b$, and, if $TI_a \cap TI_b$ intersects both I_c and I_d, $c \neq d$, then the empty word is a locally strong bispecial. For this family, not much is known; it is an open question, asked by Boshernitzan, whether all interval translation mappings have linear complexity. It is generalized to *piecewise isometries* [18] for which even less is known.

2.3   The Converse

Theorem 1. (Theorem 15 of [12]) *A language L on at least two letters is the language of a standard interval exchange transformation satisfying the i.d.o.c. condition if and only if it satisfies an order condition, is aperiodic and uniformly recurrent, and has no connection.*

Theorem 2. (Theorem 14 of [12]) *A language L is the language of a generalized, or equivalently of a standard, minimal interval exchange transformation if and only if it satisfies an order condition, is aperiodic and uniformly recurrent.*

Theorem 1 is proved in [16], and Theorem 2 uses the same method. Some results similar both to Theorem 2 and Theorem 3 below are proved in [21], using a description of the evolution of *Rauzy graphs* which is somewhat cumbersome to state, but where an order condition seems to be hidden.

Theorem 3. (Theorem 13 of [12]) *For a language L on an alphabet $\mathcal{A}$, the following are equivalent:*

- (*i*) *L satisfies an order condition and is recurrent;*
- (*ii*) *L is the language of a standard interval exchange transformation;*
- (*iii*) *L is the language of a generalized interval exchange transformation without wandering intervals.*

We want now to get rid of extra conditions besides the order condition; as the following chain of counter-examples shows, this obliges us to weaken the classical notion of standard interval exchange, thus the successive generalizations are indeed relevant.

Example 3 (Fake Sturmian). Let L be generated by the bi-infinite sequence $\dots 111222\dots$. Note that it is of complexity $n+1$ but not uniformly recurrent, and in the founding paper [27] it is not included in Sturmian languages, hence we call it a fake Sturmian language.

It satisfies the order condition with $1 <_D 2$, $1 <_A 2$, but (unsurprisingly as it is not recurrent) is *not the language of a standard interval exchange transformation* as that could only be the identity on two disjoint open intervals I_1 and I_2, and the only possible words are 1^n and 2^m. However, L is *the natural coding of an affine interval exchange transformation*: L_2 is the language of length 2 of any affine 2-interval exchange transformation, with the same orders, such that TI_1 is strictly longer than I_1, and, as L is determined by L_2 because there is no bispecial word except the empty one, L is indeed the natural coding of any of these affine interval exchange transformations.

Example 4 (Skew Sturmian). Let L be the language generated by the bi-infinite sequence $\dots 1112111\dots$, which is a skew Sturmian language as defined in [27].

It satisfies the order condition with $1 <_D 2$, $2 <_A 1$, but is *not the natural coding of any affine interval exchange transformation T*: indeed, the sequence $\dots 1111\dots$ in X_L would define a fixed point x for T, in the interior of I_1, and, if

$0 < y < x$ is the right endpoint of TI_2, T would have to send $[0, x)$ to $[y, x)$ and $[x, 1 - y)$ to $[x, 1)$, thus having a slope < 1 on a part of I_1 and a slope > 1 on another part. However, if $\tilde{L}$ is the language generated by the bi-infinite sequence $\dots 3332111\dots$, as in Example 3 $\tilde{L}$ is the natural coding of any affine interval exchange transformation T sending $I_1 = [0, x)$ to $[y, x)$, $I_3 = [x, 1 - y)$ to $[x, 1)$, $I_2 = [1 - y, 1)$ to $[0, y)$, with $0 < y < x < 1 - y$. If we now code T by the intervals $\tilde{I}_1 = I_1 \cup I_3$ and $\tilde{I}_2 = I_2$, we see that L is *a grouped coding of an affine interval exchange transformation* as in Definition 8.

Example 5 (Episkew). Let L' be the Sturmian language which is the natural coding of the unflipped standard interval exchange transformation T' sending $I_1 = [0, 1 - \alpha)$ to $[\alpha, 1)$ and $I_2 = [1 - \alpha, 1)$ to $[0, \alpha)$ for an irrational $\alpha < 1/2$. Let $y_n = i$ whenever $T^n \alpha$ is in I_i, $n \geq 0$, and $y'_n = i$ whenever $T^n(1 - 2\alpha)$ is in I_i, $n \leq 0$; when $\alpha = \frac{3-\sqrt{5}}{2}$, y is the so-called Fibonacci sequence on 1 and 2, and y' is y written backwards. Let L be the language generated by the infinite sequence $\dots y'_{-2} y'_{-1} y'_0 3 y_0 y_1 y_2 \dots$. Extending to languages the definition in [3], we can call it an episkew language. It satisfies the order condition with $1 <_D 3 <_D 2$, $2 <_A 3 <_A 1$ (note that no other order is possible, because of the way the empty bispecial is resolved).

L is *the natural coding of a generalized interval exchange transformation*, by Theorem 4 below, but it is *not the natural or grouped coding of any affine interval exchange transformation:* this will be a straightforward consequence of either one of two independent results we show below, Theorems 6 and 7.

And finally

Theorem 4. (Theorem 19 of [12]) *A language L is a natural coding of a generalized interval exchange transformation if and only if L satisfies an order condition.*

2.4   Examples and Questions

We do not have a complete characterization of the codings of affine interval exchanges. The best we can do is

Theorem 5. *(Theorem 20 of [12]) If L is a natural coding of an affine interval exchange transformation for which the absolute value of the slope is $\exp \theta_e$ on the defining interval I_e, then L satisfies an order condition and for each non recurrent sequence z in X_L, $\sum_{n \geq 0} \exp\left(\sum_{j=0}^{n} \theta_{z_j}\right) < +\infty$, and $\sum_{n > 0} \exp\left(-\sum_{j=-n}^{-1} \theta_{z_j}\right) < +\infty$.*

If L satisfies an order condition and there exist real numbers $\theta_e, e \in \mathcal{A}$, such that for each non recurrent sequence z in L, $\sum_{n \geq 0} \exp\left(\sum_{j=0}^{n} \theta_{z_j}\right) < +\infty$, and $\sum_{n > 0} \exp\left(-\sum_{j=-n}^{-1} \theta_{z_j}\right) < +\infty$, then L is a group coding of an affine interval exchange transformation.

The generalizations of standard interval exchanges have seen a recent surge in activity (see [19, 25, 26] and others) primarily centered on the conjugacy problem between these different classes of maps; in this context, standard and generalized interval exchange transformations are the extreme cases while affine interval exchange transformations constitute a fundamental middle step. The following questions and conjectures can be considered as related to this problem.

Conjecture 1. The conditions in Theorem 5 are necessary and sufficient for L to be a natural coding of an affine interval exchange transformation.

Question 1. Does there exist an aperiodic language which is a grouped coding of an affine interval exchange transformation, but not a natural coding of any affine interval exchange transformation?

Conjecture 1 and Question 1 suggest what we dare not call a conjecture.

Question 2. Is it true that L is a group coding of an affine interval exchange transformation if and only if L satisfies an order condition and there exist real numbers $\theta_e, e \in \mathcal{A}$, such that the two following conditions hold?

- For each non recurrent sequence z in L which is not ultimately periodic to the left,
  $$\sum_{n \geq 0} \exp\left(\sum_{j=0}^{n} \theta_{z_j}\right) < +\infty.$$
- For each non recurrent sequence z in L which is not ultimately periodic to the right,
  $$\sum_{n > 0} \exp\left(-\sum_{j=-n}^{-1} \theta_{z_j}\right) < +\infty.$$

There are many examples of codings of affine interval exchange transformations which are not natural codings of standard ones; they can be built by using the methods of [6, 8, 25] and others. But codings of generalized interval exchange transformations which are not codings of affine ones seem to be completely new, and we know two combinatorial ways of building them, expressed in the two following theorems.

Theorem 6. (Theorem 23 of [12]) *Let L be non recurrent, and a natural coding of a generalized interval exchange transformation T. Suppose the language L' of Lemma 5 is aperiodic, uniformly recurrent, and its arrival and departure orders are conjugate by a circular permutation. Then T cannot be of class P, class P [20] meaning that, except on a countable set of points, its derivative DT exists and $DT = h$ where h is a function with bounded variation, and $|h|$ is bounded from below by a strictly positive number.*

Theorem 7. (Theorem 24 of [12]) *Let L' be a natural coding of a non purely periodic standard interval exchange transformation. Let $w_n = a w'_n b$, $a \in \mathcal{A}$, $b \in \mathcal{A}$, be an infinite sequence of bispecial words in L'. Let u be the left-sided infinite sequence ending with w_n for all n, and v the right-sided infinite sequence beginning with w_n for all n. Let ω be a symbol which is not a letter of L', and*

L be the language generated by the union of all words in L' and the bi-infinite word uωv.

Then L is a natural coding of a generalized interval exchange transformation, but not a grouped coding of any affine interval exchange transformation.

3   Order Conditions and the Burrows-Wheeler Transform

Let $\mathcal{A} = \{a_1 < a_2 < \cdots < a_r\}$ be an ordered alphabet. For a permutation π on $\mathcal{A}$, we define the order $<_\pi$ by $x <_\pi y$ if $\pi^{-1}x < \pi^{-1}y$.

Definition 10. The *(cyclic) conjugates* of $w = w_1 \cdots w_n$ are the words $w_i \cdots w_n w_1 \cdots w_{i-1}$, $1 \leq i \leq n$. If w is primitive, w has precisely n cyclic conjugates. Let $w_{i,1} \cdots w_{i,n}$ denote the i-th conjugate of w where the n conjugates of w are ordered by ascending lexicographical order. Then the *Burrows-Wheeler transform* [7] of w, denoted by $B(w)$, is the word $w_{1,n} w_{2,n} \cdots w_{n,n}$. It depends on the given order $<$ on $\mathcal{A}$.

We say w is *clustering for the order $<$ and the permutation π* [16] if $B(w) = (\pi a_1)^{n_{\pi a_1}} \cdots (\pi a_r)^{n_{\pi a_r}}$, where π is a permutation on $\mathcal{A}$ and n_a is the number of occurrences of a in w (we allow some of the n_a to be 0, thus, given the order and w, there may be several possible π). We say w is *perfectly* clustering if it is clustering for the *symmetric* permutation $\pi a_i = a_{r+1-i}$, $1 \leq i \leq r$ ([30] though it is not named).

Non-primitive Words. As remarked in [16], the Burrows-Wheeler transform can be extended to a non-primitive word $w_1 \cdots w_n$, by ordering its n (non necessarily different) cyclic conjugates by non-strictly increasing lexicographical order and taking the word made by their last letters. Then $B(v^m)$ is deduced from $B(v)$ by replacing each of its letters x_i by x_i^m, and v^m is clustering for π if and only if v is clustering for π.

Theorem 8. *For a given order $<$ on the alphabet, a primitive word w is clustering for the order $<$ and the permutation π if and only if every bispecial word v in the language L_w generated by w^n, $n \in \mathbb{N}$, satisfies the order condition where the order $<_D$ is the order $<$ and the order $<_A$ is $<_\pi$.*
All bispecial words in L_w are factors of length at most $|w| - 2$ of ww.

Proof

We begin by the last assertion. Suppose v is a bispecial of L_w. Then v must occur at two different positions in some word w^k. If $|w| = n$ and $|v| \geq |w| - 1$, this implies in particular $w_i...w_n w_1...w_{i-2} = w_j...w_n w_1...w_{j-2}$ for $1 < j - i < n$, and we notice that each w_i is in at least one member of the equality, thus we get that w is a power of a word whose length is the GCD of n and $j - i$, which contradicts the primitivity. Thus the length of v is at most $|w| - 2$, and v occurs in ww.

We prove now that our order condition is equivalent to the following *modified order condition*: whenever $z = z_1...z_n$ and $z' = z_1'...z_n'$ are two different cyclic conjugates of w, $z < z'$ (lexicographically) if and only if $z_k <_\pi z'_k$ for the largest $k \leq n$ such that $z_k \neq z'_k$. Indeed, by definition $z < z'$ if and only if $z_j < z'_j$ for the smallest $j \geq 1$ such that $z_j \neq z'_j$. If w satifies the order condition, we apply it to the bispecial word $z_{k+1}...z_n z_1...z_{j-1}$, with k and j as defined, and get the modified order condition. Let v be a bispecial word in L_w; by the first paragraph of this proof it can be written as $z_1...z_{k-1}$ for some $1 \leq k \leq n$, with the convention that $k = 1$ whenever v is empty, and at least two different cyclic conjugates z of w, and its possible extensions are the corresponding $z_n z_1...z_k$, thus, if the modified order condition is satisfied, v does satisfy the requirement of the order condition.

The modified order condition implies clustering, as then if two cyclic conjugates of w satisfy $z < z'$, their last letters z_n and z'_n satisfy either $z_n = z'_n$ or $z_n <_\pi z'_n$. Suppose $w = w_1 \cdots w_n$ is clustering for π. Suppose two cyclic conjugates of w are such that $z_k \neq z'_k$, $z_j = z'_j$ for $k + 1 \leq j \leq n$. Then $z < z'$ is (by definition of the lexicographical order) equivalent to $z_{k+1}...z_n z_1..z_k < z'_{k+1}...z'_n z'_1..z'_k$, and, as these two words have different last letters, because of the clustering this is equivalent to $z_k <_\pi z'_k$, thus the modified order condition is satisfied.   □

Theorem 8 remains valid if $w = v^m$ is non-primitive (it can be slightly improved as there are less bispecial words to be considered, it is enough to look at factors of vv of length at most $|v| - 2$). An immediate consequence is the following, which seems to be new.

Proposition 4. *If w clusters for the order $<$ and the permutation π, its reverse clusters for the π-order, and the permutation π^{-1}.*

Proof

This follows in a straightforward way from Theorem 8.   □

The order condition can be applied to get the clustering properties of classical families of words. Thus it can be used to reprove the result of [24]: a Sturmian language contains infinitely many clustering words.

Theorem 9. *The natural coding of a standard k-interval exchange in the hyperelliptic class (this consists in all the symmetric ones, i.e. those for which the order $<_A$ is the reverse of $<_D$, and all those which can be obtained from the symmetric ones by an induction process, see [14]) satisfying the i.d.o.c. condition, or of any standard 3- or 4-interval exchange satisfying the i.d.o.c. condition, contains infinitely many clustering words.*

Proof

By Theorem 8 and Proposition 3 (or by Theorem 4 of [16]), if L is the language of a standard interval exchange satisfying the i.d.o.c. condition, w clusters if ww is in L. The fact that L contains infinitely many squares is proved in [14] for the hyperelliptic class, [13] for the other cases mentioned above.   □

Note that Theorem 9 has not yet been generalized to wider classes of interval exchanges. Also, in the forthcoming [17], we shall prove that an Arnoux-Rauzy language [2] contains finitely many clustering words, while an episturmian language [3] may contain finitely or infinitely many clustering words, with a full characterization of each case.

References

1. Arnoux, P.: Un invariant pour les échanges d'intervalles et les flots sur les surfaces, (in French) Thèse de 3e cycle: Reims (1981)
2. Arnoux, P., Rauzy, G.: Représentation géométrique de suites de complexité $2n+1$, (in French) Bull. Soc. Math. France **119**, 199–215 (1991)
3. Berstel, J.: Sturmian and Episturmian words. In: Bozapalidis, S., Rahonis, G. (eds.) CAI 2007. LNCS, vol. 4728, pp. 23–47. Springer, Heidelberg (2007). https://doi.org/10.1007/978-3-540-75414-5_2
4. Berthé, V., et al.: Acyclic, connected and tree sets, Monatsh. Math. **176**(4), 521–550 (2015)
5. Boshernitzan, M., Kornfeld, I.: Interval translation mappings. Ergodic Theor. Dynam. Syst. **15**, 821–832 (1995)
6. Bressaud, X., Hubert, P., Maass, A.: Persistence of wandering intervals in self-similar affine interval exchange transformations. Ergod. Theory Dyn. Syst. **30**, 665–686 (2010)
7. Burrows, M., Wheeler, D.J.: A block-sorting lossless data compression algorithm, Technical Report 124 (1994) Digital Equipment Corporation
8. Camelier, R., Guttierez, C.: Affine interval exchange transformations with wandering intervals. Ergod. Theory. Dyn. Syst. **17**, 1315–1338 (1997)
9. Cassaigne, J.: Complexité et facteurs spéciaux, (in French) Journées Montoises (Mons,: Bull. Belg. Math. Soc. Simon Stevin **4**(1997), 67–88 (1994)
10. Cassaigne, J., Nicolas, F.: Factor complexity, Combinatorics, automata and number theory. Encyclopedia Math. Appl. **135**, 163–247 (2010)
11. De Luca, A., Edson, M., Zamboni, L.Q.: Extremal values of semi-regular continuants and codings of interval exchange transformations, Mathematika. **69**, 432–457 (2023)
12. Ferenczi, S., Hubert, P., Zamboni, L.Q.: Languages of general interval exchange transformations, arXiv: 2212.01024
13. Ferenczi, S.: A generalization of the self-dual induction to every interval exchange transformation. Ann. Inst. Fourier (Grenoble) **64**, 1947–2002 (2014)
14. Ferenczi, S., Zamboni, L.Q.: Structure of K-interval exchange transformations: induction, trajectories, and distance theorems. J. Anal. Math. **112**, 289–328 (2010)
15. Ferenczi, S., Zamboni, L.Q.: Languages of k-interval exchange transformations. Bull. Lond. Math. Soc. **40**, 705–714 (2008)
16. Ferenczi, S., Zamboni, L.Q.: Clustering words and interval exchanges. J. Integer Seq. **16**, 9 pp. (2013). Article 13.2.1
17. Ferenczi, S., Zamboni, L.Q.: Clustering of Arnoux-Rauzy words. In: Preparation
18. Gaboriau, D., Levitt, G., Paulin, F.: Pseudogroups of isometries of R and Rips' theorem on free actions on R-trees, Israël. J. Math. **87**, 403–428 (1994)
19. Ghazouani, S., Ulcigrai, C.: A priori bounds for GIETS, affine shadows and rigidity of foliations in genus two, arXiv:2106.03529
20. Herman, M.-R.: Sur la conjugaison différentiable des difféomorphsimes du cercle à des rotations, (French). Inst. Hautes Études Sci. Publ. Math. **49**, 5–233 (1979)

21. Kanel-Belov, A.Y., Chernyat'ev, A.L.: Describing the set of words generated by interval exchange transformations, Comm. Algebra **38**, 2588–2605 (2010)
22. Keane, M.S.: Interval exchange transformations. Math. Zeitsch. **141**, 25–31 (1975)
23. Kerckhoff, S.: Simplicial systems for interval exchange maps and measured foliations. Ergod. Theory. Dyn. Syst. **5**, 257–271 (1985)
24. Mantaci, S., Restivo, A., Sciortino, M.: Burrows-Wheeler transform and Sturmian words. Inform. Process. Lett. **86**, 241–246 (2003)
25. Marmi, S., Moussa, P., Yoccoz, J.-C.: Affine interval exchange maps with a wandering interval. Proc. Lond. Math. Soc. **3**, 639–669 100 (2010)
26. Marmi, S., Moussa, P., Yoccoz, J.-C.: Linearization of generalized interval exchange maps. Ann. of Math. **176**(2), 1583–1646 (2012)
27. Morse, M., Hedlund, G.A.: Symbolic dynamics II. Sturmian trajectories, Amer. J. Math. **62**, 1–42 (1940)
28. Nogueira, A.: Nonorientable recurrence of flows and interval exchange transformations. J. Diff. Eqn. **70**, 153–166 (1987)
29. Oseledec, V.I.: The spectrum of ergodic automorphisms, (in Russian) Dokl. Akad. Nauk. SSSR **168**, 1009–1011 (1966)
30. Simpson, J., Puglisi, S.J.: Words with simple Burrows-Wheeler transforms. Electron. J. Combinatorics **15**, Research Paper 83, 17 pp. (2008)

On Sensitivity of Compact Directed Acyclic Word Graphs

Hiroto Fujimaru[1], Yuto Nakashima[2], and Shunsuke Inenaga[2(✉)]

[1] Department of Physics, Kyushu University, Fukuoka, Japan
`fujimaru.hiroto.134@s.kyushu-u.ac.jp`
[2] Department of Informatics, Kyushu University, Fukuoka, Japan
`{nakashima.yuto.003,inenaga.shunsuke.380}@m.kyushu-u.ac.jp`

Abstract. *Compact directed acyclic word graphs (CDAWGs)* [Blumer et al. 1987] are a fundamental data structure on strings with applications in text pattern searching, data compression, and pattern discovery. Intuitively, the CDAWG of a string T is obtained by merging isomorphic subtrees of the suffix tree [Weiner 1973] of the same string T, thus CDAWGs are a compact indexing structure. In this paper, we investigate the sensitivity of CDAWGs when a single character edit operation (insertion, deletion, or substitution) is performed at the left-end of the input string T, namely, we are interested in the worst-case increase in the size of the CDAWG after a left-end edit operation. We prove that if e is the number of edges of the CDAWG for string T, then the number of new edges added to the CDAWG after a left-end edit operation on T is less than e. Further, we present almost matching lower bounds on the sensitivity of CDAWGs for all cases of insertion, deletion, and substitution.

1 Introduction

Compact directed acyclic word graphs (CDAWGs) [4] are a fundamental data structure on strings that have applications in fields including text pattern searching [6,8], data compression [2,13], and pattern discovery [14]. Intuitively, the CDAWG of a string T, denoted $\mathsf{CDAWG}(T)$, is obtained by merging isomorphic subtrees of the suffix tree [15] of the same string T. Thus the size of the CDAWG is always smaller than that of the suffix tree.

It is well known that the nodes of $\mathsf{CDAWG}(T)$ correspond to *maximal repeats* in T, and the number e of right-extensions of maximal repeats in T, which is equal to the number of edges of $\mathsf{CDAWG}(T)$, has been used as one of repetitiveness measures of strings. Namely, when e is small, then the string contains a lot of repetitive substrings hence being well compressible. Indeed, it is known that e can be as small as $\Theta(\log n)$ with highly repetitive strings [11]. Further, one can obtain a *grammar-based compression* of size $O(e)$ via the CDAWG of the input string T [2]. Some relations between e and the number r of equal-letter runs in the *Burrows-Wheeler transform (BWT)* [5] have also been investigated [3].

© The Author(s), under exclusive license to Springer Nature Switzerland AG 2023
A. Frid and R. Mercaş (Eds.): WORDS 2023, LNCS 13899, pp. 168–180, 2023.
https://doi.org/10.1007/978-3-031-33180-0_13

Table 1. Our results: additive sensitivity of CDAWGs with left-end edit operations.

edit operation	upper bound	lower bound
left-end insertion $(T \Rightarrow aT)$	$e - 1$	$e - 2$
left-end deletion $(T \Rightarrow T[2..|T|])$	$e - 2$	$e - 4$
left-end substitution $(T = aS \Rightarrow bS = T')$	e	$e - 3$

Recently, Akagi et al. [1] proposed the notion of *sensitivity* of string repetitiveness measures and string compressors, including the aforementioned e and r, the smallest *string attractor* size γ [9], the *substring complexity* δ [10], and the Lempel-Ziv parse size z [16]. The sensitivity of a repetitiveness measure c asks how much the measure size increases when a single-character edit operation is performed on the input string, and thus the sensitivity allows one to evaluate the robustness of the measure/compressor against errors/edits.

This paper investigates the sensitivity of CDAWGs when a single character edit operation (insertion, deletion, or substitution) is performed at the left-end of the input string T, namely, we are interested in the worst-case increase in the size of the CDAWG after an left-end edit operation. We prove that if e is the number of edges of the CDAWG for string T, then the number of new edges which are added to the CDAWG after an left-edit operation on T is always less than e. Further, we present almost matching lower bounds on the sensitivity of CDAWGs for any left-end insertion, deletion, and substitution (see Table 1 for a summary of these results). We generalize our lower-bound instances for left-end insertion to *leftward online construction* of the CDAWG, and show that it requires $\Omega(n^2)$ time. This contrasts with the case of *rightward online CDAWG construction* for which a linear-time algorithm exists [8].

A full version of this paper can be found in [7].

Related Work. Akagi et al. [1] presented lower bounds when a new character is deleted (resp. substituted) in the middle of the string, with a series of strings for which the size e of the CDAWG additively increases by $e - 4$ (resp. $e - 2$). They also showed a lower bound when a new character is inserted at the *right-end* of the string, showing a series of strings for which the size of the CDAWG additively increases by $e-2$. While an additive $e+O(1)$ upper bound for the case of right-end insertion readily follows from the *rightward* online construction of CDAWGs [8], no non-trivial upper bounds for the other edit operations, including our case of left-end edit operations, are known. Our $\Omega(n^2)$ lower-bound for leftward online construction of the CDAWG extends the quadratic lower-bound for maintaining the CDAWG in the sliding window model [12] (remark that fixing the right-end of the sliding window is equivalent to our leftward online construction).

2   Preliminaries

Let Σ be an *alphabet* of size σ. An element of Σ^* is called a *string*. For a string $T \in \Sigma^*$, the length of T is denoted by $|T|$. The *empty string*, denoted by ε, is the string of length 0. Let $\Sigma^+ = \Sigma^* \setminus \{\varepsilon\}$. If $T = uvw$, then u, v, and w are called a *prefix*, *substring*, and *suffix* of T, respectively. The sets of prefixes, substrings, and suffixes of string T are denoted by $\mathsf{Prefix}(T)$, $\mathsf{Substr}(T)$, and $\mathsf{Suffix}(T)$, respectively. For a string T of length n, $T[i]$ denotes the ith character of T for $1 \leq i \leq n$, and $T[i..j] = T[i] \cdots T[j]$ denotes the substring of T that begins at position i and ends at position j on T for $1 \leq i \leq j \leq n$. For two strings u and T, let $\mathsf{BegPos}(u,T) = \{i \mid T[i..i + |u| - 1] = u\}$ and $\mathsf{EndPos}(u,T) = \{i \mid T[i - |u| + 1..i] = u\}$ denote the sets of beginning positions and the set of ending positions of u in T, respectively.

For any substrings $u, v \in \mathsf{Substr}(T)$ of a string T, we write $u \equiv^{\mathrm{L}}_T v$ iff $\mathsf{EndPos}(u,T) = \mathsf{EndPos}(v,T)$. Let $[\cdot]^{\mathrm{L}}_T$ denote the equivalence class of strings under $\equiv^{\mathrm{L}}_T$. For $x \in \mathsf{Substr}(T)$, let $\mathsf{long}([x]^{\mathrm{L}}_T)$ denote the longest member of $[x]^{\mathrm{L}}_T$. Let $\mathsf{LeftM}(T) = \{\mathsf{long}([x]^{\mathrm{L}}_T) \mid x \in \mathsf{Substr}(T)\}$. Any element $u \in \mathsf{LeftM}(T)$ is said to be *left-maximal* in T, since there are two distinct characters $c, d \in \Sigma$ such that $cu, du \in \mathsf{Substr}(T)$, or $u \in \mathsf{Prefix}(T)$. For any non-longest element $y \in [x]^{\mathrm{L}}_T \setminus \{\mathsf{long}([x]^{\mathrm{L}}_T)\}$ there exists a unique non-empty string α such that $\alpha y = \mathsf{long}([x]^{\mathrm{L}}_T)$, i.e. any occurrence of y in T is immediately preceded by α.

Similarly, we write $u \equiv^{\mathrm{R}}_T v$ iff $\mathsf{BegPos}(u,T) = \mathsf{BegPos}(v,T)$. Let $[\cdot]^{\mathrm{R}}_T$ denote the equivalence class of strings under $\equiv^{\mathrm{R}}_T$. For $x \in \mathsf{Substr}(T)$, let $\mathsf{long}([x]^{\mathrm{R}}_T)$ denote the longest member of $[x]^{\mathrm{R}}_T$. Let $\mathsf{RightM}(T) = \{\mathsf{long}([x]^{\mathrm{R}}_T) \mid x \in \mathsf{Substr}(T)\}$. Any element $u \in \mathsf{RightM}(T)$ is said to be *right-maximal* in T, since there are two distinct characters $c, d \in \Sigma$ such that $uc, ud \in \mathsf{Substr}(T)$, or $u \in \mathsf{Suffix}(T)$. For any non-longest element $y \in [x]^{\mathrm{R}}_T \setminus \{\mathsf{long}([x]^{\mathrm{R}}_T)\}$ there exists a unique non-empty string β such that $y\beta = \mathsf{long}([x]^{\mathrm{R}}_T)$, i.e. any occurrence of y in T is immediately followed by β. Let $\mathsf{M}(T) = \mathsf{LeftM}(T) \cap \mathsf{RightM}(T)$. Any element of $\mathsf{M}(T)$ is said to be *maximal* in T.

The *compact directed acyclic word graph* (*CDAWG*) of a string T, denoted $\mathsf{CDAWG}(T) = (\mathsf{V}, \mathsf{E})$, is an edge-labeled DAG such that

$$\mathsf{V}_T = \{[x]^{\mathrm{L}}_T \mid x \in \mathsf{RightM}(T)\},$$
$$\mathsf{E}_T = \{([x]^{\mathrm{L}}_T, \beta, [x\beta]^{\mathrm{L}}_T) \mid \beta \in \Sigma^+, x, x\beta \in \mathsf{RightM}(T), y\beta \in [x\beta]^{\mathrm{L}}_T \text{ for any } y \in [x]^{\mathrm{L}}_T\}.$$

See Fig. 1 for a concrete example of CDAWGs. Intuitively, the strings in $\mathsf{RightM}(T)$ correspond to the nodes of the suffix tree [15] of T, and the operator $[\cdot]^{\mathrm{L}}_T$ merges the isomorphic subtrees of the suffix tree. Recall that the nodes of the suffix tree for T correspond to the right-maximal substrings of T. Since $\mathsf{long}([x]^{\mathrm{L}}_T)$ is a maximal substring of T for any $x \in \mathsf{RightM}(T)$, we have the following fact:

Fact 1. *There is a one-to-one correspondence between the elements of* $\mathsf{M}(T)$ *and the nodes of* $\mathsf{CDAWG}(T)$.

We can regard each element of $\mathsf{M}(T)$ as a node of $\mathsf{CDAWG}(T)$ by Fact 1. We thus sometimes identify V_T with $\mathsf{M}(T)$ for convenience. For any $x \in \mathsf{M}(T)$, $\mathsf{d}_T(x)$ denotes the out-degree of the node x in $\mathsf{CDAWG}(T)$.

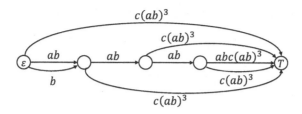

Fig. 1. Illustration for $\mathsf{CDAWG}(T)$ of string $T = (ab)^4 c(ab)^3$. Every substring of T can be spelled out from a distinct path from the source ε. There is a one-to-one correspondence between the maximal substrings in $\mathsf{M}(T) = \{\varepsilon, ab, (ab)^2, (ab)^3, (ab)^4 c(ab)^3\}$ and the nodes of $\mathsf{CDAWG}(T)$. The number of right-extensions of $\mathsf{CDAWG}(T)$ is the number $\mathsf{e}(T)$ of edges, which is 9 in this example.

A non-empty substring x of string T is called a *maximal repeat* in T if x is maximal in T and $|\mathsf{BegPos}(x,T)| = |\mathsf{EndPos}(x,T)| \geq 2$. We remark that the set of maximal repeats in T coincides with $\mathsf{M}(T) \setminus \{\varepsilon, T\}$, namely the longest elements of all internal nodes of $\mathsf{CDAWG}(T)$ are maximal repeats in T, and they are the only maximal repeats in T.

The *size* of $\mathsf{CDAWG}(T) = (V_T, E_T)$ for a string T of length n is the number $\mathsf{e}(T) = |E_T|$ of edges in $\mathsf{CDAWG}(T)$, which is also referred to as the number of right-extensions of maximal repeats in T. Using this measure e, we define the worst-case additive *sensitivity* of the CDAWG with left-end edit operations (resp. insertion, deletion, and substitution) by:

$$\mathsf{AS}_{\mathrm{LeftIns}}(\mathsf{e}, n) = \max_{T \in \Sigma^n, a \in \Sigma} \{\mathsf{e}(aT) - \mathsf{e}(T)\},$$

$$\mathsf{AS}_{\mathrm{LeftDel}}(\mathsf{e}, n) = \max_{T \in \Sigma^n} \{\mathsf{e}(T[2..n]) - \mathsf{e}(T)\},$$

$$\mathsf{AS}_{\mathrm{LeftSub}}(\mathsf{e}, n) = \max_{T \in \Sigma^n, a \in \Sigma \setminus \{T[1]\}} \{\mathsf{e}(aT[2..n]) - \mathsf{e}(T)\}.$$

For the sensitivity of CDAWGs, we first briefly describe the special case where both the original string T and an edited string T' are unary. Let $T = a^n$. Clearly, every a^i with $1 \leq i < n$ is a maximal substring of T and it is only followed by a. Thus $\mathsf{e}(T) = n - 1$. In case of insertion, i.e. $T' = aT = a^{n+1}$, we similarly have $\mathsf{e}(T') = n$. Thus $\mathsf{e}(T') - \mathsf{e}(T) = 1$ for unary strings. Symmetrically, we have $\mathsf{e}(T') - \mathsf{e}(T) = -1$ in the case of deletion with $T' = a^{n-1}$. There is no substitution when $\sigma = 1$. In what follows, we focus on the case where $\sigma \geq 2$.

3 Sensitivity of CDAWGs with Left-End Insertions

We consider the worst-case additive sensitivity $\mathsf{AS}_{\mathrm{LeftIns}}(\mathsf{e}, n)$ of $\mathsf{CDAWG}(T)$ when a new character a is prepended to input string T of length n, i.e. $T' = aT$.

3.1 Upper Bound for $AS_{LeftIns}(e, n)$ on CDAWGs

We divide the value $e(T') - e(T)$ into two components $f_{Ins}(T)$ and $g_{Ins}(T)$ s.t.

- $f_{Ins}(T)$ is the total out-degrees of new nodes that appear in $CDAWG(aT)$;
- $g_{Ins}(T)$ is the total number of new out-going edges of nodes that already exist in $CDAWG(T)$.

Clearly $e(T') - e(T) \leq f_{Ins}(T) + g_{Ins}(T)$. We first consider the above two components separately, and then we merge them to obtain the desired upper bound.

$f_{Ins}(T)$: **Total Out-Degrees of New Nodes.** Suppose u is a new node for $CDAWG(aT)$, where $u \notin M(T)$ and $u \in M(aT)$. This implies that there is a new occurrence of u in aT as a prefix. Let $u = ax$. The following is our key lemma:

Lemma 1. *If $ax \notin M(T)$ and $ax \in M(aT)$ (i.e. ax is a new node in $CDAWG(aT)$), then $x \in M(T)$. Also, $d_{aT}(ax) \leq d_T(x)$.*

Proof. Since $ax \in \mathsf{Prefix}(aT)$, $x \in \mathsf{Prefix}(T)$. Thus x is left-maximal in T. Assume on the contrary that x is not right-maximal in T. Then there exists a non-empty string $\beta \in \Sigma^+$ such that $x\beta = \mathsf{long}([x]_T^R)$, which means that any occurrence of x in T is immediately followed by β. Thus ax is also immediately followed by β in aT, however, this contradicts the precondition that $ax \in M(aT)$. Thus x is right-maximal in T. It immediately follows from $\mathsf{EndPos}(ax, aT) \subseteq \mathsf{EndPos}(x, T)$ that $d_{aT}(ax) \leq d_T(x)$.   □

It follows from Lemma 1 that the out-degree of each new node in $CDAWG(aT)$ does not exceed the maximum out-degree of $CDAWG(T)$. Also, there is an injective mapping from a new node ax in $CDAWG(aT)$ to an existing node x in $CDAWG(T)$ by Lemma 1. Thus $f_{Ins}(T) \leq e(T)$ for any string T.

In the sequel, we give a tighter bound $f_{Ins}(T) \leq e(T) - 1$. For this purpose, we pick up the case where $x = \varepsilon$, assume that $ax = a$ becomes a new node in $CDAWG(aT)$, and compare the out-degree of the source ε of $CDAWG(T)$ and the out-degree of the new node a in $CDAWG(aT)$. We consider the cases with $\sigma = 2$ and with $\sigma \geq 3$ separately:

Lemma 2. *Let $\sigma = 2$. If*

1. *$a \notin M(T)$,*
2. *$a \in M(aT)$, and*
3. *there exists a string $x \in M(T) \setminus \{\varepsilon, T\}$ such that $ax \notin M(T)$ and $ax \in M(aT)$,*

then $d_{aT}(a) < d_T(\varepsilon)$.

Proof. Let $\Sigma = \{a, b\}$. We can exclude the case where $T = b^n$ due to the following reason: Since ab^i for each $1 \leq i < n$ is not maximal in $aT = ab^n$, no new nodes are created in $CDAWG(ab^n)$ (only a new edge labeled ab^n from the source to the sink is created).

From now on, consider the case where T contains both a and b. This means that $d_T(\varepsilon) = \sigma = 2$. Since $a \in M(aT)$, a is a node of $\mathsf{CDAWG}(aT)$. Assume on the contrary that $d_{aT}(a) = d_T(\varepsilon)$. We then have $d_{aT}(a) = 2$, which means $aa, ab \in \mathsf{Substr}(aT)$. There are two cases depending on the first character of T:

- If $T[1] = a$, then let $T = aw$. Then, since $aT = aaw$, we have $ab \in \mathsf{Substr}(T)$. Since $a \notin M(T)$ (the first precondition), b is the only character that immediately follows a in T, meaning that $aa \notin \mathsf{Substr}(T)$. Recall that the new node ax must be a prefix of $aT = aaw$. Since $x \neq \varepsilon$ (the third precondition), $|ax| \geq 2$, and thus aa is a prefix of ax. However, since $aa \notin \mathsf{Substr}(T)$, aa occurs in aT exactly once as a prefix and thus ax occurs exactly once in aT. This contradicts the third precondition that ax is a new node in $\mathsf{CDAWG}(aT)$.
- If $T[1] = b$, then we have that $ab \notin \mathsf{Substr}(T)$ by similar arguments as above. Thus T must be of form $b^m a^{n-m}$ with $1 \leq m < n$. Moreover, since $a \notin M(T)$ and $a \in M(aT)$ (the first and second preconditions), we have $T = b^{n-1}a$. Then, for the edited string $aT = ab^{n-1}a$, any new internal node ax in $\mathsf{CDAWG}(aT)$ must be in form ab^i with $1 \leq i < n$. However, each $ax = ab^i$ occurs in aT exactly once, meaning that $\mathsf{long}([ab^i]^R_{aT}) = aT$. This contradicts the third precondition that ax is a new node in $\mathsf{CDAWG}(aT)$.

Consequently, $d_{aT}(a) < d_T(\varepsilon)$. □

Lemma 3. *Let $\sigma \geq 3$. If $a \notin M(T)$ and $a \in M(aT)$, then $d_{aT}(a) < d_T(\varepsilon)$.*

Proof. By similar arguments to the proof for Lemma 2, we have that T contains at least three distinct characters, one of which is a. Thus $d_T(\varepsilon) = \sigma \geq 3$.

Assume on the contrary that $d_{aT}(a) = d_T(\varepsilon) = \sigma \geq 3$. Since $a \notin M(T)$ (i.e. a is not maximal in T), we have the two following cases:

- If a is not left-maximal in T, then $T[1] \neq a$ and there is a unique character b ($\neq a$) that immediately precedes a in T, meaning that $aa \notin \mathsf{Substr}(T)$. Since $T[1] \neq a$, we also have $aa \notin \mathsf{Substr}(aT)$. Thus $d_{aT}(a) < \sigma = d_T(\varepsilon)$, a contradiction.
- If a is not right-maximal in T, then there is a unique character b that immediately follows a in T. The occurrence of a as a prefix of aT is followed by $T[1]$, and thus the number $d_{aT}(a)$ of distinct characters following a in aT is at most $2 < \sigma = d_T(\varepsilon)$, a contradiction.

Consequently, $d_{aT}(a) < d_T(\varepsilon)$. □

By Lemmas 2 and 3, even if there appear new nodes ax in $\mathsf{CDAWG}(aT)$ corresponding to all existing nodes x in $\mathsf{CDAWG}(T)$, we have a credit $d_T(\varepsilon) - d_{aT}(a) \geq 1$ in most cases. The only exception is when $\sigma = 2$ and $M(T) = \{\varepsilon, T\}$. However, in this specific case $\mathsf{CDAWG}(T)$ consists only of the two nodes (source and sink), namely $e(T) = 2$. Conversely, we have that the above arguments hold for any $e(T) \geq 3$, which leads to the following:

Lemma 4. *For any string T with $e(T) \geq 3$, $f_{\mathsf{Ins}}(T) \leq e(T) - 1$.*

$g_{Ins}(T)$: **Number of New Branches from Existing Nodes.** The following lemma states that the out-degrees of most existing nodes of $CDAWG(T)$ do not change in $CDAWG(aT)$, except for a single unique node that can obtain a single new out-going edge in $CDAWG(aT)$:

Lemma 5. *For any* $y \in Substr(T)$ *such that* $y \in M(T)$ *and* $y \in M(aT)$, $d_{aT}(y) \in \{d_T(y), d_T(y) + 1\}$. *Also, there exists at most one substring* y *with* $d_{aT}(y) = d_T(y) + 1$. *Consequently* $g_{Ins}(T) \leq 1$.

Proof. Since $y \in M(T)$ and $y \in M(aT)$, y is a node in both $CDAWG(T)$ and $CDAWG(aT)$. Then we have that:

$$d_{aT}(y) = \begin{cases} d_T(y) + 1 & \text{if } y \in Prefix(aT) \text{ and } yb \text{ occurs in } aT \text{ only as a prefix,} \\ d_T(y) & \text{otherwise,} \end{cases}$$

where b is the character that immediately follows the occurrence of y as a prefix of aT, namely $b = T[|y|]$.

Assume on the contrary that there exist two distinct substrings $x, y \in M(T) \cap M(aT)$ such that $d_{aT}(x) = d_T(x) + 1$ and $d_{aT}(y) = d_T(y) + 1$. Since both x and y must be distinct prefixes of aT, we can assume w.l.o.g. that $|x| < |y|$, which means that x is a proper prefix of y. Thus the occurrence of x as a prefix of aT is immediately followed by the character $c = y[|x| + 1]$. We recall that y occurs in T since $y \in M(T)$. Therefore there is an occurrence of x in T that is immediately followed by c, which leads to $d_{aT}(x) = d_T(x)$, a contradiction.   □

Putting All Together. Due to Lemma 4 and Lemma 5, we have an upper bound $e(T') - e(T) \leq f_{Ins}(T) + g_{Ins}(T) \leq e(T) - 1 + 1 = e(T)$ for $\sigma \geq 2$. We remark that the equality holds only if both of the following conditions are satisfied:

(a) For any $x \in M(T) \setminus \{\varepsilon\}$, $ax \notin M(T)$, $ax \in M(aT)$, and $d_{aT}(ax) = d_T(x)$;
(b) There exists a unique string $x \in Substr(T)$ such that $d_{aT}(x) = d_T(x) + 1$.

However, in the next lemma, we show that no strings x can satisfy both Conditions (a) and (b) simultaneously:

Lemma 6. *If* $ax \notin M(T)$ *and* $ax \in M(aT)$, *then* $d_{aT}(x) = d_T(x)$.

Proof. Assume on the contrary that $d_{aT}(x) \neq d_T(x)$. By Lemma 5 we have that $d_{aT}(x) = d_T(x) + 1$. Then, it also follows from the proof of Lemma 5 that x is a prefix of aT and the character $b = T[|x|]$ that immediately follows the prefix occurrence of x in aT differs from any other characters that immediately follow the occurrences of x in T. Namely, we have $b \notin \Sigma' = \{T[i+1] \mid i \in EndPos(x, T)\}$. Moreover, by Lemma 1, ax is also a prefix of aT. This means that x is a prefix of ax, and hence $ax = xb$, which means that $x = a^{|x|}$ and $a = b$. Because $\sigma \geq 2$, $T \neq x$. Since $ax \in M(aT)$ and $x \neq T$, $ax (= xb)$ occurs in T. This means that $b = c$ for some $c \in \Sigma'$, a contradiction. Thus, $d_{aT}(x) = d_T(x)$.   □

We have $e(T) \geq 3$ only if $|T| \geq 3$. By wrapping up Lemma 4, Lemma 5, and Lemma 6, we obtain the main result of this subsection:

Theorem 1. *For any $n \geq 3$ and $e \geq 3$, $\mathsf{AS}_{\mathrm{LeftIns}}(e, n) \leq e - 1$.*

3.2   Lower Bound for $\mathsf{AS}_{\mathrm{LeftIns}}(e, n)$ on CDAWGs

Theorem 2. *There exists a family of strings T such that $e(T') - e(T) = e(T) - 2$, where $T' = bT$ with $b \in \Sigma$. Therefore $\mathsf{AS}_{\mathrm{LeftIns}}(e, n) \geq e - 2$.*

The proof of the above theorem can be found in the full version [7].

4   Sensitivity of CDAWGs with Left-End Deletions

In this section we investigate the worst-case additive sensitivity $\mathsf{AS}_{\mathrm{LeftDel}}(e, n)$ of $\mathsf{CDAWG}(T)$ when $T[1]$ is deleted from the original input string T of length n.

4.1   Upper Bound for $\mathsf{AS}_{\mathrm{LeftDel}}(e, n)$ on CDAWGs

Let $a = T[1]$ be the first character of string T. Let $T = aS$ and $T' = S$, and we consider left-end deletion $aS \Rightarrow S$. Since deleting the left-end character from T never increases the right-contexts of any substring in S, it suffices for us to consider $f_{\mathrm{Del}}(T) = f_{\mathrm{Del}}(aS)$, the total out-degrees of new nodes that appear in $\mathsf{CDAWG}(T') = \mathsf{CDAWG}(S)$, namely $e(S) - e(aS) \leq f_{\mathrm{Del}}(aS)$.

Let x be a new node in $\mathsf{CDAWG}(S)$. We have the following:

Lemma 7. *If $x \notin \mathsf{M}(aS)$ and $x \in \mathsf{M}(S)$, then $x \in \mathsf{Prefix}(S)$ and $ax \in \mathsf{M}(aS)$. Also, $\mathsf{d}_S(x) = \mathsf{d}_{aS}(ax)$.*

Proof. Since $x \notin \mathsf{M}(aS)$, x is either not left-maximal or not right-maximal in aS. If x is not right-maximal in aS, then x is also not right-maximal in S, hence $x \notin \mathsf{M}(S)$. However, this contradicts the precondition $x \in \mathsf{M}(S)$. Thus x is not left-maximal in aS. Then, there exists a non-empty unique string $\alpha \in \Sigma^+$ such that $\alpha x = \mathsf{long}([x]_{aS}^L)$, which means that any occurrence of x in aS is immediately preceded by α. Assume on the contrary that $x \notin \mathsf{Prefix}(S)$. Since $x \in \mathsf{M}(S)$, $x = \mathsf{long}([x]_S^L) = \mathsf{long}([x]_{aS}^L)$, however, this contradicts that α is a non-empty string. Thus $x \in \mathsf{Prefix}(S)$, and hence $ax \in \mathsf{Prefix}(aS)$. Since $ax \in \mathsf{Prefix}(aS)$ and x is right-maximal in aS, ax is a maximal string of aS. Thus $ax \in \mathsf{M}(aS)$.

Since x is not left-maximal in aS and since $ax \in \mathsf{Prefix}(aS)$, $\mathsf{EndPos}(ax, aS) = \mathsf{EndPos}(x, aS) = \mathsf{EndPos}(x, S)$. This leads to $\mathsf{d}_{aS}(ax) = \mathsf{d}_S(x)$. □

By Lemma 7, the out-degree of each new node in $\mathsf{CDAWG}(S)$ does not exceed the maximum out-degree of $\mathsf{CDAWG}(aS)$. Also by Lemma 7, there is an injective mapping from a new node x in $\mathsf{CDAWG}(S)$ to an existing node ax in $\mathsf{M}(aS) \setminus \{\varepsilon\}$. Since $\mathsf{d}_{aS}(\varepsilon) = \sigma \geq 2$, it holds that $e(S) \leq 2(e(T) - \sigma) + \sigma \leq 2e(aS) - 2$, that is:

Theorem 3. *For any n, $\mathsf{AS}_{\mathrm{LeftDel}}(e, n) \leq e - 2$.*

4.2   Lower Bound for $\mathsf{AS}_{\mathrm{LeftDel}}(\mathsf{e}, n)$ on CDAWGs

Theorem 4. *There exists a family of strings T such that $\mathsf{e}(S) - \mathsf{e}(T) = \mathsf{e}(T) - 4$, where $T = aS$ with $a \in \Sigma$. Therefore $\mathsf{AS}_{\mathrm{LeftDel}}(\mathsf{e}, n) \geq \mathsf{e} - 4$.*

The proof of the above theorem can be found in the full version [7].

5   Sensitivity of CDAWGs with Left-End Substitutions

We consider the worst-case additive sensitivity $\mathsf{AS}_{\mathrm{LeftSub}}(\mathsf{e}, n)$ of $\mathrm{CDAWG}(T)$ when $T[1]$ is substituted by a new character $b \neq T[1]$, i.e. $T' = bT[2..n]$.

5.1   Upper Bound for $\mathsf{AS}_{\mathrm{LeftSub}}(\mathsf{e}, n)$ on CDAWGs

Similarly to the case of insertions, we separate $\mathsf{e}(T') - \mathsf{e}(T)$ into the two following components $\mathsf{f}_{\mathrm{Sub}}(T)$ and $\mathsf{g}_{\mathrm{Sub}}(T)$ such that

- $\mathsf{f}_{\mathrm{Sub}}(T)$ is the total out-degrees of new nodes that appear in $\mathrm{CDAWG}(T')$;
- $\mathsf{g}_{\mathrm{Sub}}(T)$ is the total number of new out-going edges of nodes that already exist in $\mathrm{CDAWG}(T)$.

We regard a substitution as a sequence of a deletion and an insertion, i.e. two consecutive edit operations such that $aS \ (= T) \Rightarrow S \Rightarrow bS \ (= bT[2..n] = T')$.

$\mathsf{f}_{\mathrm{Sub}}(T)$: **Total Out-Degrees of New Nodes.** Let u be a new node in $\mathrm{CDAWG}(bS)$ that does not exist in $\mathrm{CDAWG}(aS)$, namely $u \in \mathsf{M}(bS)$ and $u \notin \mathsf{M}(aS)$. We categorize each new node u to the two following types u_1 and u_2 as:

1. $u_1 \in \mathsf{M}(T)$ so that u_1 is generated by deletion $aS \Rightarrow S$;
2. $u_2 \notin \mathsf{M}(T)$ so that u_2 is generated by insertion $S \Rightarrow bS$.

Node u_1 is a new node that appears in $\mathrm{CDAWG}(S)$. Thus, it follows from Lemma 7 that node au_1 exists in $\mathrm{CDAWG}(aS)$. Since u_2 is not a node in $\mathrm{CDAWG}(S)$, it follows from Lemma 1 that $u_2 = bx$ and x is a node in $\mathrm{CDAWG}(S)$. Based on the this observation, we will show that there is an injective mapping from the new nodes in $\mathrm{CDAWG}(bS) = \mathrm{CDAWG}(T')$ to the existing nodes in $\mathrm{CDAWG}(aS) = \mathrm{CDAWG}(T)$. In so doing, we must resolve the two non-injective situations where:

(i) a new node bx is generated by insertion $S \Rightarrow bS$, where x is generated by deletion $aS \Rightarrow S$ and x remains as a node in $\mathrm{CDAWG}(bS)$;
(ii) a new node bax generated by insertion $S \Rightarrow bS$, where x is generated by deletion $aS \Rightarrow S$ and x remains as a node in $\mathrm{CDAWG}(bS)$.

Suppose (on the contrary) that Case (i) happens. Then, a new node x is generated from an existing node ax, and bx is generated from x. Therefore, two new nodes could be generated from on existed node $ax \in \mathsf{M}(aS)$. However, the next lemma shows that this situation (Case (i)) does not occur unless $x = S$:

Lemma 8. *If $x \neq S$, $x \notin \mathsf{M}(aS)$, $x \in \mathsf{M}(S)$, and $x \in \mathsf{M}(bS)$, then $bx \notin \mathsf{M}(bS)$.*

Proof. Since $x \notin M(aS)$ and $x \in M(S)$, $x \in \mathsf{Prefix}(S)$ by Lemma 7. Since $x \in \mathsf{M}(S)$ and $ax \in \mathsf{Prefix}(aS)$, $ax \equiv^{\mathsf{L}}_{aS} x$ and $ax = \mathsf{long}([x]^{\mathsf{L}}_{aS})$. This means that bx occurs exactly once in bS as a proper prefix. Thus, $bx \notin \mathsf{RightM}(bS)$ which leads to $bx \notin \mathsf{M}(bS)$. □

As for Lemma 8, the situation (Case (i)) can occur if $x = S$. However, if $x = S$, then $S \in \mathsf{M}(bS)$ which implies that S occurs in bS as prefix $bS[1..(n-1)]$. Thus, $S = b^n$, $T = aS = ab^n$ and $T' = bS = b^{n+1}$. It is clear that $\mathsf{e}(aS) = \mathsf{e}(bS) = n+1$. Therefore the size of the CDAWG does not change when $x = S$.

Now we turn our attention to Case (ii) and assume (on the contrary) that it happens. Then, two new nodes bax and x could be generated from a single existing node ax. According to the following lemma, however, this situation cannot occur:

Lemma 9. *If $ax \in \mathsf{M}(aS)$, $x \notin \mathsf{M}(aS)$, $bax \notin \mathsf{M}(aS)$, $x \in \mathsf{M}(S)$, and $bax \notin \mathsf{M}(S)$, then $bax \notin \mathsf{M}(bS)$.*

Proof. Assume on the contrary that $bax \in \mathsf{M}(bS)$. Since $x \notin M(aS)$ and $x \in M(S)$, $x \in \mathsf{Prefix}(S)$ by Lemma 7. Also, since $bax \notin M(S)$ and $bax \in M(bS)$, $ax \in \mathsf{Prefix}(S)$ by Lemma 1. This means that $x \in \mathsf{Prefix}(ax)$ and $x = a^{|x|}$. Since $ax = a^{|x|+1}$ is a maximal substring of aS, x is also a maximal substring of aS. Thus $x \in \mathsf{M}(aS)$, however, this contradicts the precondition that $x \notin \mathsf{M}(aS)$. Thus $bax \notin \mathsf{M}(bS)$. □

As a result, there is an injective mapping from the new nodes u_1 (resp. $u_2 = bx$) in $\mathsf{CDAWG}(bS)$ to the existing nodes au_1 (resp. x) in $\mathsf{CDAWG}(aS)$ by Lemmas 1, 7, 8, and 9. It also follows from these lemmas that the out-degree of each new node in $\mathsf{CDAWG}(bS)$ does not exceed the maximum out-degree of $\mathsf{CDAWG}(aS)$. Finally, we consider the source ε. By Lemmas 2, 3, and 7, if $b \in \mathsf{M}(bS)$, $b \notin \mathsf{M}(aS)$, and $\mathsf{e}(aS) \geq 3$, then $\mathsf{d}_{bS}(b) \leq \mathsf{d}_{aS}(\varepsilon)$. Thus we have:

Lemma 10. *For any string T with $\mathsf{e}(T) \geq 3$, $\mathsf{f}_{\mathsf{Sub}}(T) \leq \mathsf{e}(T) - 1$.*

$\mathsf{g}_{\mathsf{Sub}}(T)$: **Number of New Branches from Existing Nodes.** Since left-end deletions do not create new branches from existing nodes (recall Sect. 4), it is immediate from Lemma 5 that:

Lemma 11. *For any string T, $\mathsf{g}_{\mathsf{Sub}}(T) \leq 1$.*

Wrapping Up. Our main result of this section follows from Lemmas 10 and 11:

Theorem 5. *For any $n \geq 4$ and $\mathsf{e} \geq 3$, $\mathsf{AS}_{\mathsf{LeftSub}}(\mathsf{e}, n) \leq \mathsf{e}$.*

5.2   Lower Bound for $\mathsf{AS}_{\mathsf{LeftSub}}(\mathsf{e}, n)$ on CDAWGs

Theorem 6. *There exists a family of strings T such that $\mathsf{e}(T') - \mathsf{e}(T) = \mathsf{e}(T) - 3$, where $T' = bT[2..n]$ with $b \in \Sigma \setminus \{T[1]\}$. Therefore $\mathsf{AS}_{\mathsf{LeftSub}}(\mathsf{e}, n) \geq \mathsf{e} - 3$.*

The proof of the above theorem can be found in the full version [7].

6 Quadratic-Time Bound for Leftward Online Construction

The leftward online construction problem for the CDAWG is, given a string T of length n, to maintain $\mathsf{CDAWG}(T[i..n])$ for decreasing $i = n, \ldots, 1$. By extending our lower bound on the sensitivity with left-end insertions/deletions, a quadratic bound for this online CDAWG construction follows:

Theorem 7. *There exists a family of strings T_m for which the total work for building $\mathsf{CDAWG}(T_m[i..n])$ for decreasing $i = n, \ldots, 1$ is $\Omega(n^2)$, where $n = |T_m|$.*

Proof. Consider string $T_m = (ab)^{2m}cab(ab)^{2m}\$$, where $a, b, c, \$ \in \Sigma$. For $0 \le k \le m$, let $T_{k,m}$ denote a series of suffixes of T_m such that $T_{k,m} = (ab)^{m+k}cab(ab)^{2m}\$$. Notice $T_{m,m} = T_m$, $m = \Theta(n)$ with $n = |T_{m,m}|$, and $T_{k,m} = T_m[2(m-k)+1..n]$.

Now, we consider building $\mathsf{CDAWG}(T_m[i..n])$ for decreasing $i = n, \ldots, 1$, and suppose we have already built $\mathsf{CDAWG}(T_{k,m})$. For this string $T_{k,m}$, we have that $\mathsf{M}(T_{k,m}) = \{\varepsilon, ab, (ab)^2, \ldots, (ab)^{2m}, T_{k,m}\}$. For any node v of $\mathsf{CDAWG}(T_{k,m}) = (\mathsf{V}_{T_{k,m}}, \mathsf{E}_{T_{k,m}})$, let $\mathsf{d}_{T_{k,m}}(v)$ denote the out-degree of v. Then, since $\mathsf{d}_{T_{k,m}}(\varepsilon) = 4$, $\mathsf{d}_{T_{k,m}}((ab)^i) = 3$ for every $1 \le i \le m+k$, $\mathsf{d}_{T_{k,m}}((ab)^j) = 2$ for every $m+k+1 \le j \le 2m$, and $\mathsf{d}_{T_{k,m}}(T_{k,m}) = 0$. Therefore $\mathsf{e}(T_{k,m}) = 5m + k + 4$.

Let us now prepend character b to $T_{k,m}$ and obtain $T_{k+1,m} = bT_{k,m} = b(ab)^{m+k}c(ab)^{2m}\$$. It is clear that $bT_{k,m} = T_{m,m}[2(m-k)..n]$. We have that

$$\mathsf{M}(bT_{k,m}) = \{\varepsilon, ab, (ab)^2, \ldots, (ab)^{2m}, b, bab, b(ab)^2, \ldots, b(ab)^{m+k}, bT_{k,m}\}$$
$$= (\mathsf{M}(T_{k,m}) \setminus \{T_{k,m}\}) \cup \{b, bab, b(ab)^2, \ldots, b(ab)^{m+k}\} \cup \{bT_{k,m}\},$$

and that $\mathsf{d}_{bT_{k,m}}(\varepsilon) = 4$, $\mathsf{d}_{bT_{k,m}}(b) = 3$, $\mathsf{d}_{bT_{k,m}}((ab)^i) = \mathsf{d}_{bT_{k,m}}(b(ab)^i) = 3$ for every $1 \le i \le m+k$, $\mathsf{d}_{bT_{k,m}}(b(ab)^j) = 2$ for every $m+k+1 \le j \le 2m$, and $\mathsf{d}_{bT_{k,m}}(bT_{k,m}) = 0$. Thus $\mathsf{e}(bT_{k,m}) = 8m + 4k + 7$. Therefore, building $\mathsf{CDAWG}(T_{k+1,m})$ from $\mathsf{CDAWG}(T_{k,m})$ requires to *add* $|\mathsf{e}(T_{k+1,m}) - \mathsf{e}(T_{k,m})| = 3m + 3k + 3 = \Omega(m)$ new edges (see the first step of Fig. 2 for illustration).

Let us move on to the next step, where we prepend character a to $bT_{k,m}$ and obtain $T_{k+1,m} = abT_{k,m} = ab(ab)^{m+k}c(ab)^{2m}\$$. Note that $abT_{k,m} = T_{k+1,m} = T_m[2(m-k)-1..n]$, and $\mathsf{M}(T_{k+1,m}) = \{\varepsilon, ab, (ab)^2, \ldots, (ab)^{2m}, T_{k+1,m}\}$. We also have $\mathsf{d}_{T_{k+1,m}}(\varepsilon) = 4$, $\mathsf{d}_{T_{k+1,m}}((ab)^i) = 3$ for every $1 \le i \le m+k+1$, $\mathsf{d}_{T_{k+1,m}}((ab)^j) = 2$ for every $m+k+2 \le j \le 2m$, and $\mathsf{d}_{T_{k+1,m}}(T_{k+1,m}) = 0$. This leads to $\mathsf{e}(T_{k+1,m}) = 5m + k + 5$. Therefore, building $\mathsf{CDAWG}(T_{k+1,m})$ from $\mathsf{CDAWG}(bT_{k,m})$ requires to *remove* $|\mathsf{e}(T_{k+1,m}) - \mathsf{e}(bT_{k,m})| = 3m + 3k + 2 = \Omega(m)$ existing edges (see the second step of Fig. 2 for illustration).

This process of adding and removing $\Omega(m)$ edges in every two steps repeats when we update $\mathsf{CDAWG}(T_{k,m})$ to $\mathsf{CDAWG}(T_{k+1,m})$ for every increasing $k = 1, \ldots, m-1$. Since $m = \Theta(n)$, the total work for building $\mathsf{CDAWG}(T_m[i..n])$ for decreasing $i = n, \ldots, 1$ is $\Omega(m^2) = \Omega(n^2)$.   □

Remark 1. The linear-time algorithm of [8] for *rightward* online CDAWG construction maintains a slightly modified version of the CDAWG, which becomes

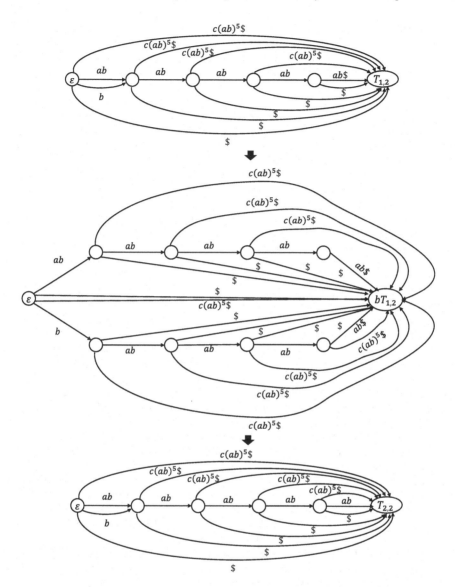

Fig. 2. Illustration for the CDAWGs of strings $T_{k,m} = (ab)^3cab(ab)^4\$$, $bT_{k,m} = b(ab)^3cab(ab)^4\$$, and $T_{k+1,m} = (ab)^4cab(ab)^4\$$ with $k = 1, m = 2$.

isomorphic to our CDAWG when a terminal symbol $ is appended to the string. Still, our lower bound instance from Theorem 7 shows that $ does not help improve the time complexity of *leftward* online CDAWG construction.

References

1. Akagi, T., Funakoshi, M., Inenaga, S.: Sensitivity of string compressors and repetitiveness measures. Inf. Comput. **291**, 104999 (2023). https://doi.org/10.1016/j.ic. 2022.104999
2. Belazzougui, D., Cunial, F.: Fast label extraction in the CDAWG. In: SPIRE 2017, pp. 161–175 (2017). https://doi.org/10.1007/978-3-319-67428-5_14
3. Belazzougui, D., Cunial, F., Gagie, T., Prezza, N., Raffinot, M.: Composite repetition-aware data structures. In: CPM 2015, pp. 26–39 (2015). https://doi. org/10.1007/978-3-319-19929-0_3
4. Blumer, A., Blumer, J., Haussler, D., McConnell, R., Ehrenfeucht, A.: Complete inverted files for efficient text retrieval and analysis. J. ACM **34**(3), 578–595 (1987). https://doi.org/10.1145/28869.28873
5. Burrows, M., Wheeler, D.J.: A block sorting lossless data compression algorithm. Tech. Rep. 124, Digital Equipment Corporation (1994)
6. Crochemore, M., Vérin, R.: On compact directed acyclic word graphs. In: Mycielski, Jan, Rozenberg, Grzegorz, Salomaa, Arto (eds.) Structures in Logic and Computer Science. LNCS, vol. 1261, pp. 192–211. Springer, Heidelberg (1997). https://doi. org/10.1007/3-540-63246-8_12
7. Fujimaru, H., Nakashima, Y., Inenaga, S.: On sensitivity of compact directed acyclic word graphs. https://doi.org/10.48550/arXiv.2303.01726 CoRR abs/ arXiv: 2303.01726 (2023)
8. Inenaga, S., et al.: On-line construction of compact directed acyclic word graphs. Discret. Appl. Math. **146**(2), 156–179 (2005). https://doi.org/10.1016/j.dam.2004. 04.012
9. Kempa, D., Prezza, N.: At the roots of dictionary compression: string attractors. In: STOC 2018, pp. 827–840 (2018). https://doi.org/10.1145/3188745.3188814
10. Kociumaka, T., Navarro, G., Prezza, N.: Towards a definitive measure of repetitiveness. In: Kohayakawa, Y., Miyazawa, F.K. (eds.) LATIN 2021. LNCS, vol. 12118, pp. 207–219. Springer, Cham (2020). https://doi.org/10.1007/978-3-030-61792-9_17
11. Radoszewski, J., Rytter, W.: On the structure of compacted subword graphs of Thue-Morse words and their applications. J. Dis. Algorithms **11**, 15–24 (2012). https://doi.org/10.1016/j.jda.2011.01.001
12. Senft, M., Dvorák, T.: Sliding CDAWG perfection. In: SPIRE 2008, pp. 109–120 (2008). https://doi.org/10.1007/978-3-540-89097-3_12
13. Takagi, T., Goto, K., Fujishige, Y., Inenaga, S., Arimura, H.: Linear-size CDAWG: New repetition-aware indexing and grammar compression. In: SPIRE 2017, pp. 304–316 (2017). https://doi.org/10.1007/978-3-319-67428-5_26
14. Takeda, M., Matsumoto, T., Fukuda, T., Nanri, I.: Discovering characteristic expressions from literary works: a new text analysis method beyond n-gram statistics and KWIC. In: Discovery Science 2000, pp. 112–126 (2000). https://doi.org/ 10.1007/3-540-44418-1_10
15. Weiner, P.: Linear pattern matching algorithms. In: Proceedings of the 14th Annual Symposium on Switching and Automata Theory, pp. 1–11. IEEE (1973). https:// doi.org/10.1109/SWAT.1973.13
16. Ziv, J., Lempel, A.: A universal algorithm for sequential data compression. IEEE Trans. Inf. Theory **23**(3), 337–343 (1977). https://doi.org/10.1109/TIT. 1977.1055714

Smallest and Largest Block Palindrome Factorizations

Daniel Gabric[1]([✉]) [ID] and Jeffrey Shallit[2] [ID]

[1] Department of Mathematics/Statistics, University of Winnipeg,
Winnipeg, MB R3B 2E9, Canada
`d.gabric@uwinnipeg.ca`
[2] School of Computer Science, University of Waterloo,
Waterloo, ON N2L 3G1, Canada
`shallit@uwaterloo.ca`

Abstract. A *palindrome* is a word that reads the same forwards and backwards. A *block palindrome factorization* (or *BP-factorization*) is a factorization of a word into blocks that becomes palindrome if each identical block is replaced by a distinct symbol. We call the number of blocks in a BP-factorization the *width* of the BP-factorization. The *largest BP-factorization* of a word w is the BP-factorization of w with the maximum width. We study words with certain BP-factorizations. First, we give a recurrence for the number of length-n words with largest BP-factorization of width t. Second, we show that the expected width of the largest BP-factorization of a word tends to a constant. Third, we give some results on another extremal variation of BP-factorization, the *smallest BP-factorization*. A *border* of a word w is a non-empty word that is both a proper prefix and suffix of w. Finally, we conclude by showing a connection between words with a unique border and words whose smallest and largest BP-factorizations coincide.

Keywords: Block palindrome · Factorization · Recurrence · Unbordered word · Unique border

1 Introduction

Let Σ_k denote the alphabet $\{0, 1, \ldots, k-1\}$. The length of a word w is denoted by $|w|$. A *border* of a word w is a non-empty word that is both a proper prefix and suffix of w. A word is said to be *bordered* if it has a border. Otherwise, the word is said to be *unbordered*. For example, the French word `entente` is bordered, and has two borders, namely `ente` and `e`.

It is well-known [1] that the number u_n of length-n unbordered words over Σ_k satisfies

$$u_n = \begin{cases} 1, & \text{if } n = 0; \\ ku_{n-1} - u_{n/2}, & \text{if } n > 0 \text{ is even}; \\ ku_{n-1}, & \text{if } n \text{ is odd}. \end{cases} \qquad (1)$$

© The Author(s), under exclusive license to Springer Nature Switzerland AG 2023
A. Frid and R. Mercaş (Eds.): WORDS 2023, LNCS 13899, pp. 181–191, 2023.
https://doi.org/10.1007/978-3-031-33180-0_14

A *palindrome* is a word that reads the same forwards as it does backwards. More formally, letting $w^R = w_n w_{n-1} \cdots w_1$ where $w = w_1 w_2 \cdots w_n$ and all w_i are symbols, a palindrome is a word w such that $w = w^R$. The definition of a palindrome is quite restrictive. The second half of a palindrome is fully determined by the first half. Thus, compared to all length-n words, the number of length-n palindromes is vanishingly small. But many words exhibit palindrome-like structure. For example, take the English word `marjoram`. It is clearly not a palindrome, but it comes close. Replacing the block `jo` with a single letter turns the word into a palindrome. In this paper, we consider a generalization of palindromes that incorporates this kind of palindromic structure.

In the 2015 British Olympiad [2], the concept of a block palindrome factorization was first introduced. Let w be a non-empty word. A *block palindrome factorization* (or *BP-factorization*) of w is a factorization $w = w_m \cdots w_1 w_0 w_1 \cdots w_m$ of a word such that w_0 is a possibly empty word, and every other factor w_i is non-empty for all i with $1 \le i \le m$. We say that a BP-factorization $w_m \cdots w_1 w_0 w_1 \cdots w_m$ is of *width* t where $t = 2m + 1$ if w_0 is non-empty and $t = 2m$ otherwise. In other words, the width of a BP-factorization is the number of non-empty blocks in the factorization. The *largest BP-factorization*[1] [3] of a word w is a BP-factorization $w = w_m \cdots w_1 w_0 w_1 \cdots w_m$ where m is maximized (i.e., where the width of the BP-factorization is maximized). See [4,5] for more on the topic of BP-factorizations and block reversals. Kolpakov and Kucherov [6] studied a special case of BP-factorizations, the *gapped palindrome*. If w_0 is non-empty and $|w_i| = 1$ for all i with $1 \le i \le m$, then w is said to be a *gapped palindrome*. Régnier [7] studied something similar to BP-factorizations, but in her paper she was concerned with borders of borders. See [8,9] for results on factoring words into palindromes.

Example 1. We use the centre dot · to denote the separation between blocks in the BP-factorization of a word.

Consider the word `abracadabra`. It has the following BP-factorizations:

$$\text{abracadabra},$$
$$\text{abra} \cdot \text{cad} \cdot \text{abra},$$
$$\text{a} \cdot \text{bracadabr} \cdot \text{a},$$
$$\text{a} \cdot \text{br} \cdot \text{acada} \cdot \text{br} \cdot \text{a},$$
$$\text{a} \cdot \text{br} \cdot \text{a} \cdot \text{cad} \cdot \text{a} \cdot \text{br} \cdot \text{a}.$$

The last BP-factorization is of width 7 and has the longest width; thus it is the largest BP-factorization of `abracadabra`.

Let w be a length-n word. Suppose $w_m \cdots w_1 w_0 w_1 \cdots w_m$ is the largest BP-factorization of w. Goto et al. [3] showed that w_i is the shortest border of $w_i \cdots w_1 w_0 w_1 \cdots w_i$ where $i \ge 1$. This means that we can compute the largest

---

[1] Largest BP-factorizations also appear in https://www.reddit.com/r/math/comments/ga2iyo/ i_just_defined_the_palindromity_function_on/.

BP-factorization of w by greedily "peeling off" the shortest borders of central factors until you hit an unbordered word or the empty word.

The rest of the paper is structured as follows. In Sect. 2 we give a recurrence for the number of length-n words with largest BP-factorization of width t. In Sect. 3 we show that the expected width of the largest BP-factorization of a length-n word tends to a constant. In Sect. 4 we consider *smallest BP-factorizations* in the sense that one "peels off" the longest non-overlapping border. We say a border u of a word w is *non-overlapping* if $|u| \leq |w|/2$; otherwise u is *overlapping*. Finally, in Sect. 5 we present some results on words with a unique border and show that they are connected to words whose smallest and largest BP-factorizations are the same.

2   Counting Largest BP-Factorizations

In this section, we prove a recurrence for the number $\mathrm{LBP}_k(n,t)$ of length-n words over Σ_k with largest BP-factorization of width t. See Table 1 for sample values of $\mathrm{LBP}_2(n,t)$ for small n, t. For the following theorem, recall the definition of u_n from Eq. 1.

Theorem 1. *Let $n, t \geq 0$, and $k \geq 2$ be integers. Then*

$$\mathrm{LBP}_k(n,t) = \begin{cases} \sum_{i=1}^{(n-t)/2+1} u_i\,\mathrm{LBP}_k(n-2i, t-2), & \text{if } n,\ t \text{ even}; \\ \sum_{i=1}^{(n-t+1)/2} u_{2i}\,\mathrm{LBP}_k(n-2i, t-1), & \text{if } n \text{ even}, t \text{ odd}; \\ 0, & \text{if } n \text{ odd, } t \text{ even}; \\ \sum_{i=1}^{(n-t)/2+1} u_{2i-1}\,\mathrm{LBP}_k(n-2i+1, t-1), & \text{if } n,\ t \text{ odd}. \end{cases}$$

where

$$\mathrm{LBP}_k(0,0) = 1,$$
$$\mathrm{LBP}_k(2n,2) = u_n,$$
$$\mathrm{LBP}_k(n,1) = u_n.$$

Proof. Let w be a length-n word whose largest BP-factorization $w_m \cdots w_1 w_0 w_1 \cdots w_m$ is of width t. Clearly $\mathrm{LBP}_k(0,0) = 1$. We know that each block in a largest BP-factorization is unbordered, since each block is a shortest border of some central factor. This immediately implies $\mathrm{LBP}_k(n,1) = u_n$ and $\mathrm{LBP}_k(2n,2) = u_n$.

Now we take care of the other cases.

– Suppose n, t are even. Then by removing both instances of w_1 from w, we get $w' = w_m \cdots w_2 w_2 \cdots w_m$, which is a length-$(n - 2|w_1|)$ word whose largest BP-factorization is of width $t - 2$. This mapping is clearly reversible, since all blocks in a largest BP-factorization are unbordered, including w_1. Thus summing over all possible w_1 and all length-$(n - 2|w_1|)$ words with largest BP-factorization of width $t - 2$ we have

$$\mathrm{LBP}_k(n,t) = \sum_{i=1}^{(n-t)/2+1} u_i\,\mathrm{LBP}_k(n-2i, t-2).$$

- Suppose n is even and t is odd. Then by removing w_0 from w, we get $w' = w_m \cdots w_1 w_1 \cdots w_m$, which is a length-$(n - |w_0|)$ word whose largest BP-factorization is of width $t - 1$. This mapping is reversible for the same reason as in the previous case. The word w' is of even length since $|w'| = 2|w_1 \cdots w_m|$. Since n is even and $|w'|$ is even, we must have that $|w_0|$ is even as well. Thus summing over all possible w_0 and all length-$(n - |w_0|)$ words with largest BP-factorization of width $t - 1$, we have

$$\mathrm{LBP}_k(n, t) = \sum_{i=1}^{(n-t+1)/2} u_{2i} \, \mathrm{LBP}_k(n - 2i, t - 1).$$

- Suppose n is odd and t is even. Then the length of w is $2|w_1 \cdots w_m|$, which is even, a contradiction. Thus $\mathrm{LBP}_k(n, t) = 0$.
- Suppose n, t are odd. Then by removing w_0 from w, we get $w' = w_m \cdots w_1 w_1 \cdots w_m$, which is a length-$(n - |w_0|)$ word whose largest BP-factorization is of width $t - 1$. This mapping is reversible for the same reasons as in the previous cases. Since n is odd and $|w'|$ is even (proved in the previous case), we must have that $|w_0|$ is odd. Thus summing over all possible w_0 and all length-$(n - |w_0|)$ words with largest BP-factorization of width $t - 1$, we have

$$\sum_{i=1}^{(n-t)/2+1} u_{2i-1} \, \mathrm{LBP}_k(n - 2i + 1, t - 1).$$

Table 1. Some values of $\mathrm{LBP}_2(n, t)$ for n, t where $10 \le n \le 20$ and $1 \le t \le 10$.

n	t									
	1	2	3	4	5	6	7	8	9	10
10	284	12	224	40	168	72	96	64	32	32
11	568	0	472	0	416	0	336	0	192	0
12	1116	20	856	88	656	176	448	224	224	160
13	2232	0	1752	0	1488	0	1248	0	896	0
14	4424	40	3328	176	2544	432	1856	640	1152	640
15	8848	0	6736	0	5440	0	4576	0	3584	0
16	17622	74	13100	372	9896	984	7408	1744	5088	2080
17	35244	0	26348	0	20536	0	16784	0	13664	0
18	70340	148	51936	760	38824	2248	29152	4416	21088	6240
19	140680	0	104168	0	79168	0	62800	0	51008	0
20	281076	284	206744	1592	153344	4992	114688	10912	84704	17312

3 Expected Width of Largest BP-Factorization

In this section, we show that the expected width $E_{n,k}$ of the largest BP-factorization of a length-n word over Σ_k is bounded by a constant. From the definition of expected value, it follows that

$$E_{n,k} = \frac{1}{k^n} \sum_{i=1}^{n} i \cdot \text{LBP}_k(n,i).$$

Table 2 shows the behaviour of $\lim\limits_{n \to \infty} E_{n,k}$ as k increases.

Lemma 1. *Let $k \geq 2$ and $n \geq t \geq 1$ be integers. Then*

$$\frac{\text{LBP}_k(n,t)}{k^n} \leq \frac{1}{k^{t/2-1}}.$$

Proof. Let w be a length-n word whose largest BP-factorization $w_m \cdots w_1 w_0 w_1 \cdots w_m$ is of width t. Since w_i is non-empty for every $1 \leq i \leq m$, we have that $\text{LBP}_k(n,t) \leq k^{n-m} \leq k^{n-t/2+1}$. So

$$\frac{\text{LBP}_k(n,t)}{k^n} \leq \frac{1}{k^{t/2-1}}$$

for all $n \geq t \geq 1$.

Theorem 2. *The limit $E_k = \lim\limits_{n \to \infty} E_{n,k}$ exists for all $k \geq 2$.*

Proof. Follows from the definition of $E_{n,k}$, Lemma 1, and the direct comparison test for convergence.

Interpreting E_k as a power series in k^{-1}, we empirically observe that E_k is approximately equal to

$$1 + \frac{2}{k} + \frac{4}{k^2} + \frac{6}{k^3} + \frac{10}{k^4} + \frac{16}{k^5} + \frac{24}{k^6} + \frac{38}{k^7} + \frac{58}{k^8} + \frac{88}{k^9} + \cdots.$$

We conjecture the following about E_k.

Conjecture 1. Let $k \geq 2$. Then

$$E_k = 1 + \sum_{i=1}^{\infty} a_i k^{-i}$$

where the sequence $(a_n/2)_{n \geq 1}$ is <u>A274199</u> in the *On-Line Encyclopedia of Integer Sequences* (OEIS) [10].

Table 2. Asymptotic expected width of a word's largest BP-factorization.

k	$\approx E_k$
2	6.4686
3	2.5908
4	1.9080
5	1.6314
6	1.4827
7	1.3902
8	1.3272
9	1.2817
10	1.2472
$\vdots$	$\vdots$
100	1.0204

Cording et al. [11] proved that the expected length of the longest unbordered factor in a word is $\Theta(n)$. Taking this into account, it is not surprising that the expected length of the largest BP-factorization of a word tends to a constant.

4   Smallest BP-Factorization

A word w, seen as a block, clearly satisfies the definition of a BP-factorization. Thus, taken literally, the smallest BP-factorization for all words is of width 1. But this is not very interesting, so we consider a different definition instead. A border u of a word w is *non-overlapping* if $|u| \leq |w|/2$; otherwise u is *overlapping*. We say that the *smallest BP-factorization* of a word w is a BP-factorization $w = w_m \cdots w_1 w_0 w_1 \cdots w_m$ where each w_i is the longest non-overlapping border of $w_i \cdots w_1 w_0 w_1 \cdots w_i$, except w_0, which is either empty or unbordered. For example, going back to Example 1, the smallest BP-factorization of abracadabra is abra·cad·abra and the smallest BP-factorization of reappear is r·ea·p·p·ea·r.

A natural question to ask is: what is the maximum possible width $f_k(n)$ of the smallest BP-factorization of a length-n word? Through empirical observation, we arrive at the following conjectures:

- We have $f_2(8n+i) = 6n+i$ for i with $0 \leq i \leq 5$ and $f_2(8n+6) = f_2(8n+7) = 6n + 5$.
- We have $f_k(n) = n$ for $k \geq 3$.

To calculate $f_k(n)$, two things are needed: an upper bound on $f_k(n)$, and words that witness the upper bound.

Theorem 3. *Let $l \geq 0$ be an integer. Then $f_2(8l+i) = 6l+i$ for i with $0 \leq i \leq 5$ and $f_2(8l + 6) = f_2(8l + 7) = 6l + 5$.*

Proof. Let $n \geq 0$ be an integer. We start by proving lower bounds on $f_2(n)$. Suppose $n = 8l$ for some $l \geq 0$. Then the width of the smallest BP-factorization of

$$(0101)^l(1001)^l$$

is $6l$, so $f_2(8l) \geq 6l$. To see this, notice that the smallest BP-factorization of 01011001 is $01 \cdot 0 \cdot 1 \cdot 1 \cdot 0 \cdot 01$, and therefore is of width 6. Suppose $n = 8l + i$ for some i with $1 \leq i \leq 7$. Then one can take $(0101)^l(1001)^l$ and insert either 0, 00, 010, 0110, 01010, 010110, or 0110110 to the middle of the word to get the desired length.

Now we prove upper bounds on $f_2(n)$. Let $t \leq n$ be a positive integer. Let w be a length-n word whose largest BP-factorization $w_m \cdots w_1 w_0 w_1 \cdots w_m$ is of width t. One can readily verify that $f_2(0) = 0$, $f_2(1) = 1$, $f_2(2) = 2$, $f_2(3) = 3$, $f_2(4) = 4$, and $f_2(5) = f_2(6) = f_2(7) = 5$ through exhaustive search of all binary words of length < 8. Suppose $m \geq 4$, so $n \geq t \geq 8$. Then we can write $w = w_m w_{m-1} w_{m-2} \cdots w_{m-2} w_{m-1} w_m$ where $|w_{m-2}|, |w_{m-1}|, |w_m| > 0$. It is easy to show that $|w_{m-2} w_{m-1} w_m| \geq 4$ by checking that all binary words of length < 8 do not admit a smallest BP-factorization of width 6. In the worst case, we can peel off prefixes and suffixes of length 4 while accounting for the 6 blocks they add to the BP-factorization until we hit the middle core of length < 8. Thus, we have $f_2(8l + i) \leq 6l + j$ where j is the width of the smallest BP-factorization of the middle core, which is of length i. We have already computed $f_2(i)$ for $0 \leq i \leq 7$, so the upper bounds follow.

Theorem 4. *Let $n \geq 0$ and $k \geq 3$ be integers. Then $f_k(n) = n$.*

Proof. Clearly $f_k(n) \leq n$. We prove $f_k(n) \geq n$. If n is divisible by 6, then consider the word $(012)^{n/6}(210)^{n/6}$. If n is not divisible by 6, then take $(012)^{\lfloor n/6 \rfloor}(210)^{\lfloor n/6 \rfloor}$ and insert either 0, 00, 010, 0110, or 01010 in the middle of the word. When calculating the smallest BP-factorization of the resulting words, it is easy to see that at each step we are removing a border of length 1. Thus, their largest BP-factorization is of width n.

5 Equal Smallest and Largest BP-Factorizations

Recall back to Example 1, that `abracadabra` has distinct smallest and largest BP-factorizations, namely `abra·cad·abra` and `a·br·a·cad·a·br·a`. However, the word `alfalfa` has the same smallest and largest BP-factorizations, namely `a·lf·a·lf·a`. Under what conditions are the smallest and largest BP-factorizations of a word the same? Looking at unique borders seems like a good place to start, since the shortest border and longest non-overlapping border coincide when a word has a unique border. However, the converse is not true—just consider the previous example `alfalfa`. The shortest border and longest non-overlapping border are both `a`, but `a` is not a unique border of `alfalfa`.

In Theorem 5 we characterize all words whose smallest and largest BP-factorization coincide.

Theorem 5. *Let* $m, m' \geq 1$ *be integers. Let* w *be a word with smallest BP-factorization* $w'_{m'} \cdots w'_1 w'_0 w'_1 \cdots w'_{m'}$ *and largest BP-factorization* $w_m \cdots w_1 w_0 w_1 \cdots w_m$. *Then* $m = m'$ *and* $w_i = w'_i$ *for all* i, $0 \leq i \leq m$ *if and only if for all* $i \neq 2$, $0 < i \leq m$, *we have that* w_i *is the unique border of* $w_i \cdots w_1 w_0 w_1 \cdots w_i$ *and for* $i = 2$ *we have that either*

1. w_2 *is the unique border of* $w_2 w_1 w_0 w_1 w_2$, *or*
2. $w_2 w_1 w_0 w_1 w_2 = w_0 w_1 w_0 w_1 w_0$ *where* w_0 *is the unique border of* $w_0 w_1 w_0$.

Proof.

$\Longrightarrow$: Let i be an integer such that $0 < i \leq m$. Let $u_i = w_i \cdots w_1 w_0 w_1 \cdots w_i$. Since w_i is both the shortest border and longest non-overlapping border of u_i (i.e., $w_i = w'_i$), we have that u_i has exactly one border of length $\leq |u_i|/2$. Thus, either w_i is the unique border of u_i, or u_i has a border of length $> |u_i|/2$. If w_i is the unique border of u_i, then we are done. So suppose that u_i has a border of length $> |u_i|/2$. Let v_i be the shortest such border. We have that w_i is both a prefix and suffix of v_i. In fact, w_i must be the unique border of v_i. Otherwise we contradict the minimality of v_i, or the assumption that w_i is both the shortest border and longest non-overlapping border of u_i. Since w_i is unbordered, it cannot overlap itself in v_i and w_i. So we can write $v_i = w_i y w_i$ for some word y where $u_i = w_i y w_i y w_i$, or $u_i = w_i x w_i x' w_i x'' w_i$ such that $y = x w_i x' = x' w_i x''$. If $u_i = w_i x w_i x' w_i x'' w_i$, then we see that $w_i x'$ is a suffix of y and $x' w_i$ is a prefix of y, implying that $w_i x' w_i$ is a new smaller border of u_i. This either contradicts the assumption that v_i is the shortest border of length $> |u_i|/2$, or the assumption that u_i has exactly one border of length $\leq |u_i|/2$. Thus, we have that $u_i = w_i y w_i y w_i$. The shortest border and longest non-overlapping border of $y w_i y$ must be y, by assumption. Additionally, w_i is unbordered, so u_i is of width 5 and $i = 2$. This implies that $w_i = w_2 = w_0$ and $y = w_1$.

$\Longleftarrow$: Let i be an integer such that $0 < i \leq m$. We omit the case when $i = 0$, since proving $w_i = w'_i$ for all other i is sufficient. Since w_i is the unique border of $u_i = w_i \cdots w_1 w_0 w_1 \cdots w_i$, we have that the shortest border and longest non-overlapping border of u_i is w_i. In other words, we have that $w_i = w'_i$. Suppose $i = 2$ and $u_2 = w_2 w_1 w_0 w_1 w_2 = w_0 w_1 w_0 w_1 w_0$ where w_0 is the unique border of $w_0 w_1 w_0$. Since w_0 is the unique border of $w_0 w_1 w_0$, it is also the shortest border of u_2. Additionally, the next longest border of u_2 is $w_0 w_1 w_0$, which is overlapping. So w_0 is also the longest non-overlapping border of u_2. Thus $w_2 = w'_2$.

Just based on this characterization, finding a recurrence for the number of words with a coinciding smallest and largest BP-factorization seems hard. So we turn to a different, related problem: counting the number of words with a unique border.

5.1   Unique Borders

Harju and Nowotka [12] counted the number $B_k(n)$ of length-n words over Σ_k with a unique border, and the number $B_k(n, t)$ of length-n words over Σ_k with a length-t unique border. However, through personal communication with the authors, a small error in one of the proofs leading up to their formula for $B_k(n,t)$ was discovered. Thus, the formula for $B_k(n,t)$ as stated in their paper is incorrect. In this section, we present the correct recurrence for the number of length-n words with a length-t unique border. We also show that the probability a length-n word has a unique border tends to a constant. See A334600 in the OEIS [10] for the sequence $(B_2(n))_{n\geq 0}$.

Suppose w is a word with a unique border u. Then u must be unbordered, and $|u|$ must not exceed half the length of w. If either of these were not true, then w would have more than one border. By combining these ideas, we get Theorem 6 and Theorem 7.

Theorem 6. *Let $n > t \geq 1$ be integers. Then the number of length-n words with a unique length-t border satisfies the recurrence*

$$B_k(n,t) = \begin{cases} 0, & \text{if } n < 2t; \\ u_t k^{n-2t} - \sum_{i=2t}^{\lfloor n/2 \rfloor} B_k(i,t) k^{n-2i}, & \text{if } n \geq 2t \text{ and } n+t \text{ odd}; \\ u_t k^{n-2t} - B_k((n+t)/2, t) - \sum_{i=2t}^{\lfloor n/2 \rfloor} B_k(i,t) k^{n-2i}, & \text{if } n \geq 2t \text{ and } n+t \text{ even}. \end{cases}$$

Proof. Let w be a length-n word with a unique length-t border u. Since u is the unique border of w, it is unbordered. Thus, we can write $w = uvu$ for some (possibly empty) word v. For $n < 2t$, we have that $B_k(n,t) = 0$ since u is unbordered and thus cannot overlap itself in w.

Suppose $n \geq 2t$. Let $\overline{B_k}(n,t)$ denote the number of length-n words that have a length-t unbordered border and have a border of length $> t$. Clearly $B_k(n,t) = u_t k^{n-2t} - \overline{B_k}(n,t)$. Suppose w has another border u' of length $> t$. Furthermore, suppose that there is no other border u'' with $|u| < |u''| < |u'|$. Then u is the largest border of u'. Since u is the shortest border, we have $|u| \leq n/2$. But we could possibly have $|u'| > n/2$. The only possible way for $|u'|$ to exceed $n/2$ is if $w = uv'uv'u$ for some (possibly empty) word v. But this is only possible if $n+t$ is even; otherwise we cannot place u in the centre of w. When $n+t$ is odd, we compute $\overline{B_k}(n,t)$ by summing over all possibilities for u' (i.e., $2t \leq |u'| \leq \lfloor n/2 \rfloor$) and the middle part of w (i.e., v'' where $w = u'v''u'$). This gives us the recurrence,

$$\overline{B_k}(n,t) = \sum_{i=2t}^{\lfloor n/2 \rfloor} B_k(i,t) k^{n-2i}.$$

When $n+t$ is even, we compute $\overline{B_k}(n,t)$ in the same fashion, except we also include the case where $|u'| = (n+t)/2$. This gives us the recurrence,

$$\overline{B_k}(n,t) = B_k((n+t)/2, t) + \sum_{i=2t}^{\lfloor n/2 \rfloor} B_k(i,t) k^{n-2i}.$$

Theorem 7. *Let $n \geq 2$ be an integer. Then the number of length-n words with a unique border is*

$$B_k(n) = \sum_{t=1}^{\lfloor n/2 \rfloor} B_k(n,t).$$

5.2   Limiting Values

We show that the probability that a random word of length n has a unique border tends to a constant. Table 3 shows the behaviour of this probability as k increases.

Let $P_{n,k}$ be the probability that a random word of length n has a unique border. Then

$$P_{n,k} = \frac{B_k(n)}{k^n} = \frac{1}{k^n} \sum_{i=1}^{\lfloor n/2 \rfloor} B_k(n,i).$$

Lemma 2. *Let $k \geq 2$ and $n \geq 2t \geq 2$ be integers. Then*

$$\frac{B_k(n,t)}{k^n} \leq \frac{1}{k^t}.$$

Proof. Let w be a length-n word. Suppose w has a unique border of length t. Since $t \leq n/2$, we can write $w = uvu$ for some words u and v where $|u| = t$. But this means that $B_k(n,t) \leq k^{n-t}$, and the lemma follows.

Theorem 8. *Let $k \geq 2$ be an integer. Then the limit $P_k = \lim_{n \to \infty} P_{n,k}$ exists.*

Proof. Follows from the definition of $P_{n,k}$, Lemma 2, and the direct comparison test for convergence.

Table 3. Probability that a word has a unique border.

k	$\approx P_k$
2	0.5155
3	0.3910
4	0.2922
5	0.2302
6	0.1890
7	0.1599
8	0.1384
9	0.1219
10	0.1089
$\vdots$	$\vdots$
100	0.0101

References

1. Nielsen, P.T.: A note on bifix-free sequences. IEEE Trans. Inform. Theory, **IT-19**, 704–706 (1973)
2. The 2015 British Informatics Olympiad (Round 1 Question 1). https://olympiad.org.uk/2015/index.html
3. Goto, K., Tomohiro, I., Bannai, H., Inenaga, S.: Block palindromes: a new generalization of palindromes. In: Gagie, T., Moffat, A., Navarro, G., Cuadros-Vargas, E. (eds.) SPIRE 2018. LNCS, vol. 11147, pp. 183–190. Springer, Cham (2018). https://doi.org/10.1007/978-3-030-00479-8_15
4. Mahalingam, K., Maity, A., Pandoh, P., Raghavan, R.: Block reversal on finite words. Theoret. Comput. Sci. **894**, 135–151 (2021)
5. Mahalingam, K., Maity, A., Pandoh, P.: Rich words in the block reversal of a word. Discrete Appl. Math. **334**, 127–138 (2023). https://doi.org/10.1016/j.dam.2023.03.013
6. Kolpakov, R., Kucherov, G.: Searching for gapped palindromes. Theoret. Comput. Sci. **410**(51), 5365–5373 (2009)
7. Régnier, M.: Enumeration of bordered words, le langage de la vache-qui-rit. RAIRO-Theor. Inf. Appl. **26**(4), 303–317 (1992)
8. Frid, A.E., Puzynina, S., Zamboni, L.Q.: On palindromic factorization of words. Adv. in Appl. Math. **50**(5), 737–748 (2013)
9. Ravsky, O.: On the palindromic decomposition of binary words. J. Autom. Lang. Comb. **8**(1), 75–83 (2003)
10. Sloane, N.J.A., et al.: OEIS Foundation Inc. The On-Line Encyclopedia of Integer Sequences (2022). https://oeis.org
11. Cording, P.H., Gagie, T., Knudsen, M.B.T., Kociumaka, T.: Maximal unbordered factors of random strings. Theoret. Comput. Sci. **852**, 78–83 (2021)
12. Harju, T., Nowotka, D.: Counting bordered and primitive words with a fixed weight. Theoret. Comput. Sci. **340**(2), 273–279 (2005)

String Attractors of Fixed Points of k-Bonacci-Like Morphisms

France Gheeraert[1]([✉]) [iD], Giuseppe Romana[2] [iD], and Manon Stipulanti[1] [iD]

[1] Department of Mathematics, University of Liège, Liège, Belgium
{france.gheeraert,m.stipulanti}@uliege.be
[2] Dipartimento di Matematica e Informatica, Università di Palermo, Palermo, Italy
giuseppe.romana01@unipa.it

Abstract. Firstly studied by Kempa and Prezza in 2018 as the cement of text compression algorithms, string attractors have become a compelling object of theoretical research within the community of combinatorics on words. In this context, they have been studied for several families of finite and infinite words. In this paper, we obtain string attractors of prefixes of particular infinite words generalizing k-bonacci words (including the famous Fibonacci word) and obtained as fixed points of k-bonacci-like morphisms. In fact, our description involves the numeration systems classically derived from the considered morphisms.

Keywords: Morphic sequences · Fibonacci word · Numeration systems · String attractors · String attractor profile function · Parry numbers

1 Introduction

Introduced in the data compression field by Kempa and Prezza [18], the concept of *string attractor* can be conceptualized as follows: it is a set of positions within a finite word that enables to catch all distinct factors. String attractors also have applications in combinatorial pattern matching [8], but the problem of finding a smallest string attractor is NP-hard [18]. However, as combinatorial properties of words yield new strategies to find a string attractor of minimum size, string attractors have become a systematic topic of research. Indeed, Sturmian words can be characterized through the structure of their smallest string attractors [22, 24], and for the ubiquitous Thue–Morse word [19,27] and of the period-doubling word [27], the corresponding *profile function* (giving the minimal size of a string attractor for each prefix) has been studied.

The story of the current work began during the international conference DLT 2022, where the three authors had the chance to meet for the first time and where they talked about the concept of string attractors. Romana's expertise

© The Author(s), under exclusive license to Springer Nature Switzerland AG 2023
A. Frid and R. Mercaş (Eds.): WORDS 2023, LNCS 13899, pp. 192–205, 2023.
https://doi.org/10.1007/978-3-031-33180-0_15

lead us to consider them for prefixes of generalized Fibonacci words to larger alphabets (on k letters, the corresponding word is called the k-*bonacci word*), as a natural extension of Sturmian words. It turned out that the string attractors that we obtained rely on the well-known k-bonacci numbers [15]. Simultaneously, Dvořáková studied string attractors of factors of episturmian words [11], which covers the case of all k-bonacci words. However, her description is less explicit.

Moreover, the fact that minimal string attractors of prefixes of the k-bonacci word can be described using the k-bonacci numbers tipped us on the probable link between string attractors and numeration systems, and lead us to believe that this bond can be adapted to other morphic sequences. More specifically, we have the following general question:

Question. Given a morphic sequence **z**, does there exist a numeration system $\mathcal{S}$ such that **z** is $\mathcal{S}$-automatic and (minimal) string attractors of the prefixes of **z** are easily described using $\mathcal{S}$?

In this paper, as a first step towards answering this question, we study a particular family of morphic words. More precisely, given parameters in the shape of a length-k word $c = c_0 \cdots c_{k-1} \in \mathbb{N}^k$, we define the morphism μ_c such that $\mu_c(i) = 0^{c_i} \cdot (i+1)$ for all $0 \le i \le k-2$ and $\mu_c(k-1) = 0^{c_{k-1}}$. When it exists, we then look at the fixed point of this morphism. This family was not randomly chosen. First, it generalizes the k-bonacci morphisms but the fixed points are not necessarily episturmian. In addition, some of these morphisms have already been studied in relation to numeration systems, in [12] for example. Indeed, if c is some β-representation of 1 for a simple Parry number β, using the terminology of [5], we can canonically associate a numeration system that is greedy and, in this case, corresponds to the sequence $(|\mu_c^n(0)|)_{n \in \mathbb{N}}$ of lengths of iterations of μ_c on 0 [3]. Under some conditions on the parameters, we show that the prefixes of the fixed point admit string attractors of size at most $k+1$ described using the associated numeration system.

This paper is organized as follows. In Sect. 2, we recall some background on combinatorics on words. We also introduce the infinite words that we will study and give some of their basic properties. Sect. 3 introduces numeration systems and explains how to associate one with a morphic word. After that, we give conditions on the parameters $c_0, \ldots, c_{k-1}$ for the numeration system to have "desirable" properties. Finally, in Sect. 4, we look at string attractors and prove the main result of this paper, namely the description of string attractors of each prefix of the studied infinite words using the associated numeration system. As a consequence, we obtain an upper bound on the string attractor profile function. We present concluding remarks and future work in Sect. 5.

2 Preliminaries

2.1 Words

We start with the bare minimum on words and introduce some notations.

Let A be an alphabet either finite or infinite (for instance, we will consider words over the set of non-negative integers $\mathbb{N}$). The length of a word is its number of letters and will be denoted with vertical bars $|\cdot|$. We let ε denote the empty word. We let A^* denote the set of finite words over A. For any integer $n \geq 0$, we let A^n be the set of length-n words over A. If $w = xyz$ for some $x, y, z \in A^*$, then x is a *prefix*, y is a *factor*, and z is a *suffix* of w. A factor of a word is *proper* if it is not equal to the initial word. A word v is a *fractional power* of a non-empty word w if there exist $\ell \in \mathbb{N}$ and x a prefix of w such that $v = w^\ell x$. We will then write $v = w^{|v|/|w|}$. Infinite words are written in bold and we start indexing them at 0. We use classical notations of intervals to denote portions of words. For a non-empty word $u \in A^*$, we let u^ω denote the concatenation of infinitely many copies of u, that is, $u^\omega = uuu \cdots$.

Let $\leq$ be a total order on A. The *lexicographic order* on A^* induced by $\leq$ is defined as follows: for $x, y \in A^*$, we say that x is *lexicographically smaller than* y, and we write $x < y$, if either x is a proper prefix of y, or $x = zax'$ and $y = zby'$ for some letters a, b with $a < b$. We write $x \leq y$ if x is lexicographically smaller than or equal to y. The *genealogical order*, also known as *radix order*, on A^* induced by $\leq$ is defined as follows: for $x, y \in A^*$, we say that x is *genealogically smaller than* y, and we write $x <_{\text{gen}} y$, if either $|x| < |y|$, or $|x| = |y|$ and $x = zax'$ and $y = zby'$ for some letters a, b with $a < b$. We write again $x \leq_{\text{gen}} y$ if x is genealogically smaller than or equal to y.

A non-empty word $w \in A^*$ is *primitive* if $w = u^n$ for $u \in A^* \setminus \{\varepsilon\}$ implies $n = 1$. Two words are *conjugates* if they are cyclic permutation of each other.

A word is *Lyndon* if it is primitive and lexicographically minimal among its conjugates for some given order. Defined in the 50's, Lyndon words are not only classical in combinatorics on words but also of utmost importance. See [21] for a presentation. A celebrated result in combinatorics on words is that London words form a so-called *complete factorization of the free monoid*.

Theorem 1 (Chen-Fox-Lyndon [7]). *For every non-empty word $w \in A^*$, there exists a unique factorization $(\ell_1, \cdots, \ell_n)$ of w into Lyndon words over A such that $\ell_1 \geq \ell_2 \geq \cdots \geq \ell_n$.*

Several variations of Lyndon words have been considered lately: generalized Lyndon [25], anti-Lyndon [14], inverse Lyndon [4], and Nyldon [6]. In this text, we will use the second.

Definition 1. *Let $(A, \leq)$ be a totally ordered alphabet. We let $\leq_-$ denote the inverse order on A, i.e., $b <_- a$ if and only if $a < b$ for all $a, b \in A$. We also let $\leq_-$ denote the inverse lexicographic order which is the lexicographic order induced by $\leq_-$. A word is* anti-Lyndon *if it is Lyndon with respect to the inverse lexicographic order.*

Otherwise stated, a word is anti-Lyndon if it is primitive and lexicographically maximal among its conjugates.

Example 1. Let $A = \{0, 1\}$ with $0 < 1$, so $1 <_- 0$. The first few anti-Lyndon words, ordered by length, are 1, 0, 10, 110, 100, 1110, 1100, and 1000.

2.2   Morphisms and Fixed Points of Interest

A *morphism* is a map $f \colon A^* \to B^*$, where A, B are alphabets, such that $f(xy) = f(x)f(y)$ for all $x, y \in A^*$. The morphism f is *prolongable* on the letter $a \in A$ if $f(a) = ax$ for some $x \in A^*$ and $f^n(x) \neq \varepsilon$ for all $n \geq 0$. In this section, we consider a specific family of morphisms defined as follows. Note that they appear under the name *generic k-bonacci* morphisms in [26, Example 2.11].

Definition 2. *Let $k \geq 2$ be an integer and let $c_0, \ldots, c_{k-1} \in \mathbb{N}$ be k parameters often summarized in the shape of a word $c = c_0 \cdots c_{k-1} \in \mathbb{N}^k$. The morphism $\mu_c \colon \{0, \ldots, k-1\}^* \to \{0, \ldots, k-1\}^*$ is given by $\mu_c(i) = 0^{c_i} \cdot (i+1)$ for all $i \in \{0, \ldots, k-2\}$ and $\mu_c(k-1) = 0^{c_{k-1}}$. For all $n \geq 0$, we then define $u_{c,n} = \mu_c^n(0)$ and $U_{c,n} = |u_{c,n}|$.*

When the context is clear, we will usually omit the subscript c in Definition 2.

Example 2. When $c = 1^k$, we recover the k-bonacci morphism and words. For $k = 3$ and $c = 102$, the first few iterations of the corresponding morphism $\mu_c \colon 0 \mapsto 01, 1 \mapsto 2, 2 \mapsto 00$ are given in Table 1. Some specific factorization of the words $(u_{c,n})_{n \geq 0}$ is highlighted in Table 1.

Table 1. Construction of the sequences $(u_n)_{n \geq 0}$ and $(U_n)_{n \geq 0}$ for $c = 102$

n	0	1	2	3	4	5
u_n	0	01	012	01200	012000101	012000101012012
fact. of u_n	0	$u_0^1 \cdot 1$	$u_1^1 \, u_0^0 \cdot 2$	$u_2^1 \, u_1^0 \, u_0^2$	$u_3^1 \, u_2^0 \, u_1^2$	$u_4^1 \, u_3^0 \, u_2^2$
U_n	1	2	3	5	9	15

The factorization presented in the previous example can be stated in general. It gives a recursive definition of the words $(u_{c,n})_{n \geq 0}$ and can be proven using a simple induction.

Proposition 1. *For all $c = c_0 \cdots c_{k-1} \in \mathbb{N}^k$ and all $n \geq 0$, we have*

$$
u_n = \begin{cases} \left(\displaystyle\prod_{i=0}^{n-1} u_{n-i-1}^{c_i} \right) \cdot n, & \text{if } n \leq k-1; \\ \displaystyle\prod_{i=0}^{k-1} u_{n-i-1}^{c_i}, & \text{if } n \geq k. \end{cases}
$$

As a consequence of Proposition 1, the sequence $(U_n)_{n \in \mathbb{N}}$ respects the following recurrence relation: if $0 \leq n \leq k-1$, then $U_n = 1 + \sum_{i=0}^{n-1} c_i U_{n-i-1}$, and if $n \geq k$, then $U_n = \sum_{i=0}^{k-1} c_i U_{n-i-1}$.

In the rest of the paper, we will assume the following working hypothesis (WH) on c:

$$
c = c_0 \cdots c_{k-1} \in \mathbb{N}^k \text{ with } c_0, c_{k-1} \geq 1. \tag{WH}
$$

The condition $c_{k-1} \geq 1$ ensures both that the recurrence relation is of order k and that the morphism μ_c is non-erasing, which is a classical assumption in combinatorics on words. Moreover, the condition $c_0 \geq 1$ guarantees that μ_c is prolongable. Under (WH), the morphism μ_c has an infinite fixed point starting with 0 denoted $\mathbf{u} := \lim_{n \to \infty} u_n$.

We make the following combinatorial observation.

Remark 1. Under (WH), using Proposition 1, a simple induction shows that the letter $1 \leq i \leq k - 1$ can only be followed by 0 and/or $i + 1$ (and only 0 in the case $i = k - 1$) in $\mathbf{u}$.

3 Fun with Numeration Systems

In this section, specific definitions will be recalled. For the reader unfamiliar with the theory of numeration systems, we refer to [2, Chapter 2] for an introduction and some advanced concepts.

A *numeration system* (for natural numbers) can be defined as a triple $\mathcal{S} = (A, \mathrm{rep}_{\mathcal{S}}, L)$, where A is an alphabet and $\mathrm{rep}_{\mathcal{S}} : \mathbb{N} \to A^*$ is an injective function such that $L = \mathrm{rep}_{\mathcal{S}}(\mathbb{N})$. The map $\mathrm{rep}_{\mathcal{S}}$ is called the *representation function* and L is the *numeration language*. If $\mathrm{rep}_{\mathcal{S}}(n) = w$ for some integer $n \in \mathbb{N}$ and some word $w \in A^*$, we say that w is the *representation (in $\mathcal{S}$)* of n and we define the *valuation (in $\mathcal{S}$)* of w by $\mathrm{val}_{\mathcal{S}}(w) = n$. Note that, when the context is clear, we omit the subscript $\mathcal{S}$ in rep and val.

Any given prolongable morphism naturally gives rise to a numeration system that we will call the *associated Dumont-Thomas numeration system* [9]. These are based on particular factorizations of the prefixes of the fixed point. Due to space limitation, we only give the definition in the particular case of the morphisms studied in this paper.

Proposition 2 (Dumont-Thomas [9]). *Let c satisfy* (WH). *For all $n \in \mathbb{N}$, there exist unique integers $N, \ell_0, \ldots, \ell_N \in \mathbb{N}$ such that $\ell_0 \geq 1$, $\mathbf{u}[0, n) = u_N^{\ell_0} \cdots u_0^{\ell_N}$, and this factorization verifies the following: u_{N+1} is not a prefix of $\mathbf{u}[0, n)$ and, for all $0 \leq i \leq N$, $u_N^{\ell_0} \cdots u_{N-i+1}^{\ell_{i-1}} u_{N-i}^{\ell_i+1}$ is not a prefix of $\mathbf{u}[0, n)$.*

Recall that a numeration system based on a suitable sequence of integers $(U_n)_{n \geq 0}$ is called *greedy* when, at each step of the decomposition of any integer, the largest possible term of the sequence $(U_n)_{n \geq 0}$ is chosen; formally, we use the Euclidean algorithm in a greedy way. As the conditions on the factorization in the previous proposition resemble that of greedy representations in numeration systems, we will refer to it as being *word-greedy*.

For a given c satisfying (WH), we then let $\mathcal{S}_c$ denote the numeration system associated with the representation function $\mathrm{rep}_{\mathcal{S}_c} : \mathbb{N} \to \mathbb{N}^*$ mapping n to $\mathrm{rep}_{\mathcal{S}_c}(n) = \ell_0 \cdots \ell_N$, where the integers $\ell_0, \ldots, \ell_N$ verify the conditions of Proposition 2 for n. By convention, we set $\mathrm{rep}_{\mathcal{S}_c}(0) = \varepsilon$.

Example 3. Using Example 2 for $c = 102$, the representations of the first few integers are given in Table 2. The word-greedy factorization of each prefix is highlighted in the second row, leading to the representation of the corresponding integer in the third row.

Table 2. Illustration of the numeration system $\mathcal{S}_c$ for $c = 102$

n	0	1	2	3	4	5	6	7	8
$\mathbf{u}[0,n)$	ε	0	01	012	$012 \cdot 0$	01200	$01200 \cdot 0$	$01200 \cdot 01$	$01200 \cdot 01 \cdot 0$
$\mathrm{rep}_{\mathcal{S}_c}(n)$	ε	1	10	100	101	1000	1001	1010	1011

Remark 2. If $\mathrm{rep}_{\mathcal{S}_c}(n) = \ell_0 \cdots \ell_N$, then $n = |u_{c,N}^{\ell_0} \cdots u_{c,0}^{\ell_N}| = \sum_{i=0}^{N} \ell_i U_{c,N-i}$. In other words, $\mathrm{val}_{\mathcal{S}_c}$ is given by the usual valuation function associated with the sequence $(U_{c,n})_{n \in \mathbb{N}}$. Such a system is sometimes called a *positional* numeration system. Note that this is not necessarily the case for the Dumont-Thomas numeration system associated with some other morphism.

The Dumont-Thomas numeration systems are a particular case of abstract numeration systems introduced in [20]. A numeration system $\mathcal{S} = (A, \mathrm{rep}, L)$ is said to be *abstract* if L is regular and $\mathrm{rep}(n)$ is the $(n+1)$st word of L in the genealogical order. We have the following result.

Theorem 2 (Rigo [26, Sect. 2.2]). *Let* $\sigma : \{\alpha_0, \ldots, \alpha_d\}^* \to \{\alpha_0, \ldots, \alpha_d\}^*$ *be a morphism prolongable on the letter* α_0. *We define the automaton* $\mathcal{A}_\sigma$ *for which* $\{\alpha_0, \ldots, \alpha_d\}$ *is the set of states,* α_0 *is the initial state, every state is final, and the (partial) transition function* δ *is such that, for each* $\alpha \in \{\alpha_0, \ldots, \alpha_d\}$ *and* $0 \leq i \leq |\sigma(\alpha)| - 1$, $\delta(\alpha, i)$ *is the* $(i+1)$st *letter of* $\sigma(\alpha)$. *If* $\mathcal{S} = (A, \mathrm{rep}, L)$ *is the Dumont-Thomas numeration system associated with* σ, *then* $L = L(\mathcal{A}_\sigma) \setminus 0\mathbb{N}^*$ *and* $\mathrm{rep}(n)$ *is the* $(n+1)$st *word of* L *in the genealogical order.*

Example 4. For $c = 102$, the automaton $\mathcal{A}_{\mu_c}$ of Theorem 2 is depicted in Fig. 1 (details are left to the reader). The first few accepted words (not starting with 0) are, in genealogical order, ε, 1, 10, 100, 101, 1000, 1001, 1010, and 1011, which indeed agree with the representations of the first few integers in Example 3.

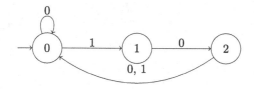

Fig. 1. The automaton $\mathcal{A}_{\mu_c}$ for $c = 102$

As the automaton in Theorem 2 can be used to produce, for all $n \geq 0$, the letter $\mathbf{u}_n$ when reading $\mathrm{rep}_{\mathcal{S}_c}(n)$ by [26, Theorem 2.24], we have the following.

Corollary 1. *Let* c *satisfy* (WH). *Then the sequence* $\mathbf{u}$ *is* $\mathcal{S}_c$-*automatic.*

Similarly to what is usually done in real base numeration systems, we will let $\mathbf{d}^\star$ denote the periodization of c, that is, $\mathbf{d}^\star = (c_0 \cdots c_{k-2}(c_{k-1} - 1))^\omega$. Using Theorem 2, we deduce the next result.

Lemma 1. *Under* (WH), *for all $n \geq 0$, we have* $\mathrm{rep}_{\mathcal{S}_c}(U_n) = 10^n$, *the numbers having a representation of length $n+1$ are those in $[U_n, U_{n+1})$, and $\mathrm{rep}_{\mathcal{S}_c}(U_{n+1} - 1) = \mathbf{d}^\star[0, n]$. In particular, $U_{n+1} - 1 = \sum_{i=0}^{n} \mathbf{d}_i^\star U_{n-i}$.*

Proof. The first claim directly follows by the definition of $\mathcal{S}_c$, and the second one by the genealogical order. The number $U_{n+1} - 1$ is then represented by the maximal length-$(n + 1)$ word accepted by the automaton $\mathcal{A}_{\mu_c}$, which is the length-$(n + 1)$ prefix of $\mathbf{d}^\star$.

Note that, if the numeration system $\mathcal{S}_c$ satisfies the greedy condition, this result follows from the characterization of numeration systems in terms of dynamical systems given by Bertrand-Mathis [3,5]. However, even though the function $\mathrm{rep}_{\mathcal{S}_c}$ is obtained using the word-greedy factorization of prefixes of $\mathbf{u}$, the numeration system $\mathcal{S}_c$ is not necessarily greedy as the following example shows.

Example 5. In Example 2 for $c = 102$, we see that $\mathbf{u}[0, 14] = 012000101 \cdot 012 \cdot 01$, so $\mathrm{rep}_{\mathcal{S}_c}(14) = 10110$, while the greedy representation of 14 associated with the sequence $(U_n)_{n \in \mathbb{N}}$ is 11000.

In fact, we have the following two characterizations.

Lemma 2. *Let c satisfy* (WH). *The numeration system $\mathcal{S}_c = (A, \mathrm{rep}_{\mathcal{S}_c}, L)$ is greedy if and only if, for all $v \in L$ and for all $i \leq |v|$, the suffix of length i of v is smaller than or equal to $\mathbf{d}^\star[0, i)$. Moreover, we then have*

$$L = \{v = v_1 \cdots v_n \in \mathbb{N}^* \setminus 0\mathbb{N}^* \mid \forall 1 \leq i \leq n, v_{n-i+1} \cdots v_n \leq \mathbf{d}^\star[0, i)\}.$$

Proof. Let us denote $\mathcal{S} = (A', \mathrm{rep}_{\mathcal{S}}, L')$ the canonical greedy numeration system associated with the sequence $(U_n)_{n \in \mathbb{N}}$. In particular, by uniqueness, $\mathcal{S}_c$ is greedy if and only if $\mathcal{S}_c = \mathcal{S}$. As $\mathcal{S}_c$ is an abstract numeration system, $\mathrm{rep}_{\mathcal{S}_c}$ respects the genealogical order, i.e., $n \leq m$ if and only if $\mathrm{rep}_{\mathcal{S}_c}(n) \leq_{\mathrm{gen}} \mathrm{rep}_{\mathcal{S}_c}(m)$. So does $\mathrm{rep}_{\mathcal{S}}$ by [2, Proposition 2.3.45]. Hence, $\mathcal{S}_c = \mathcal{S}$ if and only if $L = L'$. Moreover, for all $n \geq 0$, $\mathrm{rep}_{\mathcal{S}}(U_n) = 10^n$, so L and L' contain the same number of length-n words by Lemma 1. Thus $L = L'$ if and only if $L \subseteq L'$. The statement holds since, by [17, Lemma 5.3] and by Lemma 1, we have

$$L' = \{v = v_1 \cdots v_n \in \mathbb{N}^* \setminus 0\mathbb{N}^* \mid \forall 1 \leq i \leq n, v_{n-i+1} \cdots v_n \leq \mathbf{d}^\star[0, i)\}.$$

Theorem 3. *Let $c = c_0 \cdots c_{k-1} \in \mathbb{N}^k$ with $c_0, c_{k-1} \geq 1$. The numeration system $\mathcal{S}_c$ is greedy if and only if $c_0 \cdots c_{k-2}(c_{k-1} - 1)$ is lexicographically maximal among its conjugates.*

Proof. Using Lemma 2 and Theorem 2, $\mathcal{S}_c$ is greedy if and only if, for all $n \in \mathbb{N}$ and for all $0 \leq i \leq k - 1$, any path $\ell_0 \cdots \ell_n$ starting in State i in the automaton

$\mathcal{A}_{\mu_c}$ is such that $\ell_0 \cdots \ell_n \leq \mathbf{d}^\star[0, n]$. However, by definition of $\mathcal{A}_{\mu_c}$, the lexicographically biggest path of length n starting in state i is given by the prefix of length n of $(c_i \cdots c_{k-2}(c_{k-1} - 1)c_0 \cdots c_{i-1})^\omega$. We can therefore conclude that $\mathcal{S}_c$ is greedy if and only if $c_i \cdots c_{k-2}(c_{k-1} - 1)c_0 \cdots c_{i-1} \leq c_0 \cdots c_{k-2}(c_{k-1} - 1)$ for all $0 \leq i \leq k - 1$, i.e., $c_0 \cdots c_{k-2}(c_{k-1} - 1)$ is maximal among its conjugates.

Observe that the condition of the previous result is equivalent to the fact that $c_0 \cdots c_{k-2}(c_{k-1} - 1) = v^\ell$ for some anti-Lyndon v (in fact, v is the primitive root).

Example 6. Let $k = 4$ and $c = 1011$. In this case, $c_0 c_1 c_2 (c_3 - 1) = 1010 = v^2$ with $v = 10$, which is anti-Lyndon (see Example 1). The sequence U_n satisfies the recurrence relation $U_{n+4} = U_{n+3} + U_{n+1} + U_n$ with initial conditions $U_0 = 1$, $U_1 = 2$, $U_2 = 3$, and $U_3 = 5$. A simple induction shows that $(U_n)_{n \in \mathbb{N}}$ is in fact the sequence of Fibonacci numbers. Therefore the numeration system $\mathcal{S}_c$ corresponds to the classical Fibonacci numeration system, which can also be obtained with the parameter $c = 11$.

The observation made in the previous example is more general.

Remark 3. Let c satisfy (WH). If $c_0 \cdots c_{k-2}(c_{k-1} - 1) = v^\ell$ with v anti-Lyndon, we define the word $v' := v_1 \cdots v_{|v|-1}(v_{|v|} + 1)$ (simply put, we add 1 to the last letter of v). Then $c = v^{\ell-1}v'$ is a "partial" cyclization of v'. In particular, since $\mathbf{d}^\star_c = \mathbf{d}^\star_{v'}$ (where the dependence of $\mathbf{d}^\star$ on the chosen parameters is emphasized via a subscript), the numeration systems $\mathcal{S}_c$ and $\mathcal{S}_{v'}$ coincide by Lemma 1.

For the reader familiar with the general theory of numerations, v' satisfies $v'_i \cdots v'_{|v|} < v'$ for all indices $i \in \{2, \ldots, |v|\}$. This implies that v' is the β-expansion $d_\beta(1)$ of 1 for a simple Parry number β [23]. Therefore, c is also a representation of 1 in base β.

Example 7. We illustrate the previous remark by resuming Example 6. We have $v = 10$ and $v' = 11$. The corresponding simple Parry number is the Golden ratio φ. Observe that indeed $c = vv' = 1011$ is a representation of 1 in base φ.

4 Link to String Attractors

Using the results and concepts of the previous sections, we now turn to the concept of string attractors in relation to the fixed points of the morphisms μ_c, $c \in \mathbb{N}^k$. A *string attractor* of a finite word $y = y_1 \cdots y_n$ is a set $\Gamma \subseteq \{1, \ldots, n\}$ such that every factor of y has an occurrence crossing a position in Γ, i.e., for each factor $x \in A^m$ of y, there exists $i \in \Gamma$ and j such that $i \in \{j, \ldots, j+m-1\}$ and $x = y_j \cdots y_i \cdots y_{j+m-1}$.

Example 8. The set $\{2, 3, 4\}$ is a string attractor of the word $0\underline{1}\underline{2}\underline{0}01$. Indeed, it suffices to check that the factors 0, 1 and 01 have an occurrence crossing one of the underlined positions. No smaller string attractor exists since at least one position in the set is needed per different letter in the word.

Given an infinite word $\mathbf{x}$ and any integer $n \geq 1$, we let $s_{\mathbf{x}}(n)$ denote the size of a smallest string attractor for the length-n prefix of $\mathbf{x}$. The function $s_{\mathbf{x}} : n \mapsto s_{\mathbf{x}}(n)$ is called the *string attractor profile function* of $\mathbf{x}$ [27].

Warning. We would like to stress the following crucial point: in this paper, the letters of infinite words are indexed starting from 0 while the positions in a string attractor are counted starting at 1. This could be seen as confusing, but we use the same notation as the original paper on string attractors [18]. Where ambiguity may occur, we explicitly declare how finite words are indexed.

As we will look at prefixes of infinite words, it is natural to wonder if there is a link between the string attractors of the finite words w and wa, where a is a letter. In general, there is no trivial link although we have the following result which can be derived from the proofs of [22, Propositions 12 and 15].

Proposition 3. *Let z be a non-empty word and let $x = z^r$, $y = z^s$ be fractional powers of z with $1 \leq r \leq s$. If Γ is a string attractor of x, then $\Gamma \cup \{|z|\}$ is a string attractor of y.*

Since the considered infinite words are the limits of the sequence $(u_n)_{n \in \mathbb{N}}$, we are interested in the prefixes which are fractional powers of some u_n.

Definition 3. *Let c satisfy (WH). For all $n \geq 0$, we let q_n denote the longest prefix of $\mathbf{u}$ that is a fractional power of u_n, i.e., the longest common prefix between $\mathbf{u}$ and $(u_n)^\omega$. For all $n \geq 0$, we also let $Q_n = |q_n|$.*

The words defined above have a particular structure as stated below.

Proposition 4. *Let c satisfy (WH). Define $\mathbf{a}$ as the infinite concatenation of the longest anti-Lyndon prefix of the word $c_0 \cdots c_{k-2}$. Then for all $n \geq 0$, $q_n = u_n^{\mathbf{a}_0} u_{n-1}^{\mathbf{a}_1} \cdots u_0^{\mathbf{a}_n}$. In particular, $Q_n = \sum_{i=0}^{n} \mathbf{a}_i U_{n-i}$.*

Due to space limitation, the proof of this result can be found in [16].

Example 9. Let us pursue Example 2 for which $c = 102$. The first few words in $(q_n)_{n \geq 0}$ are $0, 01, 0120, 0120001, 0120001010120$. The longest anti-Lyndon prefix of $c_0 c_1 = 10$ is 10 itself so $\mathbf{a} = (10)^\omega$. We can easily check that the first few q_n's indeed satisfy Proposition 4.

Lemma 3. *Let c satisfy (WH). Then $c_0 \cdots c_{k-2} \geq \mathbf{a}[0, k-2]$.*

Proof. Assume the contrary and let w be the longest anti-Lyndon prefix of $c_0 \cdots c_{k-2}$. If $|w| \leq i \leq k-2$ is the smallest index such that $c_0 \cdots c_i < \mathbf{a}[0, i]$, then $c_0 \cdots c_i = w^\ell v a$ with v a proper prefix of w, a a letter, and $va < w$. So [10, Lemme 2] implies that $c_0 \cdots c_i$ is an anti-Lyndon prefix of $c_0 \cdots c_{k-2}$. As $i \geq |w|$, this contradicts the definition (maximality) of w.

We now show how the condition obtained for the greediness of the numeration system is related to the notions we have just defined.

Proposition 5. *Let c satisfy* (WH). *If $c_0 \cdots c_{k-2}(c_{k-1} - 1)$ is lexicographically maximal among its conjugates, then $\mathbf{d}^\star[0, n] \leq \mathbf{a}[0, n]$ for all $n \geq 0$.*

Proof. Let w denote the longest anti-Lyndon prefix of $c_0 \cdots c_{k-2}$. We first show that $c_0 \cdots c_{k-2}(c_{k-1} - 1) \leq \mathbf{a}[0, k - 1]$. If it is not the case, there exist $\ell \geq 1$, a proper prefix u of w, a letter a and a word v such that $c_0 \cdots c_{k-2}(c_{k-1} - 1) = w^\ell uav$ and $ua > w$. Then $uavw^\ell > c_0 \cdots c_{k-2}(c_{k-1} - 1)$, so $c_0 \cdots c_{k-2}(c_{k-1} - 1)$ is not maximal among its conjugates. This is a contradiction. Therefore we have $c_0 \cdots c_{k-2}(c_{k-1} - 1) \leq \mathbf{a}[0, k - 1]$. By Lemma 3, we get $c_0 \cdots c_{k-2} = \mathbf{a}[0, k - 2]$ and $c_{k-1} - 1 \leq \mathbf{a}_{k-1}$.

We now prove that $\mathbf{d}^\star[0, n] \leq \mathbf{a}[0, n]$ for all $n \geq 0$. If $c_{k-1} - 1 < \mathbf{a}_{k-1}$, then the conclusion is direct. If $c_{k-1} - 1 = \mathbf{a}_{k-1}$, then $c_0 \cdots c_{k-2}(c_{k-1} - 1)$ is a fractional power of w so there exist $\ell \geq 1$ and u a proper prefix of w such that $c_0 \cdots c_{k-2}(c_{k-1} - 1) = w^\ell u$. Let us write $w = uv$. If $u \neq \varepsilon$, we then have that $c_0 \cdots c_{k-2}(c_{k-1} - 1) = w^\ell u = u(vu)^\ell < uw^\ell$ as w is anti-Lyndon thus strictly greater than its conjugates. This contradicts the assumption that $c_0 \cdots c_{k-2}(c_{k-1} - 1)$ is maximal among its conjugates. Therefore, $u = \varepsilon$ and $c_0 \cdots c_{k-2}(c_{k-1} - 1)$ is a (natural) power of w. We conclude that $\mathbf{a} = \mathbf{d}^\star$, which ends the proof of the first item.

Proposition 6. *Let c satisfy* (WH). *If $c_0 \cdots c_{k-2}(c_{k-1} - 1)$ is lexicographically maximal among its conjugates, then $U_{n+1} - 1 \leq Q_n$ for all $n \geq 0$.*

Proof. Let us show the claim by contraposition. So assume that there exists an integer n such that $U_{n+1} - 1 > Q_n$. Thus $q_n = u_n^{\mathbf{a}_0} \cdots u_0^{\mathbf{a}_n}$ is a proper prefix of $\mathbf{u}[0, U_{n+1} - 1)$. By Lemma 1, $\mathrm{rep}_{\mathcal{S}_c}(U_{n+1} - 1) = \mathbf{d}^\star[0, n]$, so $\mathbf{d}_0^\star$ is the largest exponent e such that u_n^e is a prefix of $\mathbf{u}[0, U_{n+1} - 1)$. This implies that $\mathbf{d}_0^\star \geq \mathbf{a}_0$. Moreover, if $\mathbf{a}_0 = \mathbf{d}_0^\star$, the same argument implies that $\mathbf{d}_1^\star$ is the largest exponent e such that $u_n^{\mathbf{d}_0^\star} u_{n-1}^e$ is a prefix of $\mathbf{u}[0, U_{n+1} - 1)$. In both cases, we have $\mathbf{d}_0^\star \mathbf{d}_1^\star \geq \mathbf{a}_0 \mathbf{a}_1$. We may iterate the reasoning to obtain $\mathbf{d}^\star[0, n] \geq \mathbf{a}[0, n]$. As q_n is a proper prefix of $\mathbf{u}[0, U_{n+1} - 1)$, the inequality cannot be an equality. This contradicts Proposition 5, which ends the proof.

We will now prove that, under the conditions of the previous result, we can describe string attractors of every prefix of $\mathbf{u}$ using the elements of $(U_n)_{n \in \mathbb{N}}$. For $n \in \mathbb{N}$, we let Γ_n denote $\{U_0, \ldots, U_n\}$ if $0 \leq n \leq k - 1$, $\{U_{n-k+1}, \ldots, U_n\}$ otherwise. We also define P_n by U_n if $0 \leq n \leq k - 1$, $U_n + U_{n-k+1} - U_{n-k} - 1$ otherwise.

The next lemma directly follows from Proposition 6 and the definition of P_n.

Lemma 4. *Let c satisfy* (WH). *If $c_0 \cdots c_{k-2}(c_{k-1} - 1)$ is maximal among its conjugates, then $P_n \leq U_{n+1} - 1 \leq Q_n$ for all $n \in \mathbb{N}$.*

To simplify the statement of the following theorem, we set $\Gamma_{-1} = \emptyset$.

Theorem 4. *Let $c = c_0 \cdots c_{k-1} \in \mathbb{N}^k$ with $c_0, c_{k-1} \geq 1$ and $c_0 \cdots c_{k-2}(c_{k-1} - 1)$ maximal among its conjugates. Fix an integer $n \geq 0$. If $m \in [U_n, Q_n]$, then $\Gamma_{n-1} \cup \{U_n\}$ is a string attractor of $\mathbf{u}[0, m)$. Furthermore, if $m \in [P_n, Q_n]$, then Γ_n is a string attractor of $\mathbf{u}[0, m)$. In particular, $s_{\mathbf{u}}(n) \leq k + 1$ for all $n \geq 1$.*

Proof. Let us simultaneously prove the two claims by induction on n. If $n = 0$, then $1 \leq m \leq c_0$, so $\mathbf{u}[0, m) = 0^m$ and the conclusion directly follows for both claims. Assume now that the claims are satisfied for $n - 1$ and let us prove them for n. By Lemma 4 and the induction hypothesis, Γ_{n-1} is a string attractor of $\mathbf{u}[0, U_n - 1)$. This implies that $\Gamma_{n-1} \cup \{U_n\}$ is a string attractor of u_n so, by Proposition 3 and by definition of Q_n (Definition 3), of $\mathbf{u}[0, m)$ for all $m \in [U_n, Q_n]$. This ends the proof of the first claim.

Let us now prove the second claim. Observe that, using Proposition 3, it suffices to prove that Γ_n is a string attractor of $\mathbf{u}[0, P_n)$. If $0 \leq n \leq k - 1$, then $\Gamma_n = \Gamma_{n-1} \cup \{U_n\}$ so we can directly conclude using the first claim. Thus assume that $n \geq k$. Then by the first claim, $\Gamma_n \cup \{U_{n-k}\} = \Gamma_{n-1} \cup \{U_n\}$ is a string attractor of $\mathbf{u}[0, P_n)$. Therefore, it remains to show that the position U_{n-k} is not needed in the string attractor. In other words, we prove that the factors of $\mathbf{u}[0, P_n)$ that have an occurrence crossing position U_{n-k} (and no other position of $\Gamma_n \cup \{U_{n-k}\}$) have another occurrence crossing a position in Γ_n. More precisely, we show that they have an occurrence crossing position U_n. To help the reader with the proof, we illustrate the situation in Fig. 2.

As the smallest position in Γ_n is U_{n-k+1}, we need to consider the factor occurrences crossing position U_{n-k} in $\mathbf{u}[0, U_{n-k+1} - 1)$. So, if we write $\mathbf{u}[0, P_n) = u_n w$, it is sufficient to show that u_{n-k} is a suffix of u_n and that $w' := \mathbf{u}[U_{n-k}, U_{n-k+1} - 1)$ is a prefix of w. Observe that

$$|w| = P_n - U_n = U_{n-k+1} - U_{n-k} - 1 \tag{1}$$

by definition of P_n, so $|w'| = |w|$. We will actually show that $w' = w$.

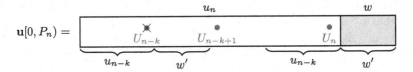

$$\mathbf{u}[0, P_n) =$$

Fig. 2. Representation of the proof of the second claim of Theorem 4. As we warned the reader before, elements in a string attractor are indexed starting at 1 (in red), while indices of letters in $\mathbf{u}$ start at 0. (Color figure online)

The fact that u_{n-k} is a suffix of u_n is a direct consequence of Proposition 1 as $c_{k-1} \geq 1$ by assumption. To prove that $w' = w$, we first make the following observation: Proposition 1 again implies that u_n is followed by $u_n^{c_0-1} u_{n-1}^{c_1} \cdots u_{n-k+1}^{c_{k-1}}$ in $\mathbf{u}$. Since u_{n-k+1} is a prefix of all the words $u_{n-k+1}, \ldots, u_{n-1}$, the word u_n is in particular followed by u_{n-k+1} in $\mathbf{u}$. As $|w| \leq U_{n-k+1}$ by Eq. (1), this implies that w is a prefix of u_{n-k+1}, so also of $\mathbf{u}$. To conclude with the claim, it is then enough to show that w' is also a prefix of $\mathbf{u}$. To prove this, we will use the numeration system S_c and consider two cases.

First, assume that $n - 2k + 1 \geq 0$. By definition of w' and by Proposition 1, w' is a prefix of $v := u_{n-k}^{c_0-1} \cdots u_{n-2k+1}^{c_{k-1}}$. Define the word

$x = (c_0 - 1)c_1 \cdots c_{k-1}0^{n-2k+1}$. If it begins with 0's, we consider instead the word obtained by removing the leading 0's. Note that x corresponds to a factorization of v into the words $u_{n-k}, \ldots, u_0$. As $c_0 \cdots c_{k-2}(c_{k-1} - 1)$ is maximal among its conjugates by assumption, x is in the numeration language by Lemma 2. By definition of $\mathcal{S}_c$, x is the Dumont-Thomas factorization of v, implying that v is a prefix of $\mathbf{u}$.

Second, if $n - 2k + 1 < 0$, then we conclude in a similar way by considering $v = u_{n-k}^{c_0-1} \cdots u_0^{c_{n-k}}$ and $x = (c_0 - 1)c_1 \cdots c_{n-k}$ instead.

The morphisms studied in Theorem 4 are associated with Parry numbers in the following way: there exists a simple Parry number β with $d_\beta(1) = c_0 \cdots c_{k-1}$ such that either $c = d_\beta(1)$, or $c = (c_0 \cdots c_{k-2}(c_{k-1} - 1))^\ell d_\beta(1)$ for some $\ell \geq 1$.

Theorem 4 implies that for some values of c (e.g., $c = 211$), we even have $s_{\mathbf{u}}(n) \leq k$ for all $n \geq 1$ and it is optimal for large n as every position in Γ_n covers a different letter (this can be proved using a simple induction).

Observe that the bounds given in the previous theorem are not necessarily tight. For example, if $c = 23$, then $\Gamma_2 = \{3, 9\}$ is a string attractor of the length-9 prefix $\mathbf{u}[0, 9) = 001001000$, while $P_2 = 10$. This is also the case for the k-bonacci morphisms ($c = 1^k$) where better bounds are provided in [15].

5 Final Comments

We end this paper by discussing the scope of use of our main result. For a given $c \in \mathbb{N}^k$ satisfying specific properties, Theorem 4 states that we can easily describe a string attractor of size at most $k + 1$ for any prefix of the fixed point $\mathbf{u}$ of μ_c defined in Sect. 2.2.

On the one hand, this result is not necessarily optimal. For example, if $c = 12$ (the corresponding fixed point is referred to as the *period-doubling* word [1]), our result sometimes yields string attractors of size 3 (e.g., for the length-8 prefix). However, string attractors of size 2 always exist as shown in [27]. More generally, we believe that, by adapting our construction, the statement of Theorem 4 may be strengthened to obtain size-k string attractors for each prefix.

On the other hand, for some $c \in \mathbb{N}^k$, the corresponding numeration system is not *addable*, meaning that the addition within the numeration system is not recognizable by a finite automaton. For example, this is the case of $c = 3203$ [13]. As a consequence, the approach from [27] does not apply; in particular, we study words outside the framework needed to use the software *Walnut* [28].

Finally, we wish to point out that this paper is a first exploration into the possible link between string attractors of prefixes of morphic words and general numeration systems. As stated in the question presented in Sect. 1, we believe that this connection can be extended to other morphisms, which is a path that we will continue exploring in the future.

Acknowledgements. France Gheeraert is a Research Fellow of the FNRS. Manon Stipulanti is supported by the FNRS Research grant 1.B.397.20F. We warmly thank M. Rigo and S. Kreczman for useful discussions on numeration systems, especially for indicating [9] and [17] respectively.

References

1. Allouche, J.P., Shallit, J.: Automatic sequences, theory, applications, generalizations Cambridge University Press, Cambridge (2003). https://doi.org/10.1017/CBO9780511546563
2. Berthé, V., Rigo, M. (eds.): Combinatorics, automata and number theory, Encyclopedia of Mathematics and its Applications, vol. 135. Cambridge University Press, Cambridge (2010). https://doi.org/10.1017/CBO9780511777653
3. Bertrand-Mathis, A.: Comment écrire les nombres entiers dans une base qui n'est pas entière. Acta Math. Hungar. **54**(3–4), 237–241 (1989). https://doi.org/10.1007/BF01952053
4. Bonizzoni, P., De Felice, C., Zaccagnino, R., Zizza, R.: Inverse Lyndon words and inverse Lyndon factorizations of words. Adv. in Appl. Math. **101**, 281–319 (2018). https://doi.org/10.1016/j.aam.2018.08.005
5. Charlier, E., Cisternino, C., Stipulanti, M.: A full characterization of Bertrand numeration systems. In: Developments in Language Theory, vol. 13257. LNCS, pp. 102–114. Springer, Cham (2022). https://doi.org/10.1007/978-3-031-05578-2_8
6. Charlier, E., Philibert, M., Stipulanti, M.: Nyldon words. J. Combin. Theory Ser. A **167**, 60–90 (2019). https://doi.org/10.1016/j.jcta.2019.04.002
7. Chen, K.T., Fox, R.H., Lyndon, R.C.: Free differential calculus. IV. The quotient groups of the lower central series. Ann. of Math. **2**(68), 81–95 (1958). https://doi.org/10.2307/1970044
8. Christiansen, A.R., Ettienne, M.B., Kociumaka, T., Navarro, G., Prezza, N.: Optimal-time dictionary-compressed indexes. ACM Trans. Algorithms **17**(1), 8:1–8:39 (2021). https://doi.org/10.1145/3426473
9. Dumont, J.M., Thomas, A.: Systèmes de numération et fonctions fractales relatifs aux substitutions. Theoret. Comput. Sci. **65**(2), 153–169 (1989). https://doi.org/10.1016/0304-3975(89)90041-8
10. Duval, J.P.: Mots de Lyndon et périodicité. RAIRO Inform. Théor. **14**(2), 181–191 (1980). https://doi.org/10.1051/ita/1980140201811
11. Dvořáková, L.: String attractors of episturmian sequence, preprint available at arXiv:2211.01660 (2022)
12. Fabre, S.: Substitutions et β-systèmes de numération. Theoret. Comput. Sci. **137**(2), 219–236 (1995). https://doi.org/10.1016/0304-3975(95)91132-A
13. Frougny, C.: On the sequentiality of the successor function. Inform. Comput. **139**(1), 17–38 (1997). https://doi.org/10.1006/inco.1997.2650
14. Gewurz, D.A., Merola, F.: Numeration and enumeration. European J. Combin. **33**(7), 1547–1556 (2012). https://doi.org/10.1016/j.ejc.2012.03.017
15. Gheeraert, F., Restivo, A., Romana, G., Sciortino, M., Stipulanti, M.: New string attractor-based complexities on infinite words (2023), work in progress
16. Gheeraert, F., Romana, G., Stipulanti, M.: String attractors of fixed points of k-bonacci-like morphisms, long version available at arXiv:2302.13647 (2023)
17. Hollander, M.: Greedy numeration systems and regularity. Theory Comput. Syst. **31**(2), 111–133 (1998). https://doi.org/10.1007/s002240000082
18. Kempa, D., Prezza, N.: At the roots of dictionary compression: string attractors. In: STOC 2018–Proceedings of the 50th Annual ACM SIGACT Symposium on Theory of Computing, pp. 827–840. ACM, New York (2018). https://doi.org/10.1145/3188745.3188814

19. Kutsukake, K., Matsumoto, T., Nakashima, Y., Inenaga, S., Bannai, H., Takeda, M.: On repetitiveness measures of Thue-Morse words. In: Boucher, C., Thankachan, S.V. (eds.) SPIRE 2020. vol. 12303. LNCS, pp. 213–220. Springer, Cham (2020). https://doi.org/10.1007/978-3-030-59212-7_15

20. Lecomte, P.B.A., Rigo, M.: Numeration systems on a regular language. Theory Comput. Syst. **34**(1), 27–44 (2001). https://doi.org/10.1007/s002240010014

21. Lothaire, M.: Combinatorics on words. Cambridge Mathematical Library. Cambridge University Press, Cambridge (1997). https://doi.org/10.1017/CBO9780511566097, corrected reprint of the 1983 original

22. Mantaci, S., Restivo, A., Romana, G., Rosone, G., Sciortino, M.: A combinatorial view on string attractors. Theoret. Comput. Sci. **850**, 236–248 (2021). https://doi.org/10.1016/j.tcs.2020.11.006

23. Parry, W.: On the β-expansions of real numbers. Acta Math. Acad. Sci. Hungar. **11**, 401–416 (1960). https://doi.org/10.1007/BF02020954

24. Restivo, A., Romana, G., Sciortino, M.: String attractors and infinite words. In: LATIN 2022: Theoretical informatics, vol. 13568. LNCS, pp. 426–442. Springer, Cham (2022). https://doi.org/10.1007/978-3-031-20624-5_26

25. Reutenauer, C.: Mots de Lyndon généralisés. Sém. Lothar. Combin. 54, Art. B54h, 16 (2005/07)

26. Rigo, M.: Formal languages, automata and numeration systems. 2. Applications to recognizability and decidability. Networks and Telecommunications Series, ISTE, London; John Wiley & Sons Inc, Hoboken, NJ (2014). https://doi.org/10.1002/9781119042853

27. Schaeffer, L., Shallit, J.: String attractor for automatic sequences, preprint available at arXiv:2012.06840 (2022)

28. Shallit, J.: The logical approach to automatic sequences. Exploring combinatorics on words with Walnut, Lond. Math. Soc, vol. 482. Lecture Notes Series. Cambridge University Press, Cambridge (2023). https://doi.org/10.1017/9781108775267

Magic Numbers in Periodic Sequences

Savinien Kreczman[1], Luca Prigioniero[2], Eric Rowland[3],
and Manon Stipulanti[1(✉)]

[1] Department of Mathematics, University of Liège, Liège, Belgium
{savinien.kreczman,m.stipulanti}@uliege.be
[2] Dipartimento di Informatica, Università degli Studi di Milano, Milan, Italy
prigioniero@di.unimi.it
[3] Department of Mathematics, Hofstra University, Hempstead, New York, USA
eric.rowland@hofstra.edu

Abstract. In formal languages and automata theory, the magic number problem can be formulated as follows: for a given integer n, is it possible to find a number d in the range $[n, 2^n]$ such that there is no minimal deterministic finite automaton with d states that can be simulated by a minimal nondeterministic finite automaton with exactly n states? If such a number d exists, it is called magic. In this paper, we consider the magic number problem in the framework of deterministic automata with output, which are known to characterize automatic sequences. More precisely, we investigate magic numbers for periodic sequences viewed as either automatic, regular, or constant-recursive.

Keywords: Magic numbers · Periodic sequences · Automatic sequences · Regular sequences · Constant-recursive sequences

1 Introduction

The magic number problem has received wide attention in formal languages and automata theory since its recent formulation [16,17]. In this setting, it is well known that, given a nondeterministic finite automaton with n states, it is always possible to obtain an equivalent deterministic finite automaton, for example by applying the *powerset construction*. The number of states of the resulting machine may vary between n and 2^n, and the upper bound cannot be reduced in the worst case [23]. Therefore, it is natural to ask whether, given a number n, it is possible to find a number d satisfying $n \leq d \leq 2^n$, such that there is no minimal deterministic finite automaton with exactly d states that can be simulated by an optimal nondeterministic finite automaton with exactly n states. If such a number d exists, it is called *magic* for n. This problem has been studied for finite automata with one- and two-letter alphabets (for which magic numbers have been found [12,16,17]), for larger alphabets (for which it has been proved that no magic number exists for each $n \geq 0$ [11,18,20]), and for some special cases [14]. The formulation of the magic number problem has been adapted to the cost of operations on regular languages [19]. In that case,

ⓒ The Author(s), under exclusive license to Springer Nature Switzerland AG 2023
A. Frid and R. Mercaş (Eds.): WORDS 2023, LNCS 13899, pp. 206–219, 2023.
https://doi.org/10.1007/978-3-031-33180-0_16

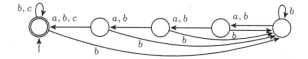

Fig. 1. A nondeterministic finite automaton with 5 states having an equivalent minimal deterministic finite automaton with 2^5 states.

the goal is to determine if the whole range of sizes between the lower and the upper bounds can be obtained when applying a given language operation.

Example 1. The nondeterministic finite automaton $\mathcal{A}$ in Fig. 1 has 5 states and its corresponding minimal deterministic finite automaton, which can be obtained by minimizing the automaton produced when applying the powerset construction to $\mathcal{A}$, has $2^5 = 32$ states. For each $d \in \{5, \ldots, 32\}$, it is possible to modify $\mathcal{A}$, by adding suitable transitions on the symbol c, to obtain a corresponding minimal deterministic finite automaton with d states. Moreover, this example can be generalized to nondeterministic finite automata with n states, for every $n \geq 0$, to prove that no magic numbers exist in the case of three-letter alphabets [20].

At first, our intention was to investigate the magic number problem in the case of *deterministic finite automata with output*. It is known that these machines, which are simple extensions of deterministic finite automata in which every state is associated with an output, generate the class of *automatic sequences*. More precisely, a sequence $s = s(n)_{n \geq 0}$ is automatic if there is a deterministic finite automaton with output that, for each integer n, outputs $s(n)$ when the digits of n are given as input [3, Chapter 5]. First systematically studied by Cobham [8], automatic sequences have become a central topic in combinatorics on words, with applications in numerous fields, especially number theory [3]. There is a well-established correspondence between a specific minimal deterministic finite automaton with output generating the sequence s and the so-called *kernel* of s, which is a set of particular subsequences extracted from s. Thus, since we are interested in the number of states, it is equivalent to study the kernel sizes of automatic sequences, which is called the *rank*. Specifically, given a period length ℓ, we want to find lower and upper bounds on the possible values of the rank for all periodic sequences with period length ℓ. Moreover, we examine the magic number problem in this context: given an integer $\ell \geq 1$, is it possible to find numbers d within the previous bounds such that there is a periodic sequence with period length ℓ generated by a deterministic finite automaton with output with exactly d states or, equivalently, having rank exactly d? In some specific case, we obtain a full characterization of magic numbers for periodic sequences.

We then extended this study to two other families of sequences, namely *constant-recursive* and *regular*. A sequence is constant-recursive if each of its elements can be obtained from a linear combination of the previous ones. The rank of such a sequence is the smallest degree of all linear recurrences producing it. Motivation to study constant-recursive sequences is that they appear in many areas of computer science and mathematics, such as theoretical biology, software verification, probabilistic model checking, quantum computing, and crystallography; for further details, see [24] or the fine survey [9], and references therein. There are still challenging open problems involving these sequences. For instance, in general, it is not known if it is decidable whether or not a constant-recursive sequence takes the value 0. This question is referred to as the *Skolem problem.* For this class of sequences, we characterize the corresponding magic numbers as sums of the Euler totient function evaluated at divisors of the period length of the sequence.

On the other hand, a sequence s is regular if the vector space generated by its kernel is finitely generated [2]. In this class, the rank of a sequence is the dimension of the vector space. Every automatic sequence, necessarily on a finite alphabet, is regular. In that sense, the notion of regular sequences generalizes that of automatic sequences to infinite alphabets and thus provides a wider framework for the study of sequences. Regular sequences naturally arise in many fields, such as analysis of algorithms, combinatorics, number theory, numerical analysis, and the theory of fractals [2]. In some specific case, we show that the regular case actually boils down to the constant-recursive one.

The paper is organized as follows. In Sect. 2, we introduce the necessary preliminaries. A specific section is then dedicated to each class of sequences: namely Sect. 3 to automatic sequences, Sect. 4 to constant-recursive sequences, and Sect. 5 to regular sequences. We chose to provide the necessary background for each class of sequences in the corresponding section[1].

2   Preliminaries

Within a family $\mathcal{F}$ of sequences, we will consider a property $P_{\mathcal{F}}$ on numbers. We say that a number is *magic* if it does not satisfy $P_{\mathcal{F}}$; otherwise, adopting the terminology from [12], it is called *muggle.* The *range* for $P_{\mathcal{F}}$ is the smallest interval containing all muggle numbers with respect to $P_{\mathcal{F}}$.

The *period length* of a periodic sequence s is the smallest integer $\ell \geq 1$ satisfying $s(n + \ell) = s(n)$ for all $n \geq 0$. Let $\mathsf{Per}(\ell)$ denote the set of such periodic sequences with period length ℓ. The *period* of a sequence $s \in \mathsf{Per}(\ell)$ is the tuple $(s(0), s(1), \ldots, s(\ell - 1))$. Note that, in the following, we discard the case $\ell = 1$, which corresponds to constant sequences and for which the three families give a corresponding rank equal to 1.

For a periodic sequence s, subsequences of the form $s(cn + r)_{n \geq 0}$ are also periodic, and their period lengths satisfy the following.

Proposition 1. *Let* $\ell \geq 2$, $c, r \geq 0$ *be integers. If* s *is a periodic sequence whose period length divides* ℓ, *then* $s(cn + r)_{n \geq 0}$ *is periodic with period length dividing* $\frac{\ell}{\gcd(c,\ell)}$.

---

[1] Proofs omitted due to space constraints may be found at https://arxiv.org/abs/2304.03268.

3   Automatic Sequences

In this section, we are interested in representations of periodic sequences as automatic sequences. We briefly mention earlier works concerning *periodicity* and *automaticity*. *Eventual periodicity* and *automatic sets* have been considered in [5, 15, 22]. For $\ell \geq 1$, Alexeev [1] and Sutner [26] studied the automata generating the sequence with period $(1, 0, 0, \ldots, 0)$, with $\ell - 1$ trailing zeroes. Sutner and Tetruashvili determined the rank of the sequence with period $(0, 1, \ldots, \ell - 1)$ for specific values of $\ell \geq 1$ [27, Claim 2.1]. Bosma studied some particular cases of several families of periodic sequences on a binary alphabet and considered upper and lower bounds on the rank [6], while Zantema gave an upper bound on the rank of all periodic sequences [28, Theorem 10].

We now introduce the definitions for this section. Let $k, \ell \geq 2$ be integers. Define $\mathrm{ord}_\ell(k)$ to be the smallest integer $m \geq 1$ such that $k^{e+m} \equiv k^e \mod \ell$ for all sufficiently large e. In the special case when k and ℓ are coprime, $\mathrm{ord}_\ell(k)$ is the usual multiplicative order of k modulo ℓ. Let $\mathrm{pre}_\ell(k)$ denote the smallest integer $e \geq 0$ such that $k^{e+\mathrm{ord}_\ell(k)} \equiv k^e \mod \ell$. In other words, $\mathrm{pre}_\ell(k)$ is the length of the preperiod of the sequence $(k^e \mod \ell)_{e \geq 0}$, and $\mathrm{ord}_\ell(k)$ is the (eventual) period length. For every integer n, we let $\mathrm{Div}(n)$ be the set of divisors of n.

Given an integer k, a sequence s is k-*automatic* if there is a deterministic finite automaton with output (DFAO) that, for each integer n, outputs $s(n)$ when fed with the base-k digits of n. To tie the automaton to the base, we call it a k-*automaton*. Here, automata read base-k representations beginning with the least significant digit (note that the other reading direction produces the same set of sequences). There is also a characterization of automatic sequences in terms of the so-called *kernel*. In particular, the k-*kernel* of a sequence s is the set of subsequences $\ker_k(s) = \{s(k^e n + j)_{n \geq 0} : e \geq 0 \text{ and } 0 \leq j \leq k^e - 1\}$. A sequence is k-automatic if and only if its k-kernel is finite. We define the k-*rank* (or simply the *rank* when the context is clear) of a k-automatic sequence s to be the quantity $\mathrm{rank}_k(s) = |\ker_k(s)|$. In fact, $\ker_k(s)$ is in bijection with the set of states in the minimal k-automaton that generates s with the property that leading 0's do not affect the output. Every periodic sequence is k-automatic for every $k \geq 2$ [7, p. 88]. Note that, in this case, the alphabet does not matter, so, without loss of generality, the sequences in this section will have nonnegative integer values. For further details on automatic sequences see, e.g., [3].

Example 2. For the sequence $s \in \mathsf{Per}(6)$ with period $(0, 1, 2, 3, 4, 5)$, we illustrate the relationship between the minimal 2-automaton for s and its 2-kernel in Fig. 2 and Table 1.

For integers $k, \ell \geq 2$ and the family of periodic sequences $\mathsf{Per}(\ell)$, we consider the following property: an integer satisfies $\mathrm{P_a}(k, \ell)$ if and only if it is equal to the

Table 1. For $s \in \mathrm{Per}(6)$ with period $(0,1,2,3,4,5)$, its 2-kernel contains 13 distinct sequences shown in the second column. Two pairs (e,j) and (e',j') are equivalent if and only if $s(k^e n + j)_{n \geq 0} = s(k^{e'} n + j')_{n \geq 0}$. Therefore, in the third column, the state in the 2-automaton for s of Fig. 2 is labeled ej where (e,j) is the lexicographically smallest representative for this equivalence relation.

(e,j)	$s((2^e \bmod 6)n + (j \bmod 6))_{n \geq 0}$	state
(0,0)	$s(n)_{n \geq 0} = 0,1,2,3,4,5,0,1,2,3,4,5,\ldots$	00
(1,0), (3,0), (3,6)	$s(2n)_{n \geq 0} = 0,2,4,0,2,4,\ldots$	10
(1,1), (3,1), (3,7)	$s(2n+1)_{n \geq 0} = 1,3,5,1,3,5,\ldots$	11
(2,0), (4,0), (4,6), (4,12)	$s(4n)_{n \geq 0} = 0,4,2,0,4,2,\ldots$	20
(2,1), (4,1), (4,7), (4,13)	$s(4n+1)_{n \geq 0} = 1,5,3,1,5,3,\ldots$	21
(2,2), (4,2), (4,8), (4,14)	$s(4n+2)_{n \geq 0} = 2,0,4,2,0,4,\ldots$	22
(2,3), (4,3), (4,9), (4,15)	$s(4n+3)_{n \geq 0} = 3,1,5,3,1,5,\ldots$	23
(3,2)	$s(2n+2)_{n \geq 0} = 2,4,0,2,4,0,\ldots$	32
(3,3)	$s(2n+3)_{n \geq 0} = 3,5,1,3,5,1,\ldots$	33
(3,4)	$s(2n+4)_{n \geq 0} = 4,0,2,4,0,2,\ldots$	34
(3,5)	$s(2n+5)_{n \geq 0} = 5,1,3,5,1,3,\ldots$	35
(4,4), (4,10)	$s(4n+4)_{n \geq 0} = 4,2,0,4,2,0,\ldots$	44
(4,5), (4,11)	$s(4n+5)_{n \geq 0} = 5,3,1,5,3,1,\ldots$	45

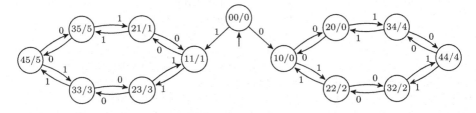

Fig. 2. The minimal 2-automaton outputting the sequence with period $(0,1,2,3,4,5)$. Each state has a label of the form α/ρ where α corresponds to a state name in the third column of Table 1 and ρ is the output.

k-rank of some sequence in $\mathrm{Per}(\ell)$. By merging the next two results, we obtain the range for $\mathrm{P_a}(k,\ell)$ as stated in Corollary 1.

Proposition 2. *Let $k, \ell \geq 2$ be two integers. The k-rank of every sequence in $\mathrm{Per}(\ell)$ is at most $B_\ell(k) := \left(\sum_{e=0}^{\mathrm{pre}_\ell(k)-1} \min(k^e, \ell)\right) + \ell \cdot \mathrm{ord}_\ell(k)$. Moreover, the k-rank of the periodic sequence with period $(0,1,\ldots,\ell-1)$ is $B_\ell(k)$.*

Proposition 3. *Let $k, \ell \geq 2$ be coprime integers. The k-rank of every sequence in $\mathrm{Per}(\ell)$ is at least ℓ. Moreover, the k-rank of the periodic sequence with period $(0,0,\ldots,0,1)$ is ℓ.*

Corollary 1. *Let $k, \ell \geq 2$ be coprime integers. The interval $[\ell, B_\ell(k)]$ is the range for $\mathrm{P_a}(k,\ell)$.*

The magic and muggle numbers with respect to $\mathrm{P}_a(k, \ell)$ when k and ℓ are coprime are characterized in the following theorem.

Theorem 1. *Let $k, \ell \geq 2$ be coprime integers. Let $A = \{d\ell : d \in \mathsf{Div}(\mathrm{ord}_\ell(k))\}$. Then A is the set of muggle numbers in the range for $\mathrm{P}_a(k, \ell)$. In particular, the set of magic numbers with respect to $\mathrm{P}_a(k, \ell)$ is $\mathbb{N} \setminus A$.*

The following lemmas are useful for the proof of Theorem 1.

Lemma 1. *Let $k, \ell \geq 2$ be coprime integers and let $s \in \mathsf{Per}(\ell)$. Then $\ker_k(s) = \{s(k^e n + j)_{n \geq 0} : e \geq 0 \text{ and } 0 \leq j \leq \ell - 1\}$.*

Lemma 2. *Let $k, \ell \geq 2$ be coprime integers. If s belongs to $\mathsf{Per}(\ell)$, then so does $s(kn)_{n \geq 0}$.*

Proof (of Theorem 1). For the sake of conciseness set $T = \{\mathrm{rank}_k(s) : s \in \mathsf{Per}(\ell)\}$. We prove $T = A$ by showing two inclusions.

First Inclusion. By Proposition 3, $\mathrm{rank}_k(s) \geq \ell$ for all $s \in \mathsf{Per}(\ell)$. Using Lemma 1, for each $e \geq 0$, let $B_{k,\ell,e}(s) = \{s(k^e n + j)_{n \geq 0} : 0 \leq j \leq \ell - 1\}$ be the set of kernel sequences arising from exponent e. Since k and ℓ are coprime, the sequences in $B_{k,\ell,e}(s)$ are precisely the ℓ shifts $s(k^e(n + i))_{n \geq 0}$ $(0 \leq i \leq \ell - 1)$ of $s(k^e n)_{n \geq 0}$, and they are distinct by Lemma 2, so $|B_{k,\ell,e}(s)| = \ell$. Therefore, if $B_{k,\ell,e}(s) \cap B_{k,\ell,e'}(s) \neq \emptyset$ for some $e, e' \geq 0$, then $B_{k,\ell,e}(s) = B_{k,\ell,e'}(s)$. The sequence $(B_{k,\ell,e}(s))_{e \geq 0}$ of sets is periodic with period length dividing $\mathrm{ord}_\ell(k)$. Therefore, $T \subseteq A$.

Second Inclusion. Let $d \in \mathsf{Div}(\mathrm{ord}_\ell(k))$. We consider the mapping $n \mapsto k^d n$ on $\mathbb{Z}/(\ell\mathbb{Z})$, which is a permutation because k^d is invertible modulo ℓ. This permutation has finite order, and the orbit of n under this mapping is the set $\{n, k^d n, k^{2d} n, \ldots\}$. In particular, the orbit of 0 is $\{0\}$. Let $X_0 = \{0\}, X_1, \ldots, X_m$ be the distinct orbits. These sets partition $\mathbb{Z}/(\ell\mathbb{Z})$ (for an illustration, see Example 3). Define the sequence s on the alphabet $\{0, 1, \ldots, m\}$ by

$$s(n) = i \quad \text{where } (n \bmod \ell) \in X_i. \tag{1}$$

Then s is periodic since $s(n) = s(n + \ell)$. Moreover, s is in $\mathsf{Per}(\ell)$ since 0 only occurs once in the period, namely at positions congruent to 0 modulo ℓ. To prove that $A \subseteq T$, it remains to show that $\mathrm{rank}_k(s) = d\ell$.

First, the sequence $(B_{k,\ell,e}(s))_{e \geq 0}$ defined previously is periodic with period length at most d since $B_{k,\ell,e}(s) = B_{k,\ell,e+d}(s)$ by definition of the orbits. We prove that the period length is exactly d. Let $e, e' \in \{0, 1, \ldots, d - 1\}$ such that $B_{k,\ell,e}(s) = B_{k,\ell,e'}(s)$; we show $e = e'$. Since 0 only occurs once in the period of s, there is precisely one sequence in $B_{k,\ell,e}(s)$ whose initial term is 0, namely $s(k^e n)_{n \geq 0}$. Therefore, $s(k^e n)_{n \geq 0} = s(k^{e'} n)_{n \geq 0}$. In particular, $s(k^e) = s(k^{e'})$, so $k^e, k^{e'} \in X_i$ for some $i \neq 0$. By definition of X_i, $k^{e+jd} \equiv k^{e'} \bmod \ell$ for some $j \geq 0$. So $e + jd \equiv e' \bmod \mathrm{ord}_\ell(k)$, which in turn implies $e \equiv e' \bmod d$, yielding in turn $e = e'$. Therefore, $\mathrm{rank}_k(s) = d\ell$, as desired.

The second part of the statement follows by Corollary 1. $\square$

Table 2. For each possible value of the rank of periodic sequences in Per(7), we provide a sequence with such rank.

d	rank	$3^d \bmod 7$	partition of $\mathbb{Z}/(7\mathbb{Z})$	period
6	42	1	{0},{1},{2},{3},{4},{5},{6}	(0,1,2,3,4,5,6)
3	21	6	{0},{1,6},{2,5},{3,4}	(0,1,2,3,3,2,1)
2	14	2	{0},{1,2,4},{3,5,6}	(0,1,1,2,1,2,2)
1	7	3	{0},{1,2,3,4,5,6}	(0,1,1,1,1,1,1)

Notice that the previous proof is constructive, that is, for each divisor d of $\mathrm{ord}_\ell(k)$, Eq. (1) defines a sequence whose rank is $d\ell$.

Example 3. We have $\mathrm{ord}_7(3) = 6$, so the muggle numbers are in the set $\{7d : d \in \mathrm{Div}(\mathrm{ord}_7(3))\} = \{7, 14, 21, 42\}$. We provide the period of a periodic sequence in Per(7) with each rank as shown in Table 2.

Remark 1. Neder [25, A217519] conjectured that if s_ℓ is the sequence with period $(0, 1, \ldots, \ell - 1)$, then "$\mathrm{rank}_2(s_\ell) \leq \ell(\ell - 1)$, with equality if and only if ℓ is a prime with primitive root 2". Recall that a *primitive root* of a prime ℓ is an integer k such that $\mathrm{ord}_\ell(k) = \ell - 1$. Neder's conjecture follows from Theorem 1 and holds even when $k \neq 2$.

4   Constant-Recursive Sequences

An integer sequence s is *constant-recursive* if there exist $c_0, c_1, \ldots, c_{d-1} \in \mathbb{Z}$, with $d \in \mathbb{N}$ and $c_0 \neq 0$ such that

$$s(n + d) = c_{d-1}s(n + d - 1) + \cdots + c_1 s(n + 1) + c_0 s(n) \tag{2}$$

for all $n \geq 0$. For a constant-recursive sequence s, we let rank(s) denote its *rank*, which is the smallest integer d such that s satisfies a recurrence relation of the form of Eq. (2). In other words, if the set $\{s(n + j)_{n \geq 0} : j \geq 0\}$ containing all shifts of s is called the 1-*kernel* of s, then the rank of s is the dimension of the $\mathbb{Q}$-vector space generated by its 1-kernel.

For an integer $\ell \geq 2$ and the family of periodic sequences Per(ℓ), we consider the following property: an integer satisfies $\mathrm{P_{cr}}(\ell)$ if and only if it is equal to the rank of some sequence in Per(ℓ). The first few possible values of rank(s) for $s \in$ Per(ℓ), with $\ell \in \{2, \ldots, 15\}$, are given in Table 3.

The strategy adopted for $\mathrm{P_{cr}}(\ell)$ is the following: we provide first the range in Proposition 4, then a full characterization of the magic and muggle numbers in Theorem 2. To state them, we need some definitions and results on *cyclotomic polynomials*. For further details on this topic we address the reader to the nice reference [10].

Definition 1. *Let $m \geq 1$. The mth cyclotomic polynomial is the monic polynomial Φ_m whose roots in $\mathbb{C}$ are the primitive mth roots of unity.*

Table 3. First few values of $\text{rank}(s)$ when $s \in \text{Per}(\ell)$ for $\ell \in \{2, \ldots, 15\}$.

ℓ	$\text{rank}(s), s \in \text{Per}(\ell)$	ℓ	$\text{rank}(s), s \in \text{Per}(\ell)$
2	1,2	9	6,7,8,9
3	2,3	10	4,5,6,8,9,10
4	2,3,4	11	10,11
5	4,5	12	4,5,6,7,8,9,10,11,12
6	2,3,4,5,6	13	12,13
7	6,7	14	6,7,8,12,13,14
8	4,5,6,7,8	15	6,7,8,9,10,11,12,13,14,15

The first few cyclotomic polynomials are $\Phi_1(x) = x - 1$, $\Phi_2(x) = x + 1$, $\Phi_3(x) = x^2 + x + 1$, $\Phi_4(x) = x^2 + 1$, $\Phi_5(x) = x^4 + x^3 + x^2 + x + 1$, and $\Phi_6(x) = x^2 - x + 1$. For all $m \geq 2$, $\Phi_m(x)$ is palindromic. We also have the following identity: $x^n - 1 = \prod_{m \in \text{Div}(n)} \Phi_m(x)$ for all $n \geq 1$. The *Euler totient function* ϕ maps every positive integer n to the number of positive integers less than n and relatively prime to n.

Lemma 3 (Garrett [10]). *For all $m \geq 1$, $\Phi_m(x)$ is monic of degree $\phi(m)$.*

The *additive version* ψ of the Euler totient function ϕ is defined as follows:
- for every prime p and every $k \geq 1$, $\psi(p^k) = \phi(p^k)$;
- if n is odd, then $\psi(2n) = \phi(n)$; if m and n are relatively prime and not equal to 2, then $\psi(mn) = \phi(m) + \phi(n)$.

Proposition 4. *Let $\ell \geq 2$. The interval $[\psi(\ell), \ell]$ is the range for $\text{P}_{cr}(\ell)$. Moreover, the sequence $s \in \text{Per}(\ell)$ with period $(0, 0, \ldots, 0, 1)$ satisfies $\text{rank}(s) = \ell$.*

Theorem 2. *Let $\ell \geq 2$. Let S be the set of non-empty sets $\{d_1, d_2, \ldots, d_j\}$ of non-negative pairwise distinct integers such that $\text{lcm}(d_1, d_2, \ldots, d_j) = \ell$. The set $R' = \{\sum_{i=1}^{j} \phi(d_i) : \{d_1, d_2, \ldots, d_j\} \in S\}$ is the set of muggle numbers in the range for $\text{P}_{cr}(\ell)$. In particular, the set of magic numbers for $\text{P}_{cr}(\ell)$ is $\mathbb{N} \setminus R'$.*

Our characterization of the magic and muggle numbers with respect to $\text{P}_{cr}(\ell)$ from Theorem 2 will follow from an intermediate result, i.e. Theorem 3 (whose proof is postponed). The latter characterizes these numbers in terms of sizes of simple matrices with order ℓ. As usual, we let $\text{GL}_d(\mathbb{Z})$ denote the *general linear group of degree d over $\mathbb{Z}$*, made of invertible matrices of size d. Note that for every matrix in $\text{GL}_d(\mathbb{Z})$, its determinant is ± 1. Given a matrix $M \in \text{GL}_d(\mathbb{Z})$, its *order* $\text{ord}\, M$ is the smallest positive integer m such that $M^m = I$, where I is the identity matrix of size d, or $+\infty$ if such an integer does not exist. We say that a matrix M is *simple* if all its eigenvalues (in $\mathbb{C}$) have algebraic multiplicity 1.

Theorem 3. *Let $\ell \geq 2$. The set*

$$R = \{d \in \mathbb{N} : \text{there exists a simple matrix } M \in \text{GL}_d(\mathbb{Z}) \text{ such that } \text{ord}\, M = \ell\}$$

is the set of muggle numbers in the range for $\text{P}_{cr}(\ell)$.

In fact, with Theorem 4, Hiller showed that the minimal size of matrices of a given order corresponds to a specific value of the function ψ. Therefore, it gives the lower bound in Proposition 4 (see [25, A080737, A152455]).

Theorem 4 (Hiller [13]). *The smallest integer $d \geq 1$ such that $\mathrm{GL}_d(\mathbb{Z})$ contains a matrix of order ℓ is $\psi(\ell)$.*

We now develop the necessary tools to prove both Theorems 2 and 3.

Let s be a constant-recursive sequence satisfying the recurrence relation in Eq. (2). The *characteristic polynomial* of the underlying recurrence is $\chi(x) = x^d - c_{d-1}x^{d-1} - \cdots - c_1 x - c_0$. For a monic polynomial $p(x) = x^\ell + a_{\ell-1}x^{\ell-1} + \cdots + a_1 x + a_0$, we let $C(p)$ denote the *companion matrix* of $p(x)$ given by the matrix of size ℓ

$$\begin{bmatrix} 0 & 0 & \cdots & 0 & -a_0 \\ 1 & 0 & \cdots & 0 & -a_1 \\ 0 & 1 & \cdots & 0 & -a_2 \\ \vdots & \vdots & \ddots & \vdots & \vdots \\ 0 & 0 & \cdots & 1 & -a_{\ell-1} \end{bmatrix}.$$

The following result links the period length of a periodic sequence satisfying Eq. (2) and the order of the companion matrix of the previous recurrence.

Proposition 5. *Let s be a constant-recursive sequence satisfying Eq. (2), let $\chi(x)$ be the characteristic polynomial of the underlying recurrence, and let $M \in \mathrm{GL}_d(\mathbb{Z})$ be the companion matrix of $\chi(x)$. Assume that $\mathrm{ord}\, M$ is finite.*

1. *The period length of s is at most $\mathrm{ord}\, M$.*
2. *Moreover, if the initial conditions are $s(0) = s(1) = \cdots = s(d-2) = 0$ and $s(d-1) = 1$, then the period length of s is exactly $\mathrm{ord}\, M$.*

The following result characterizes matrices of finite order.

Theorem 5 (Koo [21, p. 147]). *A matrix $M \in \mathrm{GL}_d(\mathbb{Q})$ has finite order if and only if there exist positive integers $r, m_1, \ldots, m_r, n_1, \ldots, n_r$ (where the m_i's are pairwise distinct), an invertible matrix $P \in \mathrm{GL}_d(\mathbb{Q})$, and a block-diagonal matrix*

$$A = \mathrm{diag}(\underbrace{C(\Phi_{m_1}), \ldots, C(\Phi_{m_1})}_{n_1\ times}, \underbrace{C(\Phi_{m_2}), \ldots, C(\Phi_{m_2})}_{n_2\ times}, \ldots, \underbrace{C(\Phi_{m_r}), \ldots, C(\Phi_{m_r})}_{n_r\ times}),$$

with $\sum_{i=1}^r n_i \deg(\Phi_{m_i}) = d$ such that $M = P^{-1}AP$. Moreover, the order of M is $\mathrm{lcm}(m_1, m_2, \ldots, m_r)$.

Observe that $r \leq \sum_{i=1}^r n_i \deg(\Phi_{m_i}) = d$ since $n_i, \deg(\Phi_{m_i}) \geq 1$ for all i.

We state two lemmas useful to prove Theorem 3, describing the characteristic polynomial of the minimal recurrence satisfied by a constant-recursive and periodic sequence.

Lemma 4 *Let $\ell \geq 2$. For every constant-recursive and periodic sequence s whose period length divides ℓ, the characteristic polynomial of the minimal recurrence satisfied by s divides $x^\ell - 1$.*

Lemma 5 *Let $\ell \geq 2$ and let $f(x)$ be a factor of $x^\ell - 1$. The companion matrix M of $f(x)$ has finite order. Moreover, $f(x) = \prod_{i=1}^{r} \Phi_{m_i}(x)$ is a product of distinct cyclotomic polynomials where each index m_i is a divisor of ℓ and $\operatorname{ord} M = \operatorname{lcm}(m_1, m_2, \ldots, m_r)$.*

We are now ready to prove Theorem 3.

Proof (of Theorem 3) For the sake of conciseness, set $T = \{\operatorname{rank}(s) : s \in \operatorname{Per}(\ell)\}$. We prove that $T = R$ by showing two inclusions.

First Inclusion. We show $T \subseteq R$. So let $s \in \operatorname{Per}(\ell)$ and let $d = \operatorname{rank}(s)$. We construct a simple matrix $M \in \operatorname{GL}_d(\mathbb{Z})$ such that $\operatorname{ord} M = \ell$. Since $\operatorname{rank}(s) = d$, the sequence s satisfies a recurrence of the form of Eq. (2) by definition. Let M be the companion matrix of the characteristic polynomial $\chi(x)$ of this recurrence. By Lemma 4, $\chi(x)$ is a factor of $x^\ell - 1$, so the matrix M is simple. By Lemma 5, M has finite order and $\operatorname{ord} M \leq \ell$. The other inequality, $\ell \leq \operatorname{ord} M$, follows from Proposition 5. Therefore, $T \subseteq R$.

Second Inclusion. We show $R \subseteq T$. Let $d \geq 1$ and let $M \in \operatorname{GL}_d(\mathbb{Z})$ be a simple matrix such that $\operatorname{ord} M = \ell$. By minimality, for all $m \in \operatorname{Div}(\ell)$ with $m < \ell$, we have $M^m \neq I$. By Theorem 5, $\chi_M(x) = \prod_{i=1}^{r} \Phi_{m_i}(x)^{n_i}$ for some integers r, m_i, n_i, where each m_i is a divisor of ℓ, the m_i's are distinct, and $\operatorname{lcm}(m_1, m_2, \ldots, m_r) = \ell$. Since M is simple, $n_i = 1$ for all i. Write $\chi_M(x) = x^d - c_{d-1}x^{d-1} - \cdots - c_1 x - c_0$ where $c_0, c_1, \ldots, c_{d-1} \in \mathbb{Z}$. Let s be the sequence satisfying the recurrence $s(n+d) = c_{d-1}s(n+d-1) + \cdots + c_1 s(n+1) + c_0 s(n)$ for all $n \geq 0$ with initial conditions $s(0) = s(1) = \cdots = s(d-2) = 0$ and $s(d-1) = 1$. By Proposition 5, the period length of s is $\operatorname{ord}(M) = \ell$, so $s \in \operatorname{Per}(\ell)$. We claim that $\operatorname{rank}(s) = d$. By definition, s has rank at most d. For all $d' < d$, the only sequence beginning with d' initial zeroes and satisfying a recurrence relation of order d' is the constant-zero sequence. Hence $\operatorname{rank}(s) = d$. Consequently, $R \subseteq T$. □

We illustrate each direction of the proof with an example.

Example 4 Let $s \in \operatorname{Per}(3)$ be the sequence with period $(-1, 0, 1)$. It satisfies $s(n+2) = -s(n+1) - s(n)$ for all $n \geq 0$. The characteristic polynomial of this recurrence is $x^2 + x + 1$, whose companion matrix is

$$M = \begin{bmatrix} 0 & -1 \\ 1 & -1 \end{bmatrix}$$

and which is a factor of $x^3 - 1$. It can be checked that $\operatorname{ord} M$ indeed equals 3.

Example 5 Let $M \in \mathrm{GL}_6(\mathbb{Z})$ be the matrix whose rows are given by the six vectors

$$\begin{bmatrix} 1 & -1 & 1 & 0 & 1 & 1 \end{bmatrix}, \quad \begin{bmatrix} 1 & -1 & 1 & 0 & 1 & 0 \end{bmatrix}, \quad \begin{bmatrix} 2 & -1 & 1 & 1 & 0 & 0 \end{bmatrix},$$
$$\begin{bmatrix} -2 & 1 & -2 & -1 & -1 & -1 \end{bmatrix}, \begin{bmatrix} -1 & 0 & 0 & 0 & -1 & 0 \end{bmatrix}, \begin{bmatrix} -1 & 1 & -1 & 0 & 0 & -1 \end{bmatrix}.$$

One can check that M has order 15 and its characteristic polynomial is equal to $\chi_M(x) = x^6 + 2x^5 + 3x^4 + 3x^3 + 3x^2 + 2x + 1 = \Phi_3(x)\Phi_5(x)$. It is not difficult to find an invertible matrix $P \in \mathrm{GL}_6(\mathbb{Q})$ such that $PMP^{-1} = \mathrm{diag}(C(\Phi_3), C(\Phi_5))$. We define s to be the sequence satisfying $s(0) = s(1) = s(2) = s(3) = s(4) = 0$, $s(5) = 1$, and, for all $n \geq 0$,

$$s(n+6) = -2s(n+5) - 3s(n+4) - 3s(n+3) - 3s(n+2) - 2s(n+1) - s(n).$$

It can be verified that s has period $(0, 0, 0, 0, 0, 1, -2, 1, 1, -2, 2, -1, -1, 2, -1)$, so $s \in \mathsf{Per}(15)$ and $\mathrm{rank}(s) = 6$.

We now turn to the proof of Theorem 2. Note that the lcm condition in the statement implies that each d_i is a divisor of ℓ.

Proof (of Theorem 2) Let $\ell \geq 2$. By Theorem 3, the sets $\{\mathrm{rank}(s) : s \in \mathsf{Per}(\ell)\}$ and R are equal. By Theorem 5 and Lemma 3, there exists a simple matrix M of size d with order ℓ if and only if $d = \sum_{i=1}^{j} \phi(d_i)$ for some set of integers $\{d_1, d_2, \ldots, d_j\} \in S$. The conclusion of the first part follows and the second one follows by Proposition 4. □

Notice that Theorem 2 directly implies the following, explaining why only two values are present on prime rows in Table 3.

Corollary 2 *For every prime ℓ, the only muggle numbers with respect to $\mathrm{P}_{cr}(\ell)$ are $\ell - 1$ and ℓ.*

5 Regular Sequences

In this section, we look at k-regular sequences [2], which are base-k analogues of constant-recursive sequences. An integer sequence s is k-*regular* if the $\mathbb{Q}$-vector space $V_k(s)$ generated by its k-kernel (defined in Sect. 3) is finitely generated. The k-*rank* (or simply the *rank* when the context is clear) of s is the dimension of $V_k(s)$. We use, again, the notation $\mathrm{rank}_k(s)$.

Every periodic sequence is k-automatic for every k and thus also k-regular. However, the corresponding k-ranks of are not necessarily equal.

Example 6 Consider the sequence $s \in \mathsf{Per}(4)$ with period $(0, 1, 1, 1)$. The 2-kernel of s consists of the sequences with periods $(0, 1, 1, 1)$, $(0, 1)$, (1), and (0). Viewed as a 2-automatic sequence, s has rank 4, while, when seen as a 2-regular sequence, it has rank 3 (the sequences with periods $(0, 1, 1, 1)$, $(0, 1)$, and (1) are linearly independent).

For integers $k, \ell \geq 2$ and the family of periodic sequences $\mathsf{Per}(\ell)$, we consider the following property: an integer satisfies $\mathrm{P_r}(k, \ell)$ if and only if it is equal to the k-rank of some sequence in $\mathsf{Per}(\ell)$. To look for magic and muggle numbers with respect to $\mathrm{P_r}(k, \ell)$, we will use the following notion. Let $n \geq 1$ be an integer and let $(a_0, a_1, \ldots, a_{n-1})$ be a sequence of integers. The *circulant matrix* $C(a_0, a_1, \ldots, a_{n-1})$ is the matrix of size n for which the ith row, $i \in \{1, \ldots, n\}$, is the $(i-1)$st shift of the first row. The polynomial $f(x) = \sum_{i=0}^{n-1} a_i x^i$ is the *associated polynomial* of $C(a_0, a_1, \ldots, a_{n-1})$.

We link the rank of the circulant matrix associated with a given periodic sequence to its constant-recursive and k-regular ranks. The proof of the first result follows from [4, Theorem 2.1.6] and the discussion in [4, Chapter 6, Sect. 1].

Proposition 6 (Berstel & Reutenauer [4]). *Let $\ell \geq 2$. Let $s \in \mathsf{Per}(\ell)$ and let $M = C(s(0), s(1), \ldots, s(\ell-1))$ be the circulant matrix of size ℓ built on the period of s. The rank of s as a constant-recursive sequence is equal to* $\mathrm{rank}\, M$.

Proposition 7. *Let $k, \ell \geq 2$ be coprime integers. Let $s \in \mathsf{Per}(\ell)$ and let $M = C(s(0), s(1), \ldots, s(\ell-1))$ be the circulant matrix of size ℓ built on the period of s. Then* $\mathrm{rank}_k(s) = \mathrm{rank}\, M$.

Theorem 6. *Let $k, \ell \geq 2$ be coprime integers. The range for $\mathrm{P_r}(k, \ell)$ is $[\psi(\ell), \ell]$. Now let S be the set of non-empty sets $\{d_1, d_2, \ldots, d_j\}$ of non-negative pairwise distinct integers with $\mathrm{lcm}(d_1, d_2, \ldots, d_j) = \ell$. The set $R' = \{\sum_{i=1}^{j} \phi(d_i) : \{d_1, d_2, \ldots, d_j\} \in S\}$ is the set of muggle numbers in the range for $\mathrm{P_r}(k, \ell)$. In particular, the set of magic numbers with respect to $\mathrm{P_r}(k, \ell)$ is $\mathbb{N} \setminus R'$.*

Proof. In this proof, for every sequence $s \in \mathsf{Per}(\ell)$, we let M_s denote the circulant matrix $C(s(0), s(1), \ldots, s(\ell-1))$ of size ℓ built on the period of s. As in Sect. 4, we also let $\mathrm{rank}(s)$ denote the rank of s viewed as constant-recursive sequence. By Propositions 6 and 7, we have that

$$\{\mathrm{rank}_k(s) : s \in \mathsf{Per}(\ell)\} = \{\mathrm{rank}\, M_s : s \in \mathsf{Per}(\ell)\} = \{\mathrm{rank}(s) : s \in \mathsf{Per}(\ell)\}.$$

In particular, the sets of muggle numbers for $\mathrm{P_r}(k, \ell)$ and $\mathrm{P_{cr}}(\ell)$ are equal. Proposition 4 and Theorem 2 then allow to conclude the proof. $\qquad\square$

6   Conclusion and Future Work

In this paper, we studied the magic number problem for three definitions of the rank of periodic sequences. In particular, in the case of constant-recursive sequences, we obtain a full characterization of magic numbers for every period length, while, for k-automatic and k-regular sequences with period length ℓ, we give a characterization provided that k and ℓ are coprime. For such sequences, the general case looks more intricate. Indeed, as Example 6 already shows, the period lengths of sequences in the k-kernel are not always equal to ℓ, but rather divide ℓ. Therefore, a deep understanding of the relationship between the sets

Per(d), where d divides ℓ, is required to characterize the possible ranks, and so in turn the magic and muggle numbers within these families of sequences. Hence, as a natural pursuance of this study, we are currently working on the generalization of the results for automatic and regular sequences to the case where no condition on k and ℓ is specified.

Acknowledgments. Savinien Kreczman and Manon Stipulanti are supported by the FNRS Research grants 1.A.789.23F and 1.B.397.20F respectively.

References

1. Alexeev, B.: Minimal DFA for testing divisibility. J. Comput. Syst. Sci. **69**(2), 235–243 (2004). https://doi.org/10.1016/j.jcss.2004.02.001
2. Allouche, J.P., Shallit, J.O.: The ring of k-regular sequences. Theor. Comput. Sci. **98**(2), 163–197 (1992). https://doi.org/10.1016/0304-3975(92)90001-V
3. Allouche, J.P., Shallit, J.O.: Automatic Sequences: Theory, Applications, Generalizations. Cambridge University Press, Cambridge (2003)
4. Berstel, J., Reutenauer, C.: Noncommutative Rational Series with Applications, Encyclopedia of Mathematics and its Applications, vol. 137. Cambridge University Press, Cambridge (2011)
5. Boigelot, B., Mainz, I., Marsault, V., Rigo, M.: An efficient algorithm to decide periodicity of b-recognisable sets using MSDF convention. In: ICALP 2017. LIPIcs, vol. 80, pp. 118:1–118:14. Schloss Dagstuhl - Leibniz-Zentrum für Informatik (2017). https://doi.org/10.4230/LIPIcs.ICALP.2017.118
6. Bosma, W.: Complexity of periodic sequences (2019). Preprint available at https://www.math.ru.nl/~bosma/pubs/periodic.pdf
7. Büchi, R.J.: Weak second-order arithmetic and finite automata. Math. Log. Q. **6**(1–6), 66–92 (1960). https://doi.org/10.1002/malq.19600060105
8. Cobham, A.: Uniform tag sequences. Math. Systems Theory **6**, 164–192 (1972). https://doi.org/10.1007/BF01706087
9. Everest, G., van der Poorten, A.J., Shparlinski, I.E., Ward, T.: Recurrence Sequences, Mathematical Surveys and Monographs, vol. 104. American Mathematical Society (2003)
10. Garrett, P.B.: Abstract Algebra. Chapman & Hall/CRC, Boca Raton (2008)
11. Geffert, V.: (Non)determinism and the size of one-way finite automata. In: DCFS 2005. Proceedings, pp. 23–37. Università degli Studi di Milano, Milan, Italy (2005)
12. Geffert, V.: Magic numbers in the state hierarchy of finite automata. Inf. Comput. **205**(11), 1652–1670 (2007). https://doi.org/10.1016/j.ic.2007.07.001
13. Hiller, H.: The crystallographic restriction in higher dimensions. Acta Crystallogr. A **41**(6), 541–544 (1985). https://doi.org/10.1107/S0108767385001180
14. Holzer, M., Jakobi, S., Kutrib, M.: The magic number problem for subregular language families. Int. J. Found. Comput. Sci. **23**(1), 115–131 (2012). https://doi.org/10.1142/S0129054112400084
15. Honkala, J.: A decision method for the recognizability of sets defined by number systems. RAIRO Theor. Inform. Appl. **20**(4), 395–403 (1986). https://doi.org/10.1051/ita/1986200403951
16. Iwama, K., Kambayashi, Y., Takaki, K.: Tight bounds on the number of states of DFAs that are equivalent to n-state NFAs. Theor. Comput. Sci. **237**(1–2), 485–494 (2000). https://doi.org/10.1016/S0304-3975(00)00029-3

17. Iwama, K., Matsuura, A., Paterson, M.: A family of NFAs which need $2^n - \alpha$ deterministic states. Theor. Comput. Sci. **301**(1–3), 451–462 (2003). https://doi.org/10.1016/S0304-3975(02)00891-5

18. Jirásková, G.: Deterministic blow-ups of minimal NFA's. RAIRO Theor. Inf. Appl. **40**(3), 485–499 (2006). https://doi.org/10.1051/ita:2006032

19. Jirásková, G.: On the state complexity of complements, stars, and reversals of regular languages. In: DLT 2008. Proceedings. LNCS, vol. 5257, pp. 431–442. Springer, Cham (2008). https://doi.org/10.1007/978-3-540-85780-8_34

20. Jirásková, G.: Magic numbers and ternary alphabet. Int. J. Found. Comput. Sci. **22**(2), 331–344 (2011). https://doi.org/10.1142/S0129054111008076

21. Koo, R.: A classification of matrices of finite order over $\mathbb{C}$, $\mathbb{R}$ and $\mathbb{Q}$. Math. Mag. **76**(2), 143–148 (2003)

22. Marsault, V.: An efficient algorithm to decide periodicity of b-recognisable sets using LSDF convention. Log. Methods Comput. Sci. **15**(3) (2019). https://doi.org/10.23638/LMCS-15(3:8)2019

23. Meyer, A.R., Fischer, M.J.: Economy of description by automata, grammars, and formal systems. In: 12th Annual Symposium on Switching and Automata Theory, East Lansing, Michigan, USA, 13–15 October 1971, pp. 188–191. IEEE Computer Society (1971). https://doi.org/10.1109/SWAT.1971.11

24. Ouaknine, J., Worrell, J.: Decision problems for linear recurrence sequences. In: RP 2012. LNCS, vol. 7550, pp. 21–28. Springer, Heidelberg (2012). https://doi.org/10.1007/978-3-642-33512-9_3

25. Sloane, N.J.A.: The On-Line Encyclopedia of Integer Sequences. http://oeis.org

26. Sutner, K.: Divisibility and state complexity. Mathematica J. **11**(3), 430–445 (2009). https://doi.org/10.3888/tmj.11.3-8

27. Sutner, K., Tetruashvili, S.: Inferring automatic sequences (2012). https://www.cs.cmu.edu/~sutner/papers/auto-seq.pdf

28. Zantema, H., Bosma, W.: Complexity of automatic sequences. Inf. Comput. **288**, 104710 (2022). https://doi.org/10.1016/j.ic.2021.104710. Special Issue: Selected Papers of the 14th International Conference on Language and Automata Theory and Applications, LATA 2020

Dyck Words, Pattern Avoidance, and Automatic Sequences

Lucas Mol[1], Narad Rampersad[2]([✉]), and Jeffrey Shallit[3]

[1] Department of Mathematics and Statistics, Thompson Rivers University,
805 TRU Way, Kamloops, BC V2C 0C8, Canada
lmol@tru.ca

[2] Department of Mathematics and Statistics, University of Winnipeg,
515 Portage Ave., Winnipeg, MB R3B 2E9, Canada
n.rampersad@uwinnipeg.ca

[3] School of Computer Science, University of Waterloo,
200 University Ave. W., Waterloo, ON N2L 3G1, Canada
shallit@uwaterloo.ca

Abstract. We study various aspects of Dyck words appearing in binary sequences, where 0 is treated as a left parenthesis and 1 as a right parenthesis. We show that binary words that are 7/3-power-free have bounded nesting level, but this no longer holds for larger repetition exponents. We give an explicit characterization of the factors of the Thue-Morse word that are Dyck, and show how to count them. We also prove tight upper and lower bounds on $f(n)$, the number of Dyck factors of Thue-Morse of length $2n$.

Keywords: Dyck word · pattern avoidance · automatic sequence

1 Introduction

We define $\Sigma_k := \{0, 1, \ldots, k - 1\}$.

Suppose $x \in \Sigma_2^*$; that is, suppose x is a finite binary word. We say it is a *Dyck word* if, considering 0 as a left parenthesis and 1 as a right parenthesis, the word represents a string of balanced parentheses [5]. For example, 010011 is Dyck, while 0110 is not. Formally, x is Dyck if x is empty, or there are Dyck words y, z such that either $x = 0y1$ or $x = yz$. The set of all Dyck words forms the *Dyck language*, denoted here by $\mathcal{D}_2$.

In this paper we are concerned with the properties of factors of infinite binary words that are Dyck words.

If x is a Dyck word, we may talk about its *nesting level* $N(x)$, which is the deepest level of parenthesis nesting in the string it represents. Formally we

Research of Lucas Mol is supported by an NSERC Grant, Grant number RGPIN-2021-04084. Research of Narad Rampersad is supported by an NSERC Grant, Grant number 2019-04111. Research of Jeffrey Shallit is supported by an NSERC Grant, Grant number 2018-04118.

© The Author(s), under exclusive license to Springer Nature Switzerland AG 2023
A. Frid and R. Mercaş (Eds.): WORDS 2023, LNCS 13899, pp. 220–232, 2023.
https://doi.org/10.1007/978-3-031-33180-0_17

have $N(\epsilon) = 0$, $N(0y1) = N(y) + 1$, and $N(yz) = \max(N(y), N(z))$ if y, z are Dyck words. The Dyck property and nesting level are intimately connected with *balance*, which is a function defined by $B(x) = |x|_0 - |x|_1$, the excess of 0's over 1's in x. It is easy to see that a word is Dyck if and only if $B(x) = 0$ and $B(x') \geq 0$ for every prefix x' of x. Furthermore, the nesting level of a Dyck word x is the maximum of $B(x')$ over all prefixes x' of x.

In this paper we will also be concerned with pattern avoidance, particularly avoidance of powers. We say a finite word $w = w[1..n]$ has period $p \geq 1$ if $w[i] = w[i + p]$ for all indices i with $1 \leq i \leq n - p$. The smallest period of w is called *the* period, and is denoted per(w). The *exponent* of a finite word w is defined to be $\exp(w) := |w|/\operatorname{per}(w)$. A word with exponent α is said to be an α-power. For example, $\exp(\texttt{alfalfa}) = 7/3$ and so $\texttt{alfalfa}$ is a 7/3-power. If a word contains no powers $\geq \alpha$, then we say it is α-*power-free*. If it contains no powers $> \alpha$, then we say it is α^+-*power-free*. A *square* is a word of the form xx, where x is a nonempty word. An *overlap* is a word of the form $axaxa$, where a is a single letter and x is a possibly empty word.

Some of our work is carried out using the $\texttt{Walnut}$ theorem prover, which can rigorously prove many results about automatic sequences. See [9,11] for more details. $\texttt{Walnut}$ is free software that can be downloaded at https://cs.uwaterloo. ca/~shallit/walnut.html .

Some details have been omitted for space considerations. For the full version of the paper, see [8].

2   Repetitions and Dyck Words

Theorem 1. *If a binary word is 7/3-power-free and Dyck, then its nesting level is at most 3.*

Proof. The 7/3-power-free Dyck words of nesting level 1 are 01 and 0101. The set of 7/3-power-free Dyck words of nesting level 2 is therefore a subset of $\{01, 0011, 001011\}^*$. Let x be a 7/3-power-free Dyck word of nesting level 3. Suppose that $x = 0y1$, where y has nesting level 2. Then to avoid the cubes 000 and 111, the word y must begin with 01 and end with 01. Furthermore, since y has nesting level 2 it must contain one of 0011 or 001011. Write $x = 001y'011$. The word y' cannot begin or end with 01, since that would imply that x contains one of the 5/2-powers 01010 or 10101. Thus y' begins with 001 and ends with 011, which means x begins with 001001 and ends with 011011. Consequently x cannot be extended to the left or to the right without creating a cube or 7/3-power. Furthermore, this implies that a 7/3-power-free Dyck word of nesting level 3 cannot be written as a concatenation of two non-empty Dyck words, nor can it be extended to a 7/3-power-free Dyck word of nesting level 4.

Theorem 2. *Define $h(0) = 01$, $h(1) = 0011$, and $h(2) = 001011$. A binary word w is an overlap-free Dyck word if and only if either*

(i) $w = h(x)$, where $x \in \Sigma_3^*$ is squarefree and contains no 212 or 20102; or

(ii) $w = 0h(x)1$, where $x \in \Sigma_3^*$ is squarefree, begins with 01 and ends with 10, and contains no 212 or 20102.

Proof. Let w be an overlap-free Dyck word. By Theorem 1, we have $N(w) \leq 3$. Suppose $N(w) \leq 2$. Then $w \in \{01, 0011, 001011\}^*$ by the proof of Theorem 1. So we have $w = h(x)$ for some $x \in \Sigma_3^*$. If $N(w) = 3$, then by the proof of Theorem 1, we have $w = 0h(x)1$. If x contains a square yy as a proper factor, then certainly w contains one of the overlaps $1h(y)h(y)$ or $h(y)h(y)0$. Furthermore, if x contains 212, then w contains the overlap 011001100 and if x contains 20102, then w contains the overlap 1101001101001. Finally, if $w = 0h(x)1$, then x must begin and end with 0 and contain at least one 1 or 2. If x begins with 02, then w contains the overlap 0010010, and if x ends with 20, then w contains the overlap 1011011. Thus x begins with 01 and ends with 10.

For the other direction, let $x \in \Sigma_3^*$ be a squarefree word that contains no 212 or 20102. First consider the word $h(x)$, which is clearly a Dyck word. We now show that $h(x)$ is overlap-free. We verify by computer that if $|x| \leq 10$, then $h(x)$ is overlap-free. So we may assume that $|x| \geq 11$. Suppose towards a contradiction that $h(x)$ contains an overlap z. Assume that $z = 0y0y0$; the case $z = 1y1y1$ is similar, and the proof is omitted. We consider several cases depending on the prefix of y.

If y starts with 0, then $h^{-1}(z0^{-1}) = h^{-1}(0y0y)$ is a square that appears as a proper factor of x.

If y starts with 100, write $y = 100y'$, so that $z = 0100y'0100y'0$. Then $h^{-1}(z0^{-1}) = h^{-1}(0100y'0100y')$ is a square that appears as a proper factor of x.

If y starts with 101, write $y = 101y'$, so that $z = 0101y'0101y'0$. Note that 00 is not a factor of x, so any occurrence of 0101 in z is as a factor of $h(2) = 001011$. Consequently, the word $h^{-1}(0z0^{-1}) = h^{-1}(00101y'00101y')$ is a square that appears as a proper factor of x.

Finally, if y starts with 11, then write $y = 11y'$, so that $z = 011y'011y'0$. Then z is a factor of $h(ax'bx'c)$, where $a, b, c \in \{1, 2\}$, and the value of b is determined by the suffix of y': if y' ends with 001 then $b = 2$ and if y' ends with 0 then $b = 1$. Clearly we have $a \neq b$ and $b \neq c$, since otherwise x contains a square as a proper factor. However, if $b = 2$ then y' ends with 001, which implies $c = 2$, a contradiction. So we have $b = 1$, and further, since $a \neq b$ and $b \neq c$, we have $a = c = 2$. We therefore have a factor $2x'1x'2$ of x. Now x' can neither begin nor end with 2 or 1, so we have $2x'1x'2 = 20x''010x''02$. Similarly, the word x'' can neither begin nor end with 0 or 1, so we have $20x''010x''02 = 202x'''20102x'''202$, whence x contains the forbidden factor 20102, a contradiction.

Thus, we conclude that $h(x)$ is an overlap-free Dyck word. Finally, assume that x begins with 01 and ends with 10, and consider the word $0h(x)1$. Again, it is clear that $0h(x)1$ is a Dyck word, and we have already shown that the word $h(x)$ is overlap-free. Now $0h(x)1$ begins with 0010011 and ends with 0011011. Note that the only occurrences of 00100 and 11011 as factors of $0h(x)1$ are as a prefix and a suffix, respectively. It follows that if $0h(x)1$ contains an overlap, then this overlap has period at most 4 and occurs as either a prefix or a suffix of

$0h(x)1$. However, one easily verifies that no such overlap exists. This completes the proof.

Corollary 3. *There are arbitrarily long overlap-free Dyck words of nesting level 2 (and 3).*

Proof. Consider the well-known word **s**, which is the infinite fixed point, starting with 0, of the morphism defined by $0 \mapsto 012$, $1 \mapsto 02$, $2 \mapsto 1$. Thue [13] proved that **s** is squarefree and contains no 010 or 212; this is also easy to verify with `Walnut` (cf. [11]). Let x be a prefix of **s** that ends in 10. Since the factor 10 appears infinitely many times in **s**, there are arbitrarily long such words x. So x is squarefree, contains no 212 or 20102, begins in 01, and ends in 10. By Theorem 2, the words $h(x)$ and $0h(x)1$ are overlap-free Dyck words. It is easy to see that $h(x)$ has nesting level 2, and $0h(x)1$ has nesting level 3, which completes the proof.

Theorem 1 says that every 7/3-power-free Dyck word has nesting level at most 3. We will see that this result is the best possible with respect to the exponent 7/3; in fact, there are $7/3^+$-power-free Dyck words of every nesting level. For a very simple construction of cube-free Dyck words of every nesting level, which serves as a preview of the main ideas in the more complicated construction of $7/3^+$-power-free Dyck words of every nesting level given below, see the arxiv version of the paper [8].

Let $g : \Sigma_3^* \to \Sigma_3^*$ be the 6-uniform morphism defined by

$$g(0) = 022012, \quad g(1) = 022112, \quad \text{and} \quad g(2) = 202101.$$

Let $f : \Sigma_3^* \to \Sigma_2^*$ be the 38-uniform morphism defined by

$$
\begin{aligned}
f(0) &= 00100110100110010110010011001011001101, \\
f(1) &= 00101100110100110110011010010110011011, \text{ and} \\
f(2) &= 00101101001101001011001101001011010011.
\end{aligned}
$$

We will show that for every $t \geq 0$, the word $f(g^t(2))$ is a $7/3^+$-power-free Dyck word of nesting level $2t + 2$. The letters f and g denote these specific morphisms throughout the remainder of this section.

Over the ternary alphabet Σ_3, we think of the letter 0 as a left parenthesis, the letter 1 as a right parenthesis, and the letter 2 as a Dyck word. So we will be particularly interested in the ternary words for which the removal of every occurrence of the letter 2 leaves a Dyck word, and we call these *ternary Dyck words*.

Definition 4. Let $\beta : \Sigma_3^* \to \Sigma_2^*$ be defined by $\beta(0) = 0$, $\beta(1) = 1$, and $\beta(2) = \varepsilon$, and let $w \in \Sigma_3^*$. If $\beta(w)$ is a Dyck word, then we say that w is a *ternary Dyck word*. In this case, the *nesting level* of w, denoted $N(w)$, is defined by $N(w) = N(\beta(w))$.

Lemma 5. *Let $w \in \Sigma_3^*$. If w is a nonempty ternary Dyck word, then $g(w)$ is a ternary Dyck word with $N(g(w)) = N(w) + 1$.*

Proof. Throughout this proof, we let $u = 01$, a Dyck word with nesting level 1. Note that $\beta(g(0)) = 001 = 0u$, $\beta(g(1)) = 011 = u1$, and $\beta(g(2)) = 0101 = u^2$.

The proof is by induction on $|\beta(w)|$. We have two base cases. If $\beta(w) = \varepsilon$, then $w = 2^i$ for some $i \geq 1$, and $N(w) = 0$. We have $\beta(g(w)) = u^{2i}$, so we see that $g(w)$ is a ternary Dyck word with $N(g(w)) = 1 = N(w) + 1$. If $\beta(w) = 01$, then $w = 2^i 0 2^j 1 2^k$ for some $i, j, k \geq 0$, and $N(w) = 1$. We have

$$\beta(g(w)) = u^{2i}(0u)u^{2j}(u1)u^{2k} = u^{2i}0u^{2j+2}1u^{2k},$$

so we see that $g(w)$ is a ternary Dyck word with $N(g(w)) = 2 = N(w) + 1$, as desired.

Now suppose that $|\beta(w)| = n$ for some $n > 2$, and that the statement holds for all ternary Dyck words w' with $|\beta(w')| < n$. We have two cases.

Case 1: We have $\beta(w) = 0y1$ for some nonempty Dyck word y.

In this case we may write $w = 2^i 0 w' 1 2^j$ for some $i, j \geq 0$, so that $\beta(w') = y$. By the induction hypothesis, the word $g(w')$ is a ternary Dyck word with $N(g(w')) = N(w') + 1$. It follows that $\beta(g(w)) = u^{2i}0u\beta(g(w'))u1u^{2j}$ is a Dyck word, so $g(w)$ is a ternary Dyck word, and

$$N(g(w)) = 1 + N(g(w')) = 1 + N(w') + 1 = N(w) + 1.$$

Case 2: We have $\beta(w) = y_1 y_2$ for some nonempty Dyck words y_1, y_2.

Write $w = w_1 w_2$ for some $w_1, w_2 \in \Sigma_3^*$ such that $\beta(w_1) = y_1$, and $\beta(w_2) = y_2$. By the induction hypothesis, the words $g(w_1)$ and $g(w_2)$ are ternary Dyck words with $N(g(w_1)) = N(w_1) + 1$, and $N(g(w_2)) = N(w_2) + 1$. Therefore, the word $g(w) = g(w_1)g(w_2)$ is a ternary Dyck word with

$$N(g(w)) = \max\left(N(g(w_1)), N(g(w_2))\right) = \max(N(w_1) + 1, N(w_2) + 1)$$
$$= \max(N(w_1), N(w_2)) + 1 = N(w) + 1.$$

$\square$

Lemma 6. *Let $w \in \Sigma_3^*$. If w is a nonempty ternary Dyck word, then $f(w)$ is a Dyck word with $N(f(w)) = 2N(w) + 2$.*

Proof. The proof is similar to the proof of Lemma 5, and is omitted here for space considerations. It can be found in the arxiv version of the paper [8]. $\square$

Theorem 7. *There are $7/3^+$-power-free Dyck words of every nesting level.*

Proof. Let $t \geq 0$. We claim that the word $f(g^t(2))$ is a $7/3^+$-free Dyck word of nesting level $2t + 2$. Since 2 is a ternary Dyck word with nesting level 0, by Lemma 5, and a straightforward induction, the word $g^t(2)$ is a ternary Dyck word with nesting level t. Thus, by Lemma 6, the word $f(g^t(2))$ is a Dyck word with nesting level $2t + 2$.

It remains only to show that $f(g^t(2))$ is $7/3^+$-power-free. We use the Walnut theorem-prover to show that $f(g^\omega(0))$ is $7/3^+$-power-free, which is equivalent (the default behaviour of some Walnut commands makes it slightly more convenient to work with fixed points starting with 0). One only need type in the following commands:

```
morphism f
"0->0010011010011001011001001100101100110
1->0010110011010011011001101001011001011
2->0010110100110100101100110100101101001011":
morphism g "0->022012 1->022112 2->202101":
promote GG g:
image DFG f GG:
eval DFGtest "?msd_6 Ei,n (n>=1) & At (3*t<=4*n) =>
DFG[i+t]=DFG[i+t+n]":
```

and Walnut returns FALSE. Here the first two morphism commands define f and g, and the next two commands create a DFAO for $f(g^\omega(0))$. Finally, the last command asserts the existence of a $7/3^+$ power in $f(g^\omega(0))$.

This was a large computation in Walnut, requiring 130G of memory and 20321 secs of CPU time. □

3   Dyck Factors of Thue-Morse

In this section we give a characterization of those factors of $\mathbf{t}$, the Thue-Morse sequence, that are Dyck.

Let $g : \Sigma_3^* \to \Sigma_2^*$ be the morphism defined by $g(0) = 011$, $g(1) = 01$, and $g(2) = 0$ and let $f : \Sigma_3^* \to \Sigma_3^*$ be the morphism defined by $f(0) = 012$, $f(1) = 02$, and $f(2) = 1$. Define $\mathbf{s} = f^\omega(0)$. It is well-known that $g(\mathbf{s}) = \mathbf{t}$. Recall the morphism $h : \Sigma_2^* \to \Sigma_2^*$ defined earlier by $h(0) = 01$, $h(1) = 0011$, and $h(2) = 001011$.

Theorem 8. *The Dyck factors of the Thue-Morse word are exactly the words $h(x)$ where x is a factor of $\mathbf{s}$.*

Proof. By considering the return words of 11 (see [2]) in $\mathbf{t}$ we see that $\mathbf{t}$ begins with 011 followed by a concatenation of the four Dyck words

$$(0(01)1), \quad (01)(0(01)1), \quad (0(01)(01)1), \quad (01)(0(01)(01)1).$$

Furthermore, these words must have the above bracketings when they occur as factors of any larger Dyck word in $\mathbf{t}$. It follows that $\mathbf{t} = 011\mathbf{t}'$, where $\mathbf{t}'$ is a concatenation of the three Dyck words $h(0) = 01$, $h(1) = 0011$, and $h(2) = 001011$.

To complete the proof, it suffices to show that $h(\mathbf{s}) = (011)^{-1}\mathbf{t} = (011)^{-1}g(\mathbf{s})$. We have

$$h(f(0)) = h(012) = g(120210) = g(0^{-1}f^2(0)0)$$
$$h(f(1)) = h(02) = g(1210) = g(0^{-1}f^2(1)0)$$
$$h(f(2)) = h(1) = g(20) = g(0^{-1}f^2(2)0),$$

so

$$h(\mathbf{s}) = h(f(\mathbf{s})) = g(0^{-1}f^2(\mathbf{s})) = g(0^{-1}\mathbf{s}) = (011)^{-1}g(\mathbf{s}),$$

as desired. □

4   Dyck Factors of Some Automatic Sequences

In this section we are concerned with Dyck factors of automatic sequences. Recall that a sequence over a finite alphabet $(s(n))_{n\geq 0}$ is k-*automatic* if there exists a DFAO (deterministic finite automaton with output) that, on input n expressed in base k, reaches a state with output $s(n)$.

Since $\mathcal{D}_2$ is not a member of the FO[+]-definable languages [4], this means that "automatic" methods (like that implemented in the Walnut system; see [9,11]) cannot always directly handle such words. However, in this section we show that if a k-automatic sequence also has a certain special property, then the number of Dyck factors of length n occurring in it is a k-regular sequence.

To explain the special property, we need the notion of synchronized sequence [10]. We say a k-automatic sequence $(v(n))_{n\geq 0}$ is *synchronized* if there is a finite automaton accepting, in parallel, the base-k representations of n and $v(n)$. Here the shorter representation is padded with leading zeros, if necessary.

Now suppose $\mathbf{s} = (s(n))_{n\geq 0}$ is a k-automatic sequence taking values in Σ_2 and define the running sum sequence $v(n) = \sum_{0\leq i<n} s(i)$. If $\mathbf{v} = (v(n))_{n\geq 0}$ is synchronized, we say that $\mathbf{s}$ is *running-sum synchronized*. For example, any fixed point of a k-uniform binary morphism such that the images of 0 and 1 have the same number of 1's is running-sum synchronized.

Theorem 9. *Suppose* $\mathbf{s} = (s(n))_{n\geq 0}$ *is a k-automatic sequence taking values in* Σ_2 *that is running-sum synchronized. Then there is an automaton accepting, in parallel, the base-k representations of those pairs (i,n) for which* $\mathbf{s}[i..i+n-1]$ *is Dyck. Furthermore, there is an automaton accepting, in parallel, the base-k representations of those triples (i,n,x) for which* $\mathbf{s}[i..i+n-1]$ *is Dyck and whose nesting level is x. In both cases, the automaton can be effectively constructed.*

Proof. We use the fact that it suffices to create first-order logical formulas for these claims (see [11, Chapter 6] for a precise description of the logical structure we are working with).

Suppose $V(n,x)$ is true if and only $v(n) = x$. Then define

$$N_1(i,n,x) : \exists y,z \; V(i,y) \wedge V(i+n,z) \wedge x+y = z$$
$$N_0(i,n,x) : \exists y \; N_1(i,n,y) \wedge n = x+y$$
$$\mathrm{Dyck}(i,n) : (\exists w \; N_0(i,n,w) \wedge N_1(i,n,w)) \wedge$$
$$(\forall t,y,z \; (t < n \wedge N_0(i,t,y) \wedge N_1(i,t,z)) \implies y \geq z).$$

Here

- $N_0(i,n,x)$ asserts that $|\mathbf{s}[i..i+n-1]|_0 = x$;
- $N_1(i,n,x)$ asserts that $|\mathbf{s}[i..i+n-1]|_1 = x$;
- $\mathrm{Dyck}(i,n)$ asserts that $\mathbf{s}[i..i+n-1]$ is Dyck.

We can now build an automaton for $\mathrm{Dyck}(i,n)$ using the methods discussed in [11].

Next we turn to nesting level. First we need a first-order formula for the balance $B(x)$ of a factor x. Since we are only interested in balance for prefixes of Dyck words, it suffices to compute $\max(0, B(x))$ for a factor x. We can do this as follows:

$$\text{Bal}(i, n, x) : \exists y, z \ N_0(i, n, y) \wedge N_1(i, n, z) \wedge ((y < z \wedge x = 0) \mid (y \geq z \wedge y = x + z)).$$

Next, we compute the nesting level of a factor, assuming it is Dyck:

$$\text{Nest}(i, n, x) : \exists m \ m < n \wedge \text{Bal}(i, m, x) \wedge \forall p, y \ (p < n \wedge \text{Bal}(i, p, y)) \implies y \leq x.$$

This completes the proof. $\qquad\square$

Corollary 10. *If* $\mathbf{s} = (s(n))_{n \geq 0}$ *is a k-automatic sequence taking values in Σ_2 that is running-sum synchronized, then it is decidable*

(a) whether $\mathbf{s}$ has arbitrarily large Dyck factors;
(b) whether Dyck factors of $\mathbf{s}$ are of unbounded nesting level.

Proof. It suffices to create first-order logical statements asserting the two properties:

(a) $\forall n \ \exists i, m \ m > n \wedge \text{Dyck}(i, m)$
(b) $\forall q \ \exists i, n, p \ \text{Dyck}(i, n) \wedge \text{Nest}(i, n, p) \wedge p > q$.

$\qquad\square$

Example 11. As an example, let us use `Walnut` to prove that there is a Dyck factor of the Thue-Morse word for all even lengths. We can use the following `Walnut` commands, which implement the ideas above. We use the fact that the sum of $T[0..n-1]$ is $n/2$ if n is even, and $(n-1)/2 + T[n-1]$ if n is odd.

```
def even "Ek n=2*k":
def odd "Ek n=2*k+1":
def V "($even(n) & 2*x=n) | ($odd(n) & 2*x+1=n & T[n-1]=@0) |
    ($odd(n) & 2*x=n+1 & T[n-1]=@1)":
# number of 1's in prefix T[0..n-1]

def N1 "Ey,z $V(i,y) & $V(i+n,z) & x+y=z":
# number of 1's in T[i..i+n-1]
def N0 "Ey $N1(i,n,y) & n=x+y":

def Dyck "(Ew $N0(i,n,w) & $N1(i,n,w)) &
    At,y,z (t<n & $N0(i,t,y) & $tmfac1(i,t,z)) => y>=z":
# is T[i..i+n-1] a Dyck word?

eval AllLengths "An $even(n) => Ei $Dyck(i,n)":
```

and `Walnut` returns TRUE.

Example 12. Continuing the previous example, let us show that the nesting level of every Dyck factor of Thue-Morse is ≤ 2. Of course, this follows from Theorem 8, but this shows how it can be done for any automatic sequence that is running-sum synchronized. We use the following `Walnut` commands:

```
def Bal "Ey,z $N0(i,n,y) & $N1(i,n,z) &
  ((y<z & x=0) | (y>=z & y=x+z))":
# computes max(0, B(T[i..i+n])) where B is balance; 14 states
def Nest "Em (m<n) & $Bal(i,m,x) &
  Ap,y (p<n & $Bal(i,p,y)) => y<=x":
# computes nesting level of factor, assuming it is Dyck
```

```
eval maxnest2 "Ai,n,x ($Dyck(i,n) & $Nest(i,n,x)) => x<=2":
```

and `Walnut` returns `TRUE` for the last assertion.

Now we turn to enumerating Dyck factors by length. Let us recall that a sequence $(s(n))_{n\geq 0}$ is k-*regular* if there is a finite set of sequences $(s_i(n))_{n\geq 0}$, $i = 1, \ldots, t$, with $s = s_1$, such that every subsequence of the form $(s(k^e n + a))_{n\geq 0}$ with $e \geq 0$ and $0 \leq a < k^e$ can be expressed as a linear combination of the s_i. See [1] for more details.

Alternatively, a sequence $(s(n))_{n\geq 0}$ is k-regular if there is a linear representation for it. If v is a row vector of dimension t, w is a column vector of dimension t, and γ is a matrix-valued morphism with domain Σ_k and range $t \times t$-matrices, then we say that the triple (v, γ, w) is a *linear representation* for a function $s(n)$, of rank t. It is defined by $s(n) = v\gamma(x)w$, where x is any base-k representation of n (i.e., possibly containing leading zeros). See [3] for more details.

It is not difficult to use the characterization of Theorem 8 to find a linear representation for $f(n)$, the number of Dyck factors of length $2n$ appearing in $\mathbf{t}$, the Thue-Morse word. However, in this section we will instead use a different approach that is more general.

Theorem 13. *Suppose* $\mathbf{s} = (s(n))_{n\geq 0}$ *is a* k-*automatic sequence that is running-sum synchronized. Then* $(f(n))_{n\geq 0}$, *the number of Dyck factors of length* $2n$ *appearing in* $\mathbf{s}$, *is* k-*regular.*

Proof. It suffices to find a linear representation for $f(n)$.

To do so, we first find a first-order formula asserting that $\mathbf{s}[i..i + n - 1]$ is *novel*; that is, it is the first occurrence of this factor in $\mathbf{s}$:

$$\text{FacEq}(i, j, n) : \forall t \ (t < n) \implies \mathbf{s}[i + t] = \mathbf{s}[j + t]$$
$$\text{Novel}(i, n) : \forall j \ \text{FacEq}(i, j, n) \implies j \geq i.$$

Then the number of i for which

$$\text{Novel}(i, 2n) \land \text{Dyck}(i, 2n) \tag{1}$$

holds is precisely the number of Dyck factors of $\mathbf{s}$ of length $2n$. Since $\mathbf{s}$ is k-automatic, and its running sum sequence $\mathbf{v}$ is synchronized, it follows that there is an automaton recognizing those i and n for which (1) evaluates to true, and from known techniques we can construct a linear representation for the number of such i.   □

Corollary 14. *Let $f(n)$ denote the number of Dyck factors of length $2n$ appearing in the Thue-Morse word. Then $(f(n))_{n \geq 0}$ is a 2-regular sequence.*

Proof. We can carry out the proof of Theorem 13 in `Walnut` for `t`, as follows:

```
def FacEq "At (t<n) => T[i+t]=T[j+t]":
def Novel "Aj $FacEq(i,j,n) => j>=i":
def NovelDyck "$Dyck(i,n) & $Novel(i,n)":
def LR n "$NovelDyck(i,2*n)":
```

The last command creates a rank-29 linear representation for the number of length-$2n$ Dyck factors.   □

Table 1 gives the first few terms of the sequence $f(n)$. It is sequence A345199 in the *On-Line Encyclopedia of Integer Sequences* [12].

Table 1. First few values of $f(n)$.

n	0	1	2	3	4	5	6	7	8	9	10	11	12	13	14	15	16	17	18	19	20
$f(n)$	1	1	2	3	2	4	6	6	4	8	8	8	12	9	12	13	8	14	16	14	16

5   Upper and Lower Bounds for $f(n)$

In this section we prove tight upper and lower bounds for $f(n)$, the number of Dyck factors of t of length $2n$.

We start with a characterization of some of the subsequences of $(f(n))_{n \geq 0}$.

Lemma 15. *We have*

$$f(2n) = 2f(n) \tag{2}$$
$$f(4n+3) = 2f(n) + f(2n+1) + q(n) \tag{3}$$
$$f(8n+1) = 2f(2n+1) + f(4n+1) - q(n) \tag{4}$$
$$f(8n+5) = 2f(n) + f(2n+1) + 2f(2n+2) \tag{5}$$

for all $n \geq 3$. Here $q(n)$ is the 2-automatic sequence computed by the DFAO in Fig. 1.

Proof. Notice that $1 \leq q(n) \leq 2$ for $n \geq 1$.

These relations can be proved using linear representations computable by `Walnut`. We only prove the most complicated one, namely Eq. (4). Substituting $n = m + 3$, we see that Eq. (4) is equivalent to the claim that $f(8m + 25) = 2f(2m + 7) + f(4m + 13) - q(m + 3)$ for $m \geq 0$. We now obtain linear representations for each of the terms, using the following `Walnut` commands.

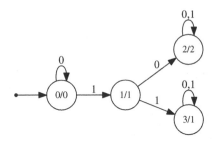

Fig. 1. DFAO computing $q(n)$. States are in the form q/a, where q is the name of the state and a is the output.

```
morphism aa "0->01 1->23 2->22 3->33":
morphism b "0->0 1->1 2->2 3->1":
promote Q1 aa:
image Q b Q1:

def term1 m "$LR(i,8*m+25)":
def term2 m "$LR(i,2*m+7)":
def term3 m "$LR(i,4*m+13)":
def term4 m "(i=0 & Q[m+3]=@1) | (i<=1 & Q[m+3]=@2)":
```

From these four linear representations, using block matrices, we can easily create a linear representation for

$$f(8m + 25) - 2f(2m + 7) - f(4m + 13) + q(m + 3).$$

It has rank 735. When we minimize it (using a `Maple` implementation of the algorithm of Schützenberger discussed in [3, Chapter 2]), we get the linear representation for the 0 function, thus proving the identity.

The other identities can be proved similarly.   □

Theorem 16. *We have $f(n) \leq n$ for all $n \geq 1$. Furthermore, this bound is tight, since $f(n) = n$ for $n = 3 \cdot 2^i$ and $i \geq 0$.*

Proof. We will actually prove the stronger bound that $f(n) \leq n - (n \bmod 2)$ for $n \geq 1$, by induction.

The base case is $1 \leq n < 29$. In this case we can verify the bound by direct computation. Otherwise assume $n \geq 29$ and the bound is true for all smaller positive $n' < n$ (the 29 comes from the fact that Eq. (5) is only valid for $n \geq 3$); we prove it for n.

There are four cases to consider: $n \equiv 0 \pmod 2$, $n \equiv 3 \pmod 4$, $n \equiv 1 \pmod 8$, and $n \equiv 5 \pmod 8$.

Suppose $n \equiv 0 \pmod 2$. By induction we have $f(n/2) \leq n/2 - (n/2 \bmod 2)$. But from Eq. (2) we have $f(n) = 2f(n/2) \leq 2(n/2) - 2(n/2 \bmod 2) \leq n$.

Suppose $n \equiv 3 \pmod 4$. By induction we have $f((n-3)/4) \leq (n-3)/4 - ((n-3)/4 \bmod 2)$ and $f((n-1)/2) \leq (n-1)/2 - ((n-1)/2 \bmod 2)$.

From Eq. (3) we have

$$f(n) = 2f((n-3)/4) + f((n-1)/2) + q((n-3)/4)$$
$$\leq (n-3)/2 - 2((n-3)/4 \bmod 2) + (n-1)/2 - ((n-1)/2 \bmod 2)$$
$$+ q((n-3)/4) \leq n-1,$$

as desired.

Suppose $n \equiv 1 \pmod 8$. By induction we have $f((n+3)/4) \leq (n+3)/4 - ((n+3)/4 \bmod 2)$ and $f((n+1)/2) \leq (n+1)/2 - ((n+1)/2 \bmod 2)$. From Eq. (4) we have

$$f(n) = 2f((n+3)/4) + f((n+1)/2) - q((n-1)/8)$$
$$\leq (n+3)/2 - 2((n+3)/4 \bmod 2) + (n+1)/2 - 2((n+1)/2 \bmod 2)$$
$$- q((n-1)/8) \leq n-1,$$

as desired.

Suppose $n \equiv 5 \pmod 8$. By induction we have

$$f((n-5)/8) \leq (n-5)/8 - ((n-5)/8 \bmod 2)$$
$$f((n-1)/4) \leq (n-1)/4 - ((n-1)/4 \bmod 2)$$
$$f((n+3)/4) \leq (n+3)/4 - ((n+3)/4 \bmod 2).$$

From Eq. (5) we have

$$f(n) = 2f((n-5)/8) + f((n-1)/4) + 2f((n+3)/4)$$
$$\leq (n-5)/4 - 2((n-5)/8 \bmod 2) + (n-1)/4 - ((n-1)/4 \bmod 2)$$
$$+ (n+3)/2 - 2((n+3)/4 \bmod 2) \leq n-1,$$

as desired. This completes the proof of the upper bound.

We can see that $f(n) = n$ for $n = 3 \cdot 2^i$ as follows. Using the linear representation for n we have $f(3 \cdot 2^i) = v_f \gamma_f(11) \gamma_f(0)^i w_f$. The minimal polynomial of $\gamma_f(0)$ is $X^2(X-1)(X+1)(X-2)$. It follows that $f(3 \cdot 2^i) = a \cdot 2^i + b + c(-1)^i$ for $i \geq 2$. Solving for the constants, we find that $a = 3$, $b = 0$, $c = 0$, and hence $f(3 \cdot 2^i) = 3 \cdot 2^i$ as claimed. $\square$

Theorem 17. *We have $f(n) \geq n/2$ for $n \geq 0$, and $f(n) \geq (n+3)/2$ for $n \geq 1$ odd. Furthermore, the bound $f(n) \geq n/2$ is attained infinitely often.*

Proof. The proof is similar to the proof of Theorem 16, and is omitted here for space considerations. It can be found in the full version of the paper [8]. $\square$

Theorem 18. *We have $\sum_{0 \leq i < 2^n} f(i) = 19 \cdot 4^n/48 - 2^n/4 + 5/3$ for $n \geq 2$.*

Proof. The summation $\sum_{0 \leq i < 2^n} f(i)$ is easily seen to equal $v_f(\gamma_f(0)+\gamma_f(1))^n w_f$. We can then apply the same techniques as above to the matrix $\gamma_f(0) + \gamma_f(1)$. $\square$

It follows that the "average" value of $f(i)$, over an interval of the form $0 \leq i < 2^n$, is asymptotically $\frac{19}{24}i$.

References

1. Allouche, J.-P., Shallit, J.O.: The ring of k-regular sequences. Theoret. Comput. Sci. **98**, 163–197 (1992)
2. Balková, L., Pelantová, E., Steiner, W.: Return words in the Thue-Morse and other sequences. https://hal.science/hal-00089863v2 2006
3. Berstel, J., Reutenauer, C.: Noncommutative Rational Series With Applications, Vol. 137 of Encyclopedia of Mathematics and Its Applications. Cambridge University Press (2011)
4. Choffrut, C., Malcher, A., Mereghetti, C., Palano, B.: First-order logics: some characterizations and closure properties. Acta Inform. **49**, 225–248 (2012)
5. Chomsky, N., Schützenberger, M.P.: The algebraic theory of context-free languages. In: Braffort, P., Hirschberg, D. (eds.) Comput. Programm. Formal Syst., pp. 118–161. North Holland, Amsterdam (1963)
6. Goč, D., Schaeffer, L., Shallit, J.: Subword complexity and k-synchronization. In: Béal, M.-P., Carton, O. (eds.) DLT 2013. LNCS, vol. 7907, pp. 252–263. Springer, Heidelberg (2013). https://doi.org/10.1007/978-3-642-38771-5_23
7. Keränen, V.: On k-repetition freeness of length uniform morphisms over a binary alphabet. Discrete Appl. Math. **9**, 297–300 (1984)
8. Mol, L., Rampersad, N., Shallit, J.: Dyck words, pattern avoidance, and automatic sequences. Arxiv preprint arXiv:2301.06145 [cs.DM], January 15 (2023). https://arxiv.org/abs/2301.06145
9. Mousavi, H.: Automatic theorem proving in Walnut. Arxiv preprint arXiv:1603.06017 [cs.FL], http://arxiv.org/abs/1603.06017 (2016)
10. Shallit, J.: Synchronized Sequences. In: Lecroq, T., Puzynina, S. (eds.) WORDS 2021. LNCS, vol. 12847, pp. 1–19. Springer, Cham (2021). https://doi.org/10.1007/978-3-030-85088-3_1
11. Shallit, J.: The Logical Approach To Automatic Sequences: Exploring Combinatorics on Words with Walnut. London Math. Soc. Lecture Notes Series, Vol. 482. Cambridge University Press (2022)
12. Sloane, N.J.A., et al.: The On-Line Encyclopedia of Integer Sequences (2022). https://oeis.org
13. Thue, A.: Über die gegenseitige Lage gleicher Teile gewisser Zeichenreihen. In: Norske vid. Selsk. Skr. Mat. Nat. Kl. **1** (1912), 1–67. Reprinted in *Selected Mathematical Papers of Axel Thue*, T. Nagell, editor, Universitetsforlaget, Oslo, pp. 413–478 (1977)

Rudin-Shapiro Sums via Automata Theory and Logic

Narad Rampersad[1] and Jeffrey Shallit[2(✉)]

[1] Department of Mathematics and Statistics, University of Winnipeg,
515 Portage Avenue, Winnipeg R3B 2E9, Canada
n.rampersad@uwinnipeg.ca
[2] School of Computer Science, University of Waterloo,
200 University Avenue W., Waterloo N2L 3G1, Canada
shallit@uwaterloo.ca

Abstract. We show how to obtain, via a unified framework provided by logic and automata theory, many classical results of Brillhart and Morton on Rudin-Shapiro sums. The techniques also facilitate easy proofs for new results.

1 Introduction

The Rudin-Shapiro coefficients $(a(n))_{n \geq 0} = (1, 1, 1, -1, 1, 1, -1, 1, 1, \ldots)$ form an infinite sequence of ± 1 defined recursively by the identities

$$a(2n) = a(n)$$
$$a(2n + 1) = (-1)^n a(n)$$

and the initial condition $a(0) = 1$. It is sequence A020985 in the On-Line Encyclopedia of Integer Sequences (OEIS) [26]. The sequence was apparently first discovered by Golay [10,11], and later studied by Shapiro [25] and Rudin [21]. The function $a(n)$ can also be defined as $a(n) = (-1)^{r_n}$, where r_n counts the number of (possibly overlapping) occurrences of 11 in the binary representation of n [3, Satz 1]. The Rudin-Shapiro coefficients have many intriguing properties and have been studied by many authors; for example, see [1,4,9,17,18]. They appear in number theory [16], analysis [13], combinatorics [14], and even optics [10,11], just to name a few places.

In a classic paper from 1978, written in German, Brillhart and Morton [3] studied sums of these coefficients, and defined the two sums[1]

$$s(n) = \sum_{0 \leq i \leq n} a(i) \qquad t(n) = \sum_{0 \leq i \leq n} (-1)^i a(i). \qquad (1)$$

---

[1] One can make the case that these definitions are "wrong", in the sense that many results become significantly simpler to state if the sums are taken over the range $0 \leq i < n$ instead. But the definitions of Brillhart-Morton are now very well-established, and using a different indexing would also make it harder to compare our results with theirs.

Research of Narad Rampersad is supported by NSERC Grant number 2019-04111.
Research of Jeffrey Shallit is supported by NSERC Grant number 2018-04118.

© The Author(s), under exclusive license to Springer Nature Switzerland AG 2023
A. Frid and R. Mercaş (Eds.): WORDS 2023, LNCS 13899, pp. 233–246, 2023.
https://doi.org/10.1007/978-3-031-33180-0_18

The first few values of the functions s and t are given in Table 1. They are sequences A020986 and A020990 respectively in the OEIS.

Table 1. First few values of $s(n)$ and $t(n)$.

n	0	1	2	3	4	5	6	7	8	9	10	11	12	13	14	15	16	17	18	19	20
$s(n)$	1	2	3	2	3	4	3	4	5	6	7	6	5	4	5	4	5	6	7	6	7
$t(n)$	1	0	1	2	3	2	1	0	1	0	1	2	1	2	3	4	5	4	5	6	7

A priori it is not even clear that these sums are always non-negative, but Brillhart and Morton proved that they are, and also proved many other properties of them. Most of these properties can be proved by induction, sometimes rather tediously.

In this paper we show how to replace nearly all of these inductions with techniques from logic and automata theory. Ultimately, almost all[2] the Brillhart-Morton results can be proved in a simple, unified manner, simply by stating them in first-order logic and applying the Walnut theorem-prover [19,23]. In fact, there are basically only three simple inductions in our entire paper. One is the brief induction used to prove Lemma 2. The other two are the inductions used in Theorem 1 to prove the correctness of our constructed automata, and in these cases the induction step itself can be proved by Walnut! We are also able to easily derive and prove new results; see Sect. 7.

Finally, another justification for this paper is that the same techniques can easily be harnessed to handle related sequences; for example, see [24].

We assume the reader is familiar with the basics of automata and regular expressions, as discussed, for example, in [12]. Regrettably, many of the details had to be omitted from this conference paper, because of space limitations. For the full paper, see [20].

2 Notation

Let $b \geq 2$ be an integer, and let Σ_b denote the alphabet $\{0, 1, \ldots, b-1\}$.

For a string $x \in \Sigma_b^*$, we let $[x]_b$ denote the integer represented by x in base b, with the more significant digits at the left. That is, if $x = c_1 c_2 \cdots c_i$, then $[x]_b := \sum_{1 \leq j \leq i} c_j b^{i-j}$. For example, $[00101011]_2 = [223]_4 = 43$.

For an integer $n \geq 0$, we let $(n)_b$ denote the canonical base-b representation of n, starting with the most significant digit, with no leading zeros. For example, $(43)_2 = 101011$. We have $(0)_b = \epsilon$, the empty string.

3 Automata and Logic

Walnut is free software, originally designed by Hamoon Mousavi, that can rigorously prove propositions about automatic sequences [19,23]. If a proposition

---

[2] The main exceptions are their results about the limit points of $s(n)/\sqrt{n}$ and $t(n)/\sqrt{n}$.

P has no free variables, one only has to state it in first-order logic and then the `Walnut` prover will prove or disprove it (subject to having enough time and space to complete the calculation).[3] If the proposition P has free variables, the program computes a finite automaton accepting exactly the values of the free variables that make P evaluate to true.

As is well-known, the Rudin-Shapiro sequence is 2-automatic and therefore, by a classic theorem of Cobham [8], also 4-automatic. This means there is a deterministic finite automaton with output (DFAO) that on input n, expressed in base 4, reaches a state with output $a(n)$. This Rudin-Shapiro automaton is illustrated below in Fig. 1. It is a simple variation on the one given in [1].

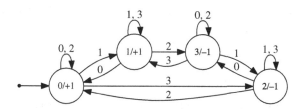

Fig. 1. DFAO computing the Rudin-Shapiro function, in base 4.

Here states are labeled a/b, where a is the state number and b is the output. The initial state is state 0, and the automaton reads the digits of the base-4 representation of n, starting with the most significant digit. Leading zeros in the inputs are allowed and do not affect the result.

Once the automaton in Fig. 1 is saved as a file named `RS4.txt`, in `Walnut` we can refer to its value at a variable n simply by writing `RS4[n]`. We would like to do the same thing for the Rudin-Shapiro summatory functions $s(n)$ and $t(n)$ defined in Eq. (1), but here we run into a fundamental limitation of `Walnut`: it can only directly deal with functions of finite range (like automatic sequences). Since $s(n)$ and $t(n)$ are unbounded, we must find another way to deal with them.

A common way to handle functions in first-order logic is to treat them as *relations:* instead of writing $f(n) = x$, we construct a relation $R_f(n, x)$ that is true iff $f(n) = x$. If the relation $R_f(n, x)$ is representable by a deterministic finite automaton (DFA) taking as input n in base b_1 and x in base b_2, in parallel, and accepting iff $R_f(n, x)$ holds, then we say that f is (b_1, b_2)-*synchronized.* For more information about synchronized functions, see [6,22].

Our first step, then, is to show that the functions $s(n)$ and $t(n)$ are $(4, 2)$-synchronized. We can obtain these automata by "guessing" them from calculated initial values of the sequences s and t, using the Myhill-Nerode theorem [12, Sect. 3.4]. However, we will see below in Remark 4 that we could have also deduced them from Satz 3 of [3] (Lemma 2 of [5]). The synchronized automaton for s is called `rss` in `Walnut`; it has 7 states. The synchronized automaton for t is called `rst` in `Walnut`; it has 8 states. Both are depicted in Fig. 2.

---

[3] In fact, all the `Walnut` code in this paper runs in a matter of milliseconds.

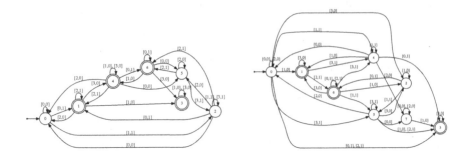

Fig. 2. Synchronized automata for $s(n)$ (left) and $t(n)$ (right).

Once we have guessed the two automata, we need to verify they are correct.

Theorem 1. *The automata in Fig. 2 correctly compute $s(n)$ and $t(n)$.*

Proof. Let $s_1(n)$ (resp., $t_1(n)$) be the functions computed by the automaton in Fig. 2. We prove that $s_1(n) = s(n)$ and $t_1(n) = t(n)$ by induction on n.

First we check that $s_1(0) = s(0) = 1$ and $t_1(0) = t(0) = 1$, which we can see simply by inspecting the automata.

Now assume that $n \geq 1$ and $s_1(n) = s(n)$ and $t_1(n) = t(n)$. We prove with Walnut that $s_1(n+1) = s_1(n) + a(n+1)$ and $t_1(n+1) = t_1(n) + (-1)^{n+1} a(n+1)$.

```
eval test1 "?msd_4 An,y ($rss(n,y) & RS4[n+1]=@1) => $rss(n+1,?msd_2 y+1)":
eval test2 "?msd_4 An,y ($rss(n,y) & RS4[n+1]=@-1) => $rss(n+1,?msd_2 y-1)":
def even4 "?msd_4 Ek n=2*k":
def odd4 "?msd_4 Ek n=2*k+1":
eval test3 "?msd_4 An,y ($rst(n,y) & ((RS4[n+1]=@1 & $even4(n+1)) |
    (RS4[n+1]=@-1 & $odd4(n+1)))) => $rst(n+1,?msd_2 y+1)":
eval test4 "?msd_4 An,y ($rst(n,y) & ((RS4[n+1]=@-1 & $even4(n+1)) |
    (RS4[n+1]=@1 & $odd4(n+1)))) => $rst(n+1, ?msd_2 y-1)":
```

and Walnut returns TRUE for all of these tests. Now the correctness of our automata follows immediately by induction.   □

Remark 1. Some Walnut syntax needs to be explained here. First, the capital A is Walnut's abbreviation for ∀ (for all); capital E is Walnut's abbreviation for ∃ (there exists); the jargon ?msd_b for a base b instructs that a parameter or expression is to be evaluated using base-b numbers, and an @ sign indicates the value of an automatic sequence (which is allowed to be negative). The symbol & is logical AND; the symbol | is logical OR; and the symbol => is logical implication.

Remark 2. There is a small technical wrinkle that we have glossed over, but it needs saying: the default domain for Walnut is ℕ, the natural numbers. But we do not know, a priori, that the functions s and t take only non-negative values. Therefore, it is conceivable that our verification might fail simply because negative numbers appear as intermediate values in a calculation. We can check that this does not happen simply by checking that `rss` and `rst` both are truly representations of functions:

```
eval test5 "?msd_4 (An Ey $rss(n,y)) &
  An ~Ex,y ($rss(n,x) & $rss(n,y) & (?msd_2 x!=y))":
eval test6 "?msd_4 (An Ey $rst(n,y)) &
  An ~Ex,y ($rst(n,x) & $rst(n,y) & (?msd_2 x!=y))":
```

These commands assert that for every n there is at least one value y such that $s(n) = y$, and there are not two different such y, and the same for t. Both evaluate to TRUE. This shows that the relations computed by `rss` and `rst` are well-defined functions, and take only non-negative values. As a result, we have already deduced Satz 11 of [3]: namely, that $t(n) \geq 0$ for all n.

The advantage of the representation of $s(n)$ and $t(n)$ as synchronized automata is that they essentially encapsulate all the needed knowledge about $s(n)$ and $t(n)$ to replace tedious inductions about them. The simple induction we used to verify them replaces, in effect, all the other needed inductions.

Remark 3. The reader may reasonably ask, as one referee did, why use base-4 for n and base-2 for the values of s and t? The reason is because the values of $s(n)$ and $t(n)$ grow like $\sqrt{n}$; if we are going to have any hope of an automaton processing, say, n and $s(n)$ in parallel, then length considerations show that the base of representation for n must be the square of that for $s(n)$ and $t(n)$. Furthermore, by a classical theorem of Cobham [7], the Rudin-Shapiro sequence itself can only be generated by an automaton using base a power of 2. This forces the base for n to be 2^{2k} for some k and the base for s and t to be 2^k.

All the code necessary to verify the results in this paper can be found at
https://cs.uwaterloo.ca/~shallit/papers.html .

4   Proofs of Results

We can now begin to *reprove*, and in some cases, *improve* some of the results of Brillhart and Morton. Let us start with their Satz 2 [3], reprised as Lemma 1 in [5]:

Theorem 2. *We have*

$$s(2n) = s(n) + t(n-1), \qquad (n \geq 1); \qquad (2)$$
$$s(2n+1) = s(n) + t(n), \qquad (n \geq 0); \qquad (3)$$
$$t(2n) = s(n) - t(n-1), \qquad (n \geq 1); \qquad (4)$$
$$t(2n+1) = s(n) - t(n), \qquad (n \geq 0). \qquad (5)$$

Proof. We use the following Walnut commands:

```
eval eq2 "?msd_4 An,x,y,z (n>=1 & $rss(2*n,x) & $rss(n,y) &
  $rst(n-1,z)) => ?msd_2 x=y+z":
eval eq3 "?msd_4 An,x,y,z ($rss(2*n+1,x) & $rss(n,y) & $rst(n,z))
  => ?msd_2 x=y+z":
eval eq4 "?msd_4 An,x,y,z (n>=1 & $rst(2*n,x) & $rss(n,y) &
  $rst(n-1,z)) => ?msd_2 x+z=y":
eval eq5 "?msd_4 An,x,y,z ($rst(2*n+1,x) & $rss(n,y) & $rst(n,z))
  => ?msd_2 x+z=y":
```

and `Walnut` returns TRUE for all of them.

For example, the `Walnut` formula `eq2` asserts that for all n, x, y, z if $n \geq 1$, $s(2n) = x$, $s(n) = y$, $t(n-1) = z$, then it must be the case that $x = y + z$. It is easily seen that this is equivalent to the statement of Eq. (2).

Note that in our `Walnut` proofs of Eqs. (3) and (4), we rearranged the statement to avoid subtractions. This is because `Walnut`'s basic domain is $\mathbb{N}$, the natural numbers, and subtractions that could potentially result in negative numbers might give anomalous results. □

Similarly, we can now verify Satz 3 of [3] (Lemma 2 of [5]):

Lemma 1. *For $n \geq 0$ we have*

$$s(4n) = 2s(n) - a(n) \tag{6}$$
$$s(4n + 1) = s(4n + 3) = 2s(n) \tag{7}$$
$$s(4n + 2) = 2s(n) + (-1)^n a(n). \tag{8}$$

Proof. We use the following `Walnut` commands:

```
eval eq6 "?msd_4 An,x,y ($rss(4*n,x) & $rss(n,y)) =>
  ((RS4[n]=@1 => ?msd_2 x+1=2*y) & (RS4[n]=@-1 =>
  ?msd_2 x=2*y+1))":

eval eq7 "?msd_4 An,x,y,z ($rss(4*n+1,x) & $rss(4*n+3,y) &
  $rss(n,z)) => ?msd_2 x=y & x=2*z":

eval eq8 "?msd_4 An,x,y ($rss(4*n+2,x) & $rss(n,y)) =>
  (((RS4[n]=@1 & $even4(n)) => ?msd_2 x=2*y+1) &
  ((RS4[n]=@-1 & $even4(n)) => ?msd_2 x+1=2*y) &
  ((RS4[n]=@1 & $odd4(n)) => ?msd_2 x+1=2*y) &
  ((RS4[n]=@-1 & $odd4(n)) => ?msd_2 x=2*y+1))":
```

and `Walnut` returns TRUE for all of them. □

Remark 4. As it turns out, Lemma 1 is more or less equivalent to our synchronized automaton for s depicted in Fig. 2. To see this, consider a synchronized automaton where the first component represents n in base 4, while the second component represents $s(n)$ in base 2, but using the nonstandard digit set

$\{-1, 0, 1\}$ instead of $\{0, 1\}$. Reading a bit i in the first component is like changing the number n read so far into $4n + i$. Theorem 1 says that $s(4n + i)$ is twice $s(n)$, plus either $-1, 0,$ or 1, depending on the value of $a(n)$ and the parity of n, both of which are (implicitly) computed by the base-4 DFAO for $a(n)$. To get the automaton in Fig. 2 from this one, we would need to combine it with a "normalizer" that can convert a nonstandard base-2 representation into a standard one.

Let us now prove Lemma 3 of [5]:

Theorem 3. *We have*

$$s(n + 2^{2k}) = s(n) + 2^k, \qquad 0 \le n \le 2^{2k-1} - 1, \ k \ge 1; \tag{9}$$
$$s(n + 2^{2k}) = -s(n) + 3 \cdot 2^k, \qquad 2^{2k-1} \le n \le 2^{2k} - 1, \ k \ge 1; \tag{10}$$
$$s(n + 2^{2k+1}) = s(n) + 2^{k+1}, \qquad 0 \le n \le 2^{2k} - 1, \ k \ge 0; \tag{11}$$
$$s(n + 2^{2k+1}) = -s(n) + 2^{k+2}, \qquad 2^{2k} \le n \le 2^{2k+1} - 1, \ k \ge 0. \tag{12}$$

Proof. We can prove these identities with `Walnut`. One small technical difficulty is that the equation $x = 2^n$ is not possible to express in the particular first-order logic that `Walnut` is built on; it cannot even multiply arbitrary variables, or raise a number to a power. Instead, we assert that x is a power of 2 without exactly specifying *which* power of 2 it is. This brings up a further difficulty, which is that we need to simultaneously express 2^{2k} and 2^k. Normally this would also not be possible in `Walnut`. However, in this case the former is expressed in base 4 and the latter in base 2, we can achieve this using the `link42` automaton:

```
reg power4 msd_4 "0*10*":
reg link42 msd_4 msd_2 "([0,0]|[1,1])*":
```

Here `power4` asserts that its argument is a power of 4; specifically, that its base-4 representation looks like 1 followed by some number of 0's, and also allowing any number of leading zeros. If this is true for x, then $x = 4^k$ for some k, and `link42` applied to the pair (x, y) asserts that $y = 2^k$ (by asserting that the base-4 representation x is the same as the base-2 representation of y).

To verify Eqs. (9)–(12), we use the following `Walnut` code:

```
eval eq9 "?msd_4 An,x,y,z ($power4(x) & x>=4 & 2*n+2<=x &
    $rss(n,y) & $link42(x,z)) => $rss(n+x,?msd_2 y+z)":
eval eq10 "?msd_4 An,x,y,z ($power4(x) & x>=4 & 2*n>=x & n<x &
    $rss(n,y) & $link42(x,z)) => $rss(n+x,?msd_2 3*z-y)":
eval eq11 "?msd_4 An,x,y,z ($power4(x) & n<x & $rss(n,y) &
    $link42(x,z)) => $rss(n+2*x,?msd_2 y+2*z)":
eval eq12 "?msd_4 An,x,y,z ($power4(x) & x<=n & n<2*x & $rss(n,y) &
    $link42(x,z)) => $rss(n+2*x,?msd_2 4*z-y)":
```

and `Walnut` returns TRUE for all of them.   □

We can also use our method to prove Lemma 4 in Brillhart and Morton [5]. It is as follows (where we have corrected a typographical error in the original statement).

Theorem 4. *Suppose* $n \in [2^{2k}, 2^{2k+1})$. *Then* $s(n) \leq 2^{k+1}$, *and furthermore, equality holds for* n *in this range iff* $n = 2^{2k+1} - 1 - \sum_{0 \leq r < k} e_r 2^{2r+1}$, *where the* $e_r \in \{0, 1\}$.

5   Inequalities

We showed in Theorem 1 that $s(n)$ and $t(n)$ are $(4, 2)$-synchronized; furthermore, $s(n)$ and $t(n)$ are both unbounded as is easily verified with Walnut. A basic result about synchronized sequences, namely Theorem 8 of [22], immediately implies that there are constants c' and c'' such that $c' \leq s(n)/\sqrt{n} \leq c''$, and similarly for $t(n)$. The main accomplishment of Brillhart and Morton's paper was to determine these constants.

Theorem 5 (Brillhart & Morton). *For* $n \geq 1$ *we have*

$$\sqrt{3n/5} \leq s(n) \leq \sqrt{6n}$$
$$0 \leq t(n) \leq \sqrt{3n}.$$

Trying to prove these results by *directly* translating the claims into Walnut leads to two difficulties: first, automata cannot compute squares or square roots. Second, our synchronized automata work with n expressed in base 4, but $s(n)$ and $t(n)$ are expressed in base 2, and Walnut cannot directly compare arbitrary integers expressed in different bases.

However, there is a way around both of these difficulties. First, we define a kind of "pseudo-square" function as follows: $m(n) = [(n)_2]_4$. In other words, m sends n to the integer obtained by interpreting the base-2 expansion of n as a number in base 4. Luckily we have already defined an automaton for m called link42; to get an automaton for m, we only have to reverse the order of the arguments in link42!

Now we need to see how far away from a real squaring function our pseudo-square function $m(n)$ is.

Lemma 2. *We have* $(n^2 + 2n)/3 \leq m(n) \leq n^2$.

Proof. We can prove the bounds by induction on n. They are clearly true for $n = 0$. Assume $n \geq 1$ and the inequalities hold for all $n' < n$; we prove them for n.

Suppose n is even. Then $n = 2k$. Clearly $m(n) = 4m(k)$. By induction we have $(k^2 + 2k)/3 \leq m(k) \leq k^2$, and multiplying through by 4 gives

$$(n^2 + 2n)/3 = (4k^2 + 4k)/3 < 4(k^2 + 2k)/3 \leq 4m(k) \leq 4k^2 = n^2.$$

Suppose n is odd. Then $n = 2k + 1$. Clearly $m(n) = 4m(k) + 1$. By induction we have $(k^2 + 2k)/3 \leq m(k) \leq k^2$. Multiplying by 4 and adding 1 gives

$$(n^2 + 2n)/3 = ((2k+1)^2 + 2(2k+1))/3 = (4k^2 + 8k + 3)/3 \leq 4m(k) + 1$$
$$\leq 4k^2 + 1 \leq (2k+1)^2 = n^2,$$

as desired.   □

We can now prove:

Lemma 3. *For $n \geq 1$ we have $\frac{3n+7}{5} \leq m(s(n)) \leq 3n + 1$, and the upper and lower bounds are tight.*

Proof. We use the `Walnut` code

```
def maps "?msd_4 Ex $rss(n,x) & $link42(y,x)":
eval ms_lowerbnd "?msd_4 An,y (n>=1 & $maps(n,y)) => y<=3*n+1":
eval ms_upperbnd "?msd_4 An,y (n>=1 & $maps(n,y)) => 3*n+7<=5*y":
```

Here maps is a synchronized automaton for the function $m(n)$. To show the bounds are tight, let us show there are infinitely many solutions to $m(s(n)) = 3n + 1$ and $m(s(n)) = (3n + 7)/5$:

```
eval lowerbnd_tight "?msd_4 Am En,y (n>m) & $maps(n,y) & y=3*n+1":
eval upperbnd_tight "?msd_4 Am En,y (n>m) & $maps(n,y) & 5*y=3*n+7":
```

□

As a consequence, we get one of the lower bounds in Theorem 5.

Corollary 1. *For $n \geq 1$ we have*

$$s(n) \geq \sqrt{\frac{3n + 7}{5}}.$$

Proof. From Lemma 2 we have $m(s(n)) \leq s(n)^2$ and from Lemma 3 we have $\frac{3n+7}{5} \leq m(s(n))$. Putting these two bounds together gives $\frac{3n+7}{5} \leq s(n)^2$.   □

Note that our lower bound is actually slightly *stronger* than that of Brillhart-Morton!

To get the upper bound $s(n) \leq \sqrt{6n}$, as in Brillhart-Morton, we need to do more work, since the results we have proved so far only suffice to show that $s(n) \leq \sqrt{9n + 3}$. To get their upper bound, Brillhart and Morton carved the various intervals for n up into three classes and proved the upper bound of $\sqrt{6n}$ for each class. We'll do the same thing, but use slightly different classes. By doing so we avoid their complicated induction entirely.

The first class is the easiest: those n for which $m(s(n)) \leq 2n$. For these n, Lemma 2 immediately gives us $s(n)^2 \leq 6n$, as desired. Furthermore, the "exceptional set" (that is, those n for which $m(s(n)) > 2n$) is calculable with `Walnut`:

```
def exceptional_set "?msd_4 Em $maps(n,m) & m>2*n":
```

The resulting automaton is quite simple (2 states!) and recognizes the set of base-4 expansions $\{0, 2\}^* \cup \{0, 2\}^* 1 \{1, 3\}^*$.

We readily see, then, that the exceptional set consists of

(a) numbers whose base-4 expansion starts with a 1 and thereafter consists of 1's and 3's, and

(b) the rest, which must start with a 2.

The numbers in group (a) are easiest to deal with, because they satisfy the inequality $M_k/2 \le n < 2^{2k+1}$ for some $k \ge 0$, where $M_k = (2^{2k+3} - 2)/3$. Now for all n (not just those in the exceptional set) in the half-open interval $I_k := [M_k/2,\, 2^{2k+1})$ we can show with Walnut that $s(n) \le 2^{k+1}$, as follows:

```
eval maxcheck "?msd_4 An,x,y,z ($power4(x) & 3*n+1>=4*x & n<2*x
  & $rss(n,y) & $link42(x,z)) => ?msd_2 y<=2*z":
```

So for all $n \in I_k$ we have

$$\frac{s(n)^2}{n} \le \frac{(\max_{n \in I_k} s(n))^2}{\min_{n \in I_k} n} = \frac{(2^{k+1})^2}{M_k/2} = 3\frac{2^{2k+2}}{2^{2k+2} - 1} \le 4.$$

This handles the numbers in group (a).

Finally, we turn to group (b), which are the hardest to deal with. These numbers lie in the interval $I_k' = [2^{2k+1}, M_k]$. We will split these numbers into the following intervals: $J_{k,i} := [M_k - M_i, M_k - M_{i-1})$ for $0 \le i < k$. Since $M_k - M_{k-1} = 2^{2k+1}$, the union

$$J_{k,0} \cup J_{k,1} \cup \cdots \cup J_{k,k-1} \cup \{M_k\}$$

forms a disjoint partition of the interval I_k'.

Now with Walnut we can prove that for $n \in J_{k,i}$ we have $s(n) \le 2^{k+2} - 2^{i+1}$.

```
eval J_inequality "?msd_4 An,x,y,z,w,m ($rss(n,m) & $power4(x) &
  $power4(y) & x>y & $link42(x,w) & $link42(y,z) & 8*x<=3*n+8*y
  & 3*n+2*y<8*x) => ?msd_2 m+2*z<=4*w":
```

It now follows that for $n \in J_{k,i}$, $k \ge 1$, and $0 \le i < k$, we have

$$\frac{s(n)^2}{n} \le \frac{(\max_{n \in J_{k,i}} s(n))^2}{\min_{n \in J_{k,i}} n} \le \frac{(2^{k+2} - 2^{i+1})^2}{M_k - M_i}$$

and a routine manipulation shows this is less than 6. The only remaining case is M_k. But then $s(M_k) = 2^{k+2} - 1$, and then $s(M_k)^2 < 6M_k$ by another routine calculation.

Finally, we should verify that we have really covered all the possible n:

```
def left_endpoint "?msd_4 3*z+8*y=8*x":
def right_endpoint "?msd_4 3*z+2*y=8*x":
eval check_all "?msd_4 An (n>=1) => ((~$exceptional_set(n)) |
  (Ex $power4(x) & 4*x<=3*n+1 & n<2*x) |
  (Ex,y,z,w $power4(x) & $power4(y) & x>y &
  $left_endpoint(x,y,z) & $right_endpoint(x,y,w) & n>=z & n<w) |
  (Ex $power4(x) & 3*n+2=8*x))":
```

which evaluates to TRUE.

Thus we have proved one upper bound from Theorem 5:

Theorem 6. $s(n) \le \sqrt{6n}$ *for* $n \ge 1$.

Using exactly the same techniques we can prove

Lemma 4. *For all $n \geq 0$ we have $m(t(n)) \leq n+1$, with equality iff $(n)_4 \in (0 \cup 11^*0)^*3^*$.*

From the first claim of Lemma 4 we see that $m(t(n)) \leq n+1$, and by Lemma 2 we have $t(n)^2/3 \leq m(t(n))$. Putting these bounds together gives us $t(n) \leq \sqrt{3(n+1)}$, which is very close to the Brillhart-Morton upper bound for $t(n)$.

To get the other Brillhart-Morton upper bound of Theorem 5, we just use Eqs. (2) and (3), just as Brillhart and Morton did. This gives us $t(n) = s(n/2) - t(n/2-1) \leq s(n/2) \leq \sqrt{3n}$ for $n \geq 2$ even and $t(n) = s((n-1)/2) - t((n-1)/2) \leq s((n-1)/2) \leq \sqrt{3(n-1)}$ for n odd. Thus we have proved

Theorem 7. *We have $t(n) \leq \sqrt{3n}$ for $n \geq 1$.*

6 Counting the k for Which $s(k) = n$

One of the most fun properties of the Rudin-Shapiro summation function $s(n)$ is Satz 22 of [3]:

Theorem 8. *There are exactly n values of k for which $s(k) = n$.*

Proof. We can prove this theorem "purely mechanically" by using another capability of Walnut: the fact that it can create base-b linear representations for values of synchronized sequences. By a *base-b linear representation* for a function $f(n)$ we mean vectors v, w, and a matrix-valued morphism γ such that $f(n) = v\gamma(x)w$ for all strings x representing n in base b. (The dimension of v is called the *rank* of the representation.) So let us find a base-2 linear representation for the number of such k for which $s(k) = n$:

```
eval satz22 n "$rss(?msd_4 k,n)":
```

This gives us a base-2 linear representation of rank 7 computing some function $f(n)$. Next we use Walnut to compute a base-2 linear representation for the function $g(n) = n$:

```
eval gfunc n "i<n":
```

From this, we can easily compute a base-2 linear representation for $f(n) - g(n)$, and minimize it using an algorithm[4] of Schützenberger [2, §2.3]. When we do so, we get the representation for the 0 function, so $f(n) = n$.   □

---

[4] **Maple** code implementing this algorithm is available from the second author.

7   New Results

One big advantage to the synchronized representation of the Rudin-Shapiro sum functions is that it becomes almost trivial to explore and rigorously prove new properties. As a new result, let's consider the analogue of Theorem 8, but for the function t. Here we run into the problem that every natural number k appears as a value of $t(n)$ infinitely often:

```
eval tvalues "?msd_4 An,k Em (m>n) & $rst(m,k)":
```

and Walnut returns TRUE.

So it makes sense to count the number of times k appears as a value of $t(n)$ in some initial segment, say the first $0, 1, \ldots, 2^r - 1$. Some empirical calculations suggest the following, which we can easily prove using Walnut.

Theorem 9.

(a) For $n \in [0, 4^m/2)$, 0 appears as a value of $t(n)$ exactly 2^{m-1} times, and k appears exactly $2^m - k$ times for $1 \leq k < 2^m$.

(b) For $n \in [0, 4^m)$, 0 appears as a value of $t(n)$ exactly $2^m - 1$ times, 2^m appears exactly once, and k appears exactly $2(2^m - k)$ times for $1 \leq k < 2^m$.

Proof. We use the following Walnut commands.

```
def counta1 k x "?msd_4 $rst(n,k) & $power4(x) & x>1 & 2*n<x ":
def counta2 k x "?msd_4 Ey $power4(x) & x>1 & $link42(x,y) &
    (?msd_2 (k=0 & 2*n<y)|(1<=k & k<y & n+k<y))":
def countb1 k x "?msd_4 $rst(n,k) & $power4(x) & n<x":
def countb2 k x "?msd_4 Ey $power4(x) & $link42(x,y) &
    (?msd_2 (k=0 & n+1<y)|(k=y & n=0)|(1<=k & k<y & n+2*k<2*y))":
```

The first two statements are used for part (a). The code counta1 asserts that $x = 4^m$ for some m, and that $t(n) = k$ for some $n < x$. It returns a linear representation for the number of n for which this holds, as a function of k and x. The code counta2 creates a formula that says that the number of n fulfills the conclusion of the theorem. From these linear representations we can create a linear representation for their difference. When we minimize it, we get the linear representation for the 0 function, so they compute the same function.

The same approach can be used for (b).   □

Acknowledgments. We are grateful to Jean-Paul Allouche and to the referees for several helpful suggestions.

References

1. Allouche, J.-P.: Automates finis en théorie des nombres. Exposition. Math. **5**, 239–266 (1987)
2. Berstel, J., Reutenauer, C.: Noncommutative Rational Series With Applications, vol. 137. Encyclopedia of Mathematics and Its Applications. Cambridge University Press, Cambridge (2011)
3. Brillhart, J., Morton, P.: Über Summen von Rudin-Shapiroschen Koeffizienten. Illinois J. Math. **22**, 126–148 (1978)
4. Brillhart, J., Erdős, P., Morton, P.: On sums of Rudin-Shapiro coefficients. II. Pacific J. Math. **107**, 39–69 (1983)
5. Brillhart, J., Morton, P.: A case study in mathematical research: the Golay-Rudin-Shapiro sequence. Amer. Math. Monthly **103**, 854–869 (1996)
6. Carpi, A., Maggi, C.: On synchronized sequences and their separators. RAIRO Inform. Théor. App. **35**, 513–524 (2001)
7. Cobham, A.: On the base-dependence of sets of numbers recognizable by finite automata. Math. Systems Theory **3**, 186–192 (1969)
8. Cobham, A.: Uniform tag sequences. Math. Systems Theory **6**, 164–192 (1972)
9. Dekking, F.M., Mendès France, M., van der Poorten, A.J.: Folds! Math. Intelligencer **4**, 130–138, 173–181, 190–195 (1982). Erratum **5**, 5 (1983)
10. Golay, M.J.E.: Multi-slit spectrometry. J. Optical Soc. Amer. **39**, 437–444 (1949)
11. Golay, M.J.E.: Static multislit spectrometry and its application to the panoramic display of infrared spectra. J. Opt. Soc. Am. **41**, 468–472 (1951)
12. Hopcroft, J.E., Ullman, J.D.: Introduction to Automata Theory, Languages, and Computation. Addison-Wesley, Boston (1979)
13. Kahane, J.-P.: Some Random Series of Functions. Cambridge Studies in Advanced Mathematics, vol. 5, 2nd edn. Cambridge University Press, Cambridge (1994)
14. Konieczny, J.: Gowers norms for the Thue-Morse and Rudin-Shapiro sequences. Annales de l'Institut Fourier **69**, 1897–1913 (2019)
15. Lafrance, P., Rampersad, N., Yee, R.: Some properties of a Rudin-Shapiro-like sequence. Adv. Appl. Math. **63**, 19–40 (2015)
16. Mauduit, C., Rivat, J.: Prime numbers along Rudin-Shapiro sequences. J. Eur. Math. Soc. **17**, 2595–2642 (2015)
17. Mendès France, M., Tenenbaum, G.: Dimension des courbes planes, papiers pliés et suites de Rudin-Shapiro. Bull. Soc. Math. France **109**, 207–215 (1981)
18. Mendès France, M.: Paper folding, space-filling curves and Rudin-Shapiro sequences. In: Papers in Algebra, Analysis and Statistics, vol. 9. Contemporary Mathematics, American Mathematical Society, pp. 85–95 (1982)
19. Mousavi, H.: Automatic theorem proving in Walnut. Arxiv preprint arXiv:1603.06017 [cs.FL] (2016)
20. Rampersad, N., Shallit, J.: Rudin-Shapiro sums via automata theory and logic. Arxiv preprint arXiv:2302.00405 [math.NT], 1 February 2023
21. Rudin, W.: Some theorems on Fourier coefficients. Proc. Am. Math. Soc. **10**, 855–859 (1959)
22. Shallit, J.: Synchronized sequences. In: Lecroq, T., Puzynina, S. (eds.) WORDS 2021. LNCS, vol. 12847, pp. 1–19. Springer, Cham (2021). https://doi.org/10.1007/978-3-030-85088-3_1
23. Shallit, J.: The Logical Approach To Automatic Sequences: Exploring Combinatorics on Words with Walnut, vol. 482. London Mathematical Society Lecture Note Series. Cambridge University Press, Cambridge (2022)

24. Shallit, J.: Rarefied Thue-Morse sums via automata theory and logic. Arxiv preprint ArXiv:2302.09436 [math.NT], 18 February 2023. https://arxiv.org/abs/2302.09436
25. Shapiro, H.S.: Extremal problems for polynomials and power series. Master's thesis, MIT (1951)
26. Sloane, N.J.A., et al.: The On-line Encyclopedia of Integer Sequences (2023). https://oeis.org

Automaticity and Parikh-Collinear Morphisms

Michel Rigo⬤, Manon Stipulanti⬤, and Markus A. Whiteland(✉)⬤

Department of Mathematics, University of Liège, Liège, Belgium
{m.rigo,m.stipulanti,mwhiteland}@uliege.be

Abstract. Parikh-collinear morphisms have recently received a lot of attention. They are defined by the property that the Parikh vectors of the images of letters are collinear. We first show that any fixed point of such a morphism is automatic. Consequently, we get under some mild technical assumption that the abelian complexity of a binary fixed point of a Parikh-collinear morphism is also automatic, and we discuss a generalization to arbitrary alphabets. Then, we consider the abelian complexity function of the fixed point of the Parikh-collinear morphism $0 \mapsto 010011$, $1 \mapsto 1001$. This 5-automatic sequence is shown to be aperiodic, answering a question of Salo and Sportiello.

Keywords: Automatic sequences · Morphic words · Abelian complexity · Automated theorem proving · Walnut

1 Introduction

Let us briefly introduce the main concept of this paper. Details and precise definitions are given in Sect. 2. Let A be a finite alphabet. A morphism $f \colon A^* \to B^*$ is *Parikh-collinear* if the Parikh vectors $\Psi(f(a))$, $a \in A$, are collinear (or pairwise $\mathbb{Z}$-linearly dependent).

Example 1. The morphism $f \colon \{0,1\}^* \to \{0,1\}^*$ with $0 \mapsto 010011$, $1 \mapsto 1001$ is Parikh-collinear, as is the morphism $g \colon \{0,1,2\}^* \to \{0,1,2\}^*$ defined by $0 \mapsto 012$, $1 \mapsto 102201$, $2 \mapsto \varepsilon$. Any *Parikh-constant* morphism [26] (i.e., the Parikh vectors of images of letters are equal) is Parikh-collinear.

Parikh-collinear morphisms have received some attention in recent years. Cassaigne *et al.* characterized Parikh-collinear morphisms as those morphisms that map all words to words with bounded abelian complexity [8]. These morphisms also provide infinite words with interesting properties with respect to the so-called k-binomial equivalence $\sim_k$. Two words $u, v \in A^*$ are *k-binomially equivalent* if $\binom{u}{x} = \binom{v}{x}$, for all $x \in A^*$ with $|x| \leq k$. Recall that a binomial coefficient $\binom{u}{x}$ counts the number of times x occurs as a subword of u. The k-*binomial complexity function* of an infinite word $\mathbf{x}$ introduced in [26] is defined as $\mathsf{b}_{\mathbf{x}}^{(k)} \colon \mathbb{N} \to \mathbb{N}$, $n \mapsto \#(\mathcal{L}_n(\mathbf{x})/\sim_k)$, i.e., length-$n$ factors in $\mathbf{x}$ are counted up

© The Author(s), under exclusive license to Springer Nature Switzerland AG 2023
A. Frid and R. Mercaş (Eds.): WORDS 2023, LNCS 13899, pp. 247–260, 2023.
https://doi.org/10.1007/978-3-031-33180-0_19

to k-binomial equivalence. (Here $b_x^{(1)}$ is the usual abelian complexity function [14].) For a survey on abelian properties of words, see [15]. In a recent work[1], we showed that a morphism is Parikh-collinear iff it maps all words with bounded k-binomial complexity to words with bounded $(k+1)$-binomial complexity (for all k) [27]. Thus any fixed point of a Parikh-collinear morphism has a bounded k-binomial complexity for all k (and thus a bounded abelian complexity).

In our computer experiments and research presentations, the first few values of these bounded complexities on various examples suggested that the abelian complexity of a fixed point of a Parikh-collinear morphism might be ultimately periodic. This question was asked independently by Ville Salo when the third author visited Turku University and by Andrea Sportiello when the second author gave a presentation at "Journées Combinatoires de Bordeaux" 2023.

Our Contributions. Even though Parikh-collinear morphisms are generally non-uniform, we show in Sect. 3 that their fixed points are k-automatic for $k = \sum_{b \in A} |f(b)|_b$. The result itself, in fact, can be considered folklore. The (constructive) proof given here was inspired by [7,9] but, as pointed out by the referees, it can be seen as a consequence of [1, Thm. 2.2 or 4.2], the former of which is itself a reformulation of a result of Dekking [12] (we note however, that the statements speak of non-erasing morphisms). It is well known that there exist infinite sequences that are the fixed points of non-uniform morphisms, but not k-automatic for any k, and that every k-automatic sequence is the image of a fixed point of a non-uniform morphism [3]. A recent preprint [18] completely characterizes those uniformly recurrent (i.e., every factor occurs infinitely often and with bounded gaps) morphic words that are automatic.

Making use of Büchi's theorem and first-order logic, we prove in Sect. 4 that under some mild assumptions the abelian complexity of a binary fixed point of a Parikh-collinear morphism is automatic. This result supports the expectation of the abelian complexity of a k-automatic sequence to exhibit regular behavior in base-k (cf. Rigo's conjecture [22], named after the first author of this paper). We however recall that abelian properties in general cannot be handled with such a formalism (for instance, Schaeffer showed that the set of occurrences of abelian squares in the paperfolding word is not k-automatic for any k [28]).

Coming back to Salo and Sportiello's question; a positive answer to it would suggest that our second main result is trivial in the sense that any ultimately periodic sequence is automatic over any base. In Sect. 5, we propose an answer to their question by considering the abelian complexity of the fixed point $\mathbf{w} = 0100111001\cdots$ of the morphism $f\colon \{0,1\}^* \to \{0,1\}^*$ given in Example 1: we show that its abelian complexity is aperiodic. We provide two proofs of this result: we make use of, on the one hand, classical combinatorial arguments, and on the other, the software `Walnut` [21,30] to illustrate, in this case, that the study of abelian complexities is amenable to the first-order logic formalism in practice.

Finally, we explain in Sect. 6 how the mild assumptions considered here may be alleviated and the result generalized to arbitrary alphabets.

---

[1] A long version is available at https://doi.org/10.48550/arXiv.2201.04603.

2   Preliminaries

We recall some basics needed for the paper. More can be found in [2,24,25]. We let A^* (resp., $A^{\mathbb{N}}$) denote the set of finite (resp., infinite) words over A equipped with concatenation. Infinite words are written in bold unless otherwise stated. We let ε denote the empty word. The length of the word w is denoted by $|w|$ and the number of occurrences of a letter a in w is denoted by $|w|_a$. The *Parikh vector* of a word $w \in A^*$ is defined as the vector $\Psi(w) = (|w|_a)_{a \in A} \in \mathbb{N}^A$. An infinite word is *ultimately periodic* if it can be written as $uvvv \cdots$ where $u, v \in A^*$ and $v \neq \varepsilon$. If it is not the case, it is said to be *aperiodic*. For an infinite word $\mathbf{x}$ and an integer $n \geq 0$, we let $\mathcal{L}(\mathbf{x})$ and $\mathcal{L}_n(\mathbf{x})$ respectively denote the set of factors of $\mathbf{x}$ and that of length-n factors of $\mathbf{x}$.

A *morphism* is a map $f \colon A^* \to B^*$, where A, B are alphabets, such that $f(xy) = f(x)f(y)$ for all $x, y \in A^*$. The morphism f is *prolongable* on the letter $a \in A$ if $f(a) = ax$ for some $x \in A^*$ and $\lim_{n \to \infty} |f^n(x)| = \infty$. We let $f^{\omega}(a) := \lim_{n \to \infty} f^n(a)$ denote the fixed point of f starting with a; an infinite word $\mathbf{x}$ is called *pure morphic* if $\mathbf{x} = f^{\omega}(a)$ for some such f and a. An infinite word is *morphic* if it can be written as $g(f^{\omega}(a))$, where g and f are morphisms such that f is prolongable on a. For a given morphism $f \colon A^* \to A^*$, a letter $a \in A$ is called *mortal* if $f^n(a) = \varepsilon$ for some $n \geq 1$. If a is not mortal, we call it *immortal*. For an integer $k \geq 1$, a morphism $f \colon A^* \to B^*$ is *k-uniform* if $|f(a)| = k$ for all letters $a \in A$. A 1-uniform morphism is called a *coding*.

For a morphism $f \colon A^* \to A^*$ let $M_f \in \mathbb{N}^{A \times A}$ denote the associated matrix defined by $(M_f)_{a,b} = |f(b)|_a$ for $a, b \in A$. Then we have $\Psi(f(w)) = M_f \Psi(w)$ for all words $w \in A^*$. A morphism f is *primitive* if the corresponding matrix M_f is primitive, that is, there exists a power of M_f having only positive entries.

Two words $u, v \in A^*$ are *abelian equivalent* if they are obtained as permutations of each other, and we write $u \sim_1 v$. The latter relation is called the *abelian equivalence* already introduced by Erdős [14]. The *abelian complexity function* of an infinite word $\mathbf{x}$ is defined as $\mathsf{a_x} \colon \mathbb{N} \to \mathbb{N}$, $n \mapsto \#(\mathcal{L}_n(\mathbf{x})/\sim_1)$.

Introduced by Cobham [11], automatic words have several equivalent definitions. See [2] for a comprehensive presentation. Let $k \geq 2$ be an integer. An infinite word $\mathbf{x}$ is *k-automatic* if it is the image, under a coding, of a fixed point of a k-uniform morphism; they form a subclass of morphic words. They can also be characterized by means of first-order logic. Consider the structure $\langle \mathbb{N}, +, V_k \rangle$, where $V_k(0) := 1$ and, for all $n \geq 1$, $V_k(n)$ is the largest power of k dividing n. A set $X \subseteq \mathbb{N}^d$ is *k-definable* if it can be defined by a first-order formula with d free variables within $\langle \mathbb{N}, +, V_k \rangle$. As a consequence of a theorem of Büchi [5], an infinite word $\mathbf{x}$ is k-automatic iff for every letter a, the set of positions where a occurs in $\mathbf{x}$ is k-definable. Again, we refer the reader to [2,4,10,25].

3   Automaticity of Parikh-Collinear Fixed Points

We focus on Parikh-collinear morphisms f that are prolongable on some letter. For any mortal letter $m \in A$ of f we have that $f(m) = \varepsilon$. Indeed, if $f(b) \neq \varepsilon$

for a letter $b \in A$ and f is prolongable on a, then $f(b)$ contains an occurrence of a by Parikh-collinearity. Therefore b cannot be mortal. We shall always assume that the underlying alphabet is minimal. More precisely, we assume that each letter of A appears in $f(a)$ for any immortal letter a. Again, Parikh-collinearity implies that the minimal alphabet is well-defined. For an immortal letter $a \in A$, for each $b \in A$ there exists $r_b \in \mathbb{Q}$ such that $\Psi(f(b)) = r_b \Psi(f(a))$.

Lemma 2. Let $f \colon A^* \to A^*$ be Parikh-collinear and $a \in A$ be immortal. Then $\Psi(f(a))$ is an eigenvector of M_f associated with the eigenvalue $\sum_{b \in A} |f(b)|_b$.

Proof. For any word $w \in A^*$, we have

$$M_f \Psi(w) = \sum_{b \in A} |w|_b M_f \Psi(b) = \sum_{b \in A} |w|_b \Psi(f(b)) = \sum_{b \in A} (|w|_b r_b) \cdot \Psi(f(a)).$$

With the choice $w = f(a)$, we find that $\Psi(f(a))$ is an eigenvector of M_f associated with the eigenvalue $\sum_{b \in A} |f(a)|_b r_b = \sum_{b \in A} |f(b)|_b$. □

When speaking of *the eigenvalue* of a Parikh-collinear morphism f, we mean the eigenvalue $\sum_{b \in A} |f(b)|_b$ of M_f. As M_f has rank 1, the only other eigenvalue is 0 (with multiplicity $\#A - 1$).

Remark 3. If f is prolongable on a letter a, then the eigenvalue k of f is at least 2. Indeed, $f(a)$ must contain at least two occurrences of immortal letters (the first letter a and another one, say b). If $b = a$ then $k \geq |f(a)|_a \geq 2$, otherwise $|f(a)|_a, |f(b)|_b \geq 1$ by Parikh-collinearity and again $k \geq 2$.

In what follows, for an infinite word $\mathbf{x}$, a letter $a \in A$ is called *left deterministic* (resp., *right deterministic*) if it is always preceded (resp., followed) by a unique letter $b \in A$ in $\mathbf{x}$. In particular, the first letter of $\mathbf{x}$ is not left deterministic.

Lemma 4. Let $\mathbf{x} \in A^{\mathbb{N}}$ be a fixed point of the morphism $f \colon A^* \to A^*$. Assume further that, for distinct letters $a_1, \ldots, a_\ell$, such that $a_1 \cdots a_\ell \in \mathcal{L}(\mathbf{x})$, a_i is right deterministic for $i < \ell$, and a_j is left deterministic for $j \geq 2$. Factorize $f(a_1 \cdots a_\ell) = u_1 \cdots u_\ell$, with $u_i \in A^*$. Then $g(\mathbf{x}) = \mathbf{x}$, where g is the morphism defined by $a_i \mapsto u_i$ for all $i \in \{1, \ldots, \ell\}$, and $c \mapsto f(c)$ for all other $c \in A$.

Proof. Writing $w = a_1 \cdots a_\ell$, we may factorize $\mathbf{x} = x_0 w x_1 \cdots w x_n \cdots$, where $x_i \in (A \setminus \{a_1, \ldots, a_\ell\})^*$ (and if w appears only a finite number n of times, then $x_n \in (A \setminus \{a_1, \ldots, a_\ell\})^{\mathbb{N}}$). But we now have $g(w) = f(w)$ and $g(x_i) = f(x_i)$ for all i by construction, whence $g(\mathbf{x}) = f(\mathbf{x}) = \mathbf{x}$. □

Notice that if g above is prolongable on the first letter a of $\mathbf{x}$, then $\mathbf{x} = g^\omega(a)$.

The reader may wish to consult Example 6 for illustrations of the constructions provided the proof the following theorem.

Theorem 5. Let $f \colon A^* \to A^*$ be a Parikh-collinear morphism prolongable on a letter $a \in A$. Then $f^\omega(a)$ is k-automatic for the eigenvalue k of f.

Proof. Let us write $\mathbf{x} := f^{\omega}(a)$. Recall from Lemma 2 that $\Psi(f(b)) = r_b \Psi(f(a))$ is either the zero vector or an eigenvector of the adjacency matrix M_f associated with the eigenvalue k. It therefore follows that $|f^2(b)| = k \cdot |f(b)|$ for each $b \in A$.

As an intermediate step, we construct a pure morphic word $\mathbf{y}$ which can be mapped, by a coding, to $\mathbf{x}$. To this end, for each immortal letter $b \in A$ define the letters $\widehat{b}_1, \ldots, \widehat{b}_{|f(b)|}$ and let $B = \{\widehat{b}_i : b \in A \text{ is immortal}, i = 1, \ldots, |f(b)|\}$. Let $\tau: B^* \to A^*$ be the coding defined by $\tau(\widehat{b}_i) = c$ if the ith letter of $f(b)$ equals c. Set, for each $c \in A$, the word $\mathfrak{w}_c := \widehat{c}_1 \cdots \widehat{c}_{|f(c)|}$ if c is immortal, otherwise set $\mathfrak{w}_c = \varepsilon$. It is now evident that $\tau(\mathfrak{w}_c) = f(c)$. Define then the morphism $\varphi: B^* \to B^*$ as follows: for each immortal $b \in A$ and $i \in \{1, \ldots, |f(b)|\}$, we set $\varphi(\widehat{b}_i) = \mathfrak{w}_{\tau(\widehat{b}_i)}$. Notice now that $\tau(\varphi(\widehat{b}_i)) = \tau(\mathfrak{w}_{\tau(\widehat{b}_i)}) = f(\tau(\widehat{b}_i))$, and so $\tau \circ \varphi = f \circ \tau$ as morphisms $B^* \to A^*$. Moreover $\tau \circ \varphi^n = f \circ \tau \circ \varphi^{n-1} = \ldots = f^n \circ \tau$. Since a is immortal we get $\tau(\varphi^n(\widehat{a}_1)) = f^n(a)$ for all $n \geq 0$, yielding $\tau(\varphi^{\omega}(\widehat{a}_1)) = \mathbf{x}$. We set $\mathbf{y} = \varphi^{\omega}(\widehat{a}_1)$.

Next we define a k-uniform morphism $g: B^* \to B^*$ for which $g^{\omega}(\widehat{a}_1) = \mathbf{y}$ (recall that $k \geq 2$ by Remark 3). This implies that $\mathbf{x}$ is k-automatic, as then $\mathbf{x} = \tau(g^{\omega}(\widehat{a}_1))$. Fix an immortal letter $b \in A$ (we will proceed iteratively for each of them). The letters $\widehat{b}_i$, with $i \in \{1, \ldots, |f(b)|\}$, satisfy the assumptions of Lemma 4 in $\mathbf{y}$. Notice that $\tau(\varphi(\mathfrak{w}_b)) = \tau(\varphi^2(\widehat{b}_1)) = f^2(\tau(\widehat{b}_1)) = f^2(b)$, whence $|\varphi(\mathfrak{w}_b)| = k \cdot |f(b)|$. Factorize $\varphi(\mathfrak{w}_b) = u_1 \cdots u_{|\mathfrak{w}_b|}$, each of the words u_i having length k. By Lemma 4, we have $\mathbf{y} = g_b(\mathbf{y})$, where $g_b: B^* \to B^*$ is defined by $g_b(\widehat{b}_i) = u_i$ and $g_b(\mathfrak{a}) = \varphi(\mathfrak{a})$ for $\mathfrak{a} \in B \setminus \{\widehat{b}_i : i = 1, \ldots, |f(b)|\}$ (note that $|g_b(\widehat{b}_i)| = k$ for each $i \in \{1, \ldots, |f(b)|\}$). We may repeat this operation on g_b (sequentially) for all the other immortal letters $c \in A$, and the resulting morphism g is k-uniform for which $g(\mathbf{y}) = \mathbf{y}$. Clearly g is prolongable on $\widehat{a}_1$, which suffices for the claim. $\square$

Example 6. Let f be defined by $0 \mapsto 012;\ 1 \mapsto 112002;\ 2 \mapsto \varepsilon$, and let $\mathbf{x} = f^{\omega}(0)$. Here $k = 3$. We thus have $B = \{\widehat{0}_i, \widehat{1}_j : i = 1, 2, 3,\ j = 1, \ldots, 6\}$, and τ is defined by $\widehat{0}_1, \widehat{1}_4, \widehat{1}_5 \mapsto 0;\ \widehat{0}_2, \widehat{1}_1, \widehat{1}_2 \mapsto 1;\ \widehat{0}_3, \widehat{1}_3, \widehat{1}_6 \mapsto 2$. We then define φ by

$$\widehat{0}_1, \widehat{1}_4, \widehat{1}_5 \mapsto \mathfrak{w}_0 = \widehat{0}_1 \widehat{0}_2 \widehat{0}_3;\quad \widehat{0}_2, \widehat{1}_1, \widehat{1}_2 \mapsto \mathfrak{w}_1 = \widehat{1}_1 \widehat{1}_2 \widehat{1}_3 \widehat{1}_4 \widehat{1}_5 \widehat{1}_6;\quad \widehat{0}_3, \widehat{1}_3, \widehat{1}_6 \mapsto \varepsilon.$$

Factorizing $\varphi(\mathfrak{w}_0)$ and $\varphi(\mathfrak{w}_1)$, respectively, as

$$\varphi(\mathfrak{w}_0) = \widehat{0}_1 \widehat{0}_2 \widehat{0}_3 \cdot \widehat{1}_1 \widehat{1}_2 \widehat{1}_3 \cdot \widehat{1}_4 \widehat{1}_5 \widehat{1}_6$$
$$\varphi(\mathfrak{w}_1) = \widehat{1}_1 \widehat{1}_2 \widehat{1}_3 \cdot \widehat{1}_4 \widehat{1}_5 \widehat{1}_6 \cdot \widehat{1}_1 \widehat{1}_2 \widehat{1}_3 \cdot \widehat{1}_4 \widehat{1}_5 \widehat{1}_6 \cdot \widehat{0}_1 \widehat{0}_2 \widehat{0}_3 \cdot \widehat{0}_1 \widehat{0}_2 \widehat{0}_3,$$

we define g by $\widehat{0}_1, \widehat{1}_5, \widehat{1}_6 \mapsto \widehat{0}_1 \widehat{0}_2 \widehat{0}_3;\ \widehat{0}_2, \widehat{1}_1, \widehat{1}_3 \mapsto \widehat{1}_1 \widehat{1}_2 \widehat{1}_3;$ and $\widehat{0}_3, \widehat{1}_2, \widehat{1}_4 \mapsto \widehat{1}_4 \widehat{1}_5 \widehat{1}_6$, which gives $\tau(g^{\omega}(\widehat{0}_1)) = \mathbf{x}$.

One notes that there are redundant letters (i.e., they have equal images under both τ and $g \circ \tau$); we find a simpler morphism h by identifying them: $0 \mapsto 012;\ 1 \mapsto 134;\ 2 \mapsto 506;\ 3 \mapsto 506;\ 4 \mapsto 134;\ 5 \mapsto 506;\ 6 \mapsto 012$, with which $\tau'(h^{\omega}(0)) = \mathbf{x}$, where τ' is defined by $0, 5 \mapsto 0;\ 1, 3 \mapsto 1;\ 2, 4, 6 \mapsto 2$.

4 Automaticity of the Abelian Complexity

In this section, we consider a Parikh-collinear morphism $f \colon A \to A^*$ prolongable on a letter $a \in A$ and its fixed point $\mathbf{x} = f^\omega(a)$. We set $k = \sum_{b \in A} |f(b)|_b$ to be the eigenvalue of f. For all $n \geq 0$, we let $\mathrm{pref}_n(\mathbf{x})$ be the length-n prefix of $\mathbf{x}$. The corresponding *cutting set* is defined by

$$\mathrm{CS}_{f,a} := \{|f(\mathrm{pref}_n(\mathbf{x}))| \colon n \geq 0\}. \tag{1}$$

This set simply provides the indices where blocks $f(b)$, with $b \in A$, start in a factorization of $\mathbf{x}$ of the form $f(x_0)f(x_1)f(x_2)\cdots$.

To help the reader, we now start a running example throughout the section.

Running Example 7. Consider $f \colon 0 \mapsto 010011,\ 1 \mapsto 1001$ and $\mathbf{w} = f^\omega(0)$. The first five elements in $\mathrm{CS}_{f,0}$ are 0, 6, 10, 16, and 22.

Given any integer i, we look for two consecutive integers: the next and previous elements found in C around i.

Lemma 8. *Let $C = \{0 = c_0 < c_1 < c_2 < \cdots\}$ be an infinite k-definable subset of $\mathbb{N}$. The functions $\mathrm{ne} \colon \mathbb{N} \to \mathbb{N}$ mapping i to the least element in C greater than or equal to i and $\mathrm{pr} \colon \mathbb{N} \to \mathbb{N}$ mapping i to the greatest element in C less than i, are k-definable. (We set $\mathrm{pr}(0) = 0$.)*

Proof. Since C is k-definable by some formula φ_C, i.e., $\varphi_C(j)$ holds iff $j \in C$. The functions of the statement are then defined by

$$\mathrm{ne}(i) = j \equiv \varphi_C(j) \wedge (i \leq j) \wedge (\forall k)(\varphi_C(k) \wedge i \leq k) \to (j \leq k),$$
$$\mathrm{pr}(i) = j \equiv \varphi_C(j) \wedge (j < i) \wedge (\forall k)(\varphi_C(k) \wedge k < i) \to (k \leq j). \qquad \square$$

It is easy to see that the abelian complexity of a binary word $\mathbf{x}$, fixed point of a (non-erasing) Parikh-collinear morphism, is given by

$$\mathsf{a}_{\mathbf{x}}(n) = \frac{1}{r + r'}\left(\max_{x \in \mathcal{L}_n(\mathbf{x})}(r'|x|_1 - r|x|_0) - \min_{x \in \mathcal{L}_n(\mathbf{x})}(r'|x|_1 - r|x|_0)\right) + 1, \tag{2}$$

where $|f(a)|_1 = \frac{r}{q}|f(a)|$ and $|f(a)|_0 = \frac{r'}{q}|f(a)|$ for both $a \in \{0, 1\}$. Notice that we have $r + r' = q$ and $r'|f(a)|_1 = r|f(a)|_0$.

Observe that, since f is Parikh-collinear, each full block $f(b)$ occurring in a factor x has no contribution to the value of $r'|x|_1 - r|x|_0$ in the above formula. This observation is at the core of our reasoning. A similar strategy can be found in [16] where for factors of the Thue–Morse word, one can disregard full images.

Let x be a length-n factor of $\mathbf{x}$. Then there exist two letters $b, b' \in A$, a proper suffix s of $f(b)$, a factor u, and a proper prefix p of $f(b')$ such that $x = sf(u)p$. Due to the previous observation what matters to compute the abelian classes is therefore the total contribution of both p and s. Note that there are only finitely many such proper prefixes and suffixes, which is enough to be encoded into a formula. In addition, empty prefixes or suffixes have no contribution.

Running Example 9. The word $11|1001|010011|010 = 11f(10)010$ is a factor of length 15 of **w** occurring at position 4. For this morphism f, we have $r = r' = 1$ and $q = 2$ with the notation of Eq. (2). We have $\mathrm{pr}(4) = 0$ and $\mathrm{ne}(4) = 6$. The prefix 11 has a contribution of 2 to $r'|x|_1 - r|x|_0$ and the suffix 010 a contribution of -1. In Table 1, the contribution (symbolized by c) of each suffix (symbolized by s) and each prefix (symbolized by p) of $f(0)$ and $f(1)$ is given.

Table 1. Contributions (c) to $r' | \cdot |_1 - r | \cdot |_0$ of the suffixes (s) and prefixes (p) of $f(0)$ and $f(1)$ respectively.

s	c	s	c	p	c	p	c
1	1	1	1	01001	−1	100	−1
11	2	01	0	0100	−2	10	0
011	1	001	−1	010	−1	1	1
0011	0			01	0		
10011	1			0	−1		

For the sake of presentation, we give the next result for binary words under a mild assumption on $\mathsf{CS}_{f,a}$. In Sect. 6 we discuss how to generalize it.

Theorem 10. *Let* $\mathbf{x} = f^\omega(0) \in \{0,1\}^\mathbb{N}$ *be a binary fixed point of a Parikh-collinear morphism. If* $\mathsf{CS}_{f,a}$ *is* k-*automatic then* $\mathsf{a_x}(n)$ *is* k-*automatic.*

Proof. We can assume that $|f(0)| \neq |f(1)|$. Because otherwise f is Parikh-constant and Guo *et al.* [17, Thm. 3] showed that the Parikh-constant image of a k-automatic sequence has k-automatic abelian complexity.

Let $x = sf(u)p$ be the factor of length n occurring in position i where, as usual, s is a proper suffix of $f(b)$, u is a factor, and p is a proper prefix of $f(b')$ for some letters b, b'. Let r, r' be the constants as in Eq. (2).

If s is non-empty, i.e., if $i \neq \mathrm{ne}(i)$, the letter b is uniquely determined by $\mathrm{ne}(i) - \mathrm{pr}(i)$. This difference is equal to $|f(b)|$ and by assumption, distinct letters have images with distinct length. The length of s is given by $\mathrm{ne}(i) - i$. Consequently, $i, \mathrm{ne}(i), \mathrm{pr}(i)$ determine if $s = \varepsilon$ or, a unique suffix s with a specific contribution to $r'|x|_1 - r|x|_0$.

Similarly, the length of p is zero if $i + n - 1 = \mathrm{ne}(i + n - 1)$, i.e., $i + n - 1$ belongs to $\mathsf{CS}_{f,a}$. If p is non-empty, the letter b' is uniquely determined by $\mathrm{ne}(i + n - 1) - \mathrm{pr}(i + n - 1)$. Otherwise, the length of p is given by $i + n - 1 - \mathrm{pr}(i + n - 1)$. Consequently, $i + n - 1, \mathrm{ne}(i + n - 1), \mathrm{pr}(i + n - 1)$ determine if $p = \varepsilon$ or, a unique prefix p with a specific contribution to $r'|x|_1 - r|x|_0$. Since there are finitely many prefixes and suffixes, we may define a function $\mathrm{contr} : (i, n) \mapsto r'|x|_1 - r|x|_0$ where x is the length-n factor occurring at position i in **x**. It is a finite disjunction of terms of the form

$$(\text{ne}(i) - \text{pr}(i) = z_1 \wedge \text{ne}(i) - i = z_2 \wedge \text{ne}(i + n - 1) - \text{pr}(i + n - 1) = z_3$$
$$\wedge \, i + n - 1 - \text{pr}(i + n - 1) = z_4) \rightarrow \text{contr}(i, n) = \lambda,$$

where λ depends on the 4-tuple (z_1, z_2, z_3, z_4). The complete formula should also take into account the test for empty p or s. We illustrate this on our running example below.

The conclusion follows: the maximum in Eq. (2) is defined as a function of n by $\max_{x \in \mathcal{L}_n(\mathbf{x})}(r'|x|_1 - r|x|_0) = M \equiv (\exists i)(\text{contr}(i, n) = M) \wedge (\forall j)(\text{contr}(i, n) \leq M)$. One proceeds similarly with the minimum and we thus see that $a_{\mathbf{x}}(n)$ is k-definable. Simply recall that multiplication by a constant is definable in Presburger arithmetic and by Büchi's theorem, the claim follows.   □

Running Example 11. With $i = 4$ and $n = 15$, we find $\text{pr}(i) = 0$, $\text{ne}(i) = 6$, so $z_1 = 6 = |f(0)|$, and $i + n - 1 = 18$, $\text{pr}(18) = 15$, $\text{ne}(18) = 21$, so $z_3 = 6 = |f(0)|$. Since $z_2 = \text{ne}(i) - i = 2$, we know that we have the suffix of $f(0)$ of length 2 with contribution 2 (recall Table 1). Similarly, since $z_4 = i + n - 1 - \text{pr}(i + n - 1) = 3$, we known that we have the prefix of $f(0)$ of length 3 with contribution -1. So the 4-tuple $(6, 2, 6, 3)$ is associated with the total contribution $\frac{3 - (-1)}{2} - 1 = 1$.

We now show on our running example that $\mathsf{CS}_{f,0}$ is k-definable.

Running Example 12. The factor 0100 occurs in $\mathbf{w}$ only in position corresponding to prefixes of blocks $f(0)$. Since $\mathbf{w}$ is k-automatic for $k = 5$, there are two k-definable unary relations $\varphi_0(i)$ and $\varphi_1(i)$ which are true iff 0 and 1 respectively occur in $\mathbf{w}$ at position i. So the index i is the starting position of a block $f(0)$ iff $\Psi_0(i) \equiv \varphi_0(i) \wedge \varphi_1(i + 1) \wedge \varphi_0(i + 2) \wedge \varphi_0(i + 3)$ holds. In a similar way, every block $f(1)$ is preceded in $\mathbf{w}$ by a letter 1, so the factor 1100 occurs in position $i - 1$ iff an occurrence of $f(1)$ starts in position i. Therefore the index i is the starting position of a block $f(1)$ iff $\Psi_1(i) \equiv \varphi_1(i - 1) \wedge \varphi_1(i) \wedge \varphi_0(i + 1) \wedge \varphi_0(i + 2)$ holds. From this, we deduce that $\mathsf{CS}_{f,0}$ is k-definable by $\{i : \Psi_0(i) \vee \Psi_1(i)\}$ and we can therefore apply the above theorem.

5   The Fixed Point of $0 \mapsto 010011$, $1 \mapsto 1001$

We consider the pure morphic word $\mathbf{w} = f^{\omega}(0)$, where $f : 0 \mapsto 010011$, $1 \mapsto 1001$ (see Running Example 7). We give two different proofs of the following result.

Proposition 13. *The abelian complexity function $a_{\mathbf{w}}$ of $\mathbf{w}$ is aperiodic.*

5.1   Proving Aperiodicity the Old-Fashioned Way

It is straightforward to see that $\mathbf{w}$ is aperiodic and uniformly recurrent. Furthermore, the frequency of 0 exists and equals $1/2$ as this is the case for the images of both letters. Hence, for each $n \geq 0$, there exist length-n factors u, v for which $|u|_0 > n/2$ and $|v|_0 < n/2$ (see, for example, [23, Lem. 4.3]). The next lemma shows that $a_{\mathbf{w}}$ is bounded by 4.

Lemma 14. *We have* $\mathsf{a_w}(2n + 1) \leq 4$ *and* $\mathsf{a_w}(2n) = 3$ *for all* $n \geq 0$. *More precisely, for a factor* $x \in \mathcal{L}(\mathbf{w})$, *we have* $\lfloor |x|/2 \rfloor - 1 \leq |x|_0 \leq \lceil |x|/2 \rceil + 1$.

Proof. The claim can be verified for factors of length at most 4 straightforwardly. For a factor x of length $n \geq 5$ we may write $x = sf(u)p$, where s is a proper suffix and p is a proper prefix of the image of a letter. Notice that $|f(u)|_0 = |f(u)|/2$, hence $|x|_0 = (|x| - |ps|)/2 + |ps|_0$. Notice here that $|ps|$ and $|x|$ have the same parity, as $|f(u)|$ has even length. By inspecting all combinations of p and s (with total length odd and even separately), we find the more precise claim.

To see that $\mathsf{a_w}(2n) = 3$, all three values n and $n \pm 1$, for the number of 0s, are attained, as was asserted in the beginning of this section. $\square$

The proof of the following lemma is a tedious exercise using properties established above. We omit the proof due to space constraints.

Lemma 15. *The abelian complexity function of* $\mathbf{w}$ *satisfies the following.*

1. *We have* $\mathsf{a_w}(5n) = \mathsf{a_w}(n)$ *for all* $n \geq 0$.
2. *For each* $j \in \{1, 2, 3, 4\}$ *and for all* $n \geq 1$, *we have* $\mathsf{a_w}(5n + j) \geq 3$.

Proof sketch. 1. The claim essentially follows from two correspondences: We have $\min_{u \in \mathcal{L}_n(\mathbf{w})}\{|u|_0\} < \lfloor n/2 \rfloor$ iff $\max_{x \in \mathcal{L}_{5n}(\mathbf{w})}\{|x|_0\} > \lceil 5n/2 \rceil$. Similarly, $\max_{u \in \mathcal{L}_n(\mathbf{w})}\{|u|_0\} > \lceil n/2 \rceil$ iff $\min_{x \in \mathcal{L}_n(\mathbf{w})}\{|x|_0\} < \lfloor 5n/2 \rfloor$. For the proof one takes a factor $x = sf(u)p$, and computes bounds on $|u|$. Inspecting all possibilities (utilizing Lemma 14) gives the claimed properties.

2. When $5n + j$ is even, Lemma 14 shows that $\mathsf{a_w}(5n + j) = 3$. For each of the cases $j \in \{1, 2, 3, 4\}$ and for each n for which $5n + j$ is odd, we exhibit a factor x of length $5n + j$ for which $|x|_0 \notin \frac{5n+j\pm1}{2}$. For example, let $u \in \mathcal{L}(\mathbf{w})$ be of odd length $n - 1$ such that $1u0 \in \mathcal{L}(\mathbf{w})$ and $|u|_0 = \lfloor \frac{n-1}{2} \rfloor$. Such a factor always exists: there exist two factors v, v' and an index i such that $|v|_0 = \frac{n-2}{2}$, $|v'|_0 = \frac{n}{2}$, v begins at position i in $\mathbf{w}$, and v' begins at position $i + 1$ in $\mathbf{w}$. The only possibility is that v begins with 1 and v' ends with 0; u may be then taken to be the overlap of v and v'. We have $|f(u)| = 4 \cdot |u| + 2 \cdot |u|_0 = 5n - 6$ since n is even. Take $x = 001f(u)0100$. Then $|x| = 5n + 1$ and $|x|_0 = \frac{|f(u)|}{2} + 5 = \frac{5n-6}{2} + 5 = \lceil \frac{5n+1}{2} \rceil + 1$. $\square$

Proof of Proposition 13. We show that $\mathsf{a_w}(n) = 2$ iff $n = 5^m$ for some $m \geq 0$. The value 2 is unattainable for all even n by Lemma 14. Let n be odd and write $n = 5^m \cdot k$ with $\gcd(k, 5) = 1$; then $\mathsf{a_w}(5^m \cdot k) = \mathsf{a_w}(k)$ by repeated application of Lemma 15(1). Write $k = 5\ell + j$ for some $1 \leq j \leq 4$ and $\ell \geq 0$. If $\ell = 0$ and $j = 1$ (so $n = 5^m$) we find $\mathsf{a_w}(n) = 2$, and if $j \geq 2$ we find $\mathsf{a_w}(n) = 3$ by inspection. For $\ell \geq 1$ Lemma 15(2) shows $\mathsf{a_x}(n) \geq 3$. $\square$

5.2 Proof of Proposition 13 via `Walnut in a Shell`

Originally designed by Mousavi [21], `Walnut` is a free and publicly available software that allows to prove theorems and properties for the particular family of automatic words through the lens of first-order logic; see Shallit's recent book [30] for a comprehensive presentation. We shall use `Walnut` to prove Proposition 13.

The procedure associated to Theorem 5 (complemented with identifying redundant letters and renaming) gives the 5-uniform morphism g: $0 \mapsto$ 01023, $1 \mapsto$ 14501, $2 \mapsto$ 10102, $3 \mapsto$ 31450, $4 \mapsto$ 45010, and $5 \mapsto$ 10231, and the coding τ: $0, 2, 5 \mapsto 0$; $1, 3, 4 \mapsto 1$, for which $\mathbf{w} = \tau(g^\omega(0))$, and which can be used to define $\mathbf{w}$ in Walnut conveniently by:

morphism g "0->01023 1->14501 2->10102 3->31450 4->45010 5->10231";
morphism tau "0->0 1->1 2->0 3->1 4->1 5->0";
promote G g;
image W tau G;

Following the procedure described in Sect. 4, we construct a DFAO generating $a_\mathbf{w}$. We first recall basic syntax for Walnut. The letters A and E are abbreviations for ∀ ("for all") and ∃ ("there exists") respectively. The letters ?msd_5 indicates that an expression is to be evaluated using base-5 representations. The symbol @ specifies the value of an automatic sequence. The symbols &, |, ~, and => are logical "and" , "or", negation and implication respectively.

As in Eq. (1), we define the cutting set (details to understand the formula can be found in Running Example 12):

def CS "?msd_5 (W[i]=@0 & W[i+1]=@1 & W[i+2]=@0 & W[i+3]=@0)
             | (i>0 & W[i-1]=@1 & W[i]=@1 & W[i+1]=@0 & W[i+2]=@0)";

As in Lemma 8, we define the maps ne and pr:

def ne "?msd_5 k >= i & $CS(k) & (A j (j >= i & $CS(j)) => j>=k)";

def pr "?msd_5 k < i & $CS(k) & (A j (j < i & $CS(j)) => j<=k)";

Next we describe the contribution of factors of $\mathbf{w}$ starting at position i for all $i \geq 0$. Using the notation from Sect. 4, for the length-n factor x of $\mathbf{w}$ starting at position i, we write $x = sf(u)p$ where $b, b' \in \{0, 1\}$, s is a proper suffix of $f(b)$, u is a word, and p is a proper prefix of $f(b')$. We know that x only contributes to the abelian complexity $a_\mathbf{w}$ via s and p. Based on Table 1, the following two binary predicates correspond to integer pairs (c, i) such that the contribution from the suffix s of $f(b)$ (resp., prefix p of $f(b')$) starting at position i (resp., ending at position $i - 1$) equals $c - 2$. We notably shift true value by 2 as the variable domain in Walnut is $\mathbb{N}$.

def suffContr "?msd_5 (c=1 & ($ne(i,i+3) & $pr(i,i-1)))
    | (c=2 & ($ne(i,i) | $ne(i,i+4)) | ($ne(i,i+2) & $pr(i,i-2)))
    | (c=3 & ($ne(i,i+1) | ($ne(i,i+3) & $pr(i,i-3)) | $ne(i,i+5)))
    | (c=4 & ($ne(i,i+2) & $pr(i,i-4)))";
def prefContr "?msd_5 (c=0 & ($pr(i,i-4) & $ne(i,i+2)))
    | (c=1 & ($pr(i,i-3) | ($pr(i,i-1) & $ne(i,i+5)) | $pr(i,i-5)))
    | (c=2 & ($ne(i,i) | $pr(i,i-2))) | (c=3 & ($pr(i,i-1) & $ne(i,i+3)))";

The ternary predicate $\mathrm{contr}(c, i, n)$ is satisfied when the prefix contribution at position i and the suffix contribution at position $i + n$ total up to $c - 4$.

def contr "?msd_5 Ed,e $suffContr(d,i) & $prefContr(e,i+n) & (e+d=c)";

Here recall that $d - 2$ (resp., $e - 2$) is the contribution of p at position i (resp., s ending at position $i + n - 1$). (In the end we will call on this predicate only for $n > 4$, so one does not need to worry what happens when, e.g., $n = 0$.)

The following predicates accept the pairs (c, n), where $c - 4$ is the max. (resp., min.) contribution for the prefixes and suffixes p, s for length-n factors.

```
def MaxContr "?msd_5 (Ei $contr(c,i,n)) & (Ai,d $contr(d,i,n) => d<=c)";
def MinContr "?msd_5 (Ei $contr(c,i,n)) & (Ai,d $contr(d,i,n) => d>=c)";
```

We may now define the abelian complexity function as a binary predicate $abComp(a, n)$. We have $\mathbf{a_w}(n) = a$ iff $2 \cdot a = \max(|u|_1 - |u|_0) - \min(|u|_1 - |u|_0) + 2$. Adding 4 to both min and max simultaneously does not change the sum, so we may use `MaxContr` and `MinContr` in their places (interpreted as functions).

```
def abComp "?msd_5 (n=0 & a=1) | (n=1 & a=2) | ((n>=2 & n<=4) & a =3)
    | (n>4 & (Ec,d $MaxContr(c,n) & $MinContr(d,n)  & 2*a = c-d + 2))";
```

Next we generate $\mathbf{a_w}$ as a 5-automatic word, so we define the following predicates, for $z = 1, 2, 3, 4$, which recognize the base-5 representations of integers n such that $\mathbf{a_w}(n) = z$:

```
def abCompz "?msd_5 $abComp(z,n)";
```

To express $\mathbf{a_w}$ as an automatic sequence, we combine the predicates `abCompz` into one; in `Walnut`, this is done with the command

```
combine abCompW abComp1=1 abComp2=2 abComp3=3 abComp4=4;
```

`Walnut` then returns the 9-state deterministic finite automaton with output reading base-5 representations generating $\mathbf{a_w}$ in Fig. 1.

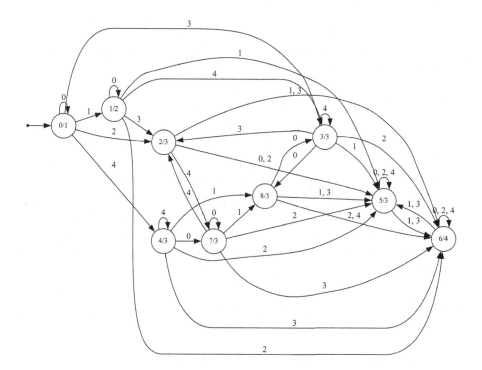

Fig. 1. The minimal deterministic finite automaton with output reading base-5 representations generating the abelian complexity $\mathbf{a_w}$.

We now have an automatic way of proving Proposition 13. One may perform the following query about the aperiodicity of $\mathbf{a_w}$, for which `Walnut` replies TRUE:

```
eval isAper "?msd_5 ~(Ei,p (p>0) & Aj ((j>=i)
                        => (abCompW[j] = abCompW[j+p])))";
```

6   Final Remarks

Our first main result Theorem 5 is constructive, given the morphism f and letter $a \in A$ for which $\mathbf{x} = f^\omega(a)$. Our second main result Theorem 10 holds for those binary $\mathbf{x}$ for which the set $CS_{f,a}$ is k-definable. We briefly sketch a plan for relaxing the assumptions of Theorem 10 to obtain: *Assuming f above is non-erasing, $\mathsf{a_x}$ is k-automatic (for the eigenvalue k of f), and a DFAO defining it can be effectively constructed.* The main idea is the following: we show that, for the sequence $(p_n)_{n\geq 0}$ of length-n prefixes $p_n = \mathrm{pref}_n(\mathbf{x})$ and any letter $b \in A$, the sequence $(|p_n|_b)_{n\geq 0}$ is k-synchronized, namely, there is an automaton which accepts the tuples $(\mathrm{rep}_k(n), \mathrm{rep}_k(|p_n|_b))$ accordingly padded (where $\mathrm{rep}_k(n)$ denote the base-k expansion of n). See again [30, §10] for an excellent introduction. We may then invoke the result and methods of Shallit [29] to conclude.

As mentioned above, with k the eigenvalue of f, Theorem 5 shows that $\mathbf{x}$ is effectively k-automatic. Since f is assumed non-erasing, it is primitive. Therefore, by a result of Mossé [19,20], there exists a constant L such that, from any position i of $\mathbf{x}$, one can determine the indices $\mathrm{pr}(i)$ and $\mathrm{ne}(i)$ of $CS_{f,a}$ by inspecting the factor $\mathbf{x}[i - L...i + L]$. Moreover, given f and a, the constant L is effectively computable by [13]. Therefore, $CS_{f,a}$ is effectively k-definable.

For a prefix p_n of the form $f(x_n)t_n$ where x_n is a prefix of $\mathbf{x}$ such that $\mathrm{pr}(n) = |f(x_n)|$, we have that $|f(x_n)|_b = \frac{r}{q}|f(x)|$, where $r = |f(a)|_b$ and $q = |f(a)|$. Hence $q|p_n|_b = r|f(x_n)| + q|t_n|_b$. Define the function $F(n) = |p_n|_b$. Then

$$y = F(n) \equiv \exists m, z : (\mathrm{pr}(n) = m) \wedge (q \cdot (y - z) = r \cdot m) \wedge (|\mathbf{x}[m...n]|_b = z).$$

Recall that r and q are constants, and notice that $|\mathbf{x}[\mathrm{pr}(n)...n]|$ attains finitely many values, whence the last check ($|\mathbf{x}[m...n]|_b = z$) can be expressed by a first-order logical formula with indexing into $\mathbf{x}$. Hence the function $F(n)$ is defined by a first-order logical formula for which [30, Thm. 10.2.3] applies, and is therefore synchronized.

We remark that Mossé's recognizability result referred to in the above has recently been extended to deal with erasing morphisms [6]. One could hope that the results therein are useful to obtain the automaticity of the abelian complexity of any infinite word generated by a Parikh-collinear morphism.

Acknowledgments. M. Stipulanti and M. Whiteland are supported by the FNRS Research grants 1.B.397.20F and 1.B.466.21F, respectively. We thank Julien Leroy for fruitful discussions on morphic words and pointing out useful references. We thank A. Sportiello and V. Salo for asking the question leading to this paper. We also thank the reviewers for their suggestions.

References

1. Allouche, J.P., Dekking, M., Queffélec, M.: Hidden automatic sequences. Comb. Theory **1**(#20) (2021). https://doi.org/10.5070/C61055386
2. Allouche, J.P., Shallit, J.: Automatic Sequences: Theory, Applications, Generalizations. Cambridge University Press, Cambridge (2003). https://doi.org/10.1017/CBO9780511546563
3. Allouche, J.-P., Shallit, J.: Automatic sequences are also non-uniformly morphic. In: Raigorodskii, A.M., Rassias, M.T. (eds.) Discrete Mathematics and Applications. SOIA, vol. 165, pp. 1–6. Springer, Cham (2020). https://doi.org/10.1007/978-3-030-55857-4_1
4. Bruyère, V., Hansel, G., Michaux, C., Villemaire, R.: Logic and p-recognizable sets of integers. Bull. Belg. Math. Soc. Simon Stevin **1**(2), 191–238 (1994). http://projecteuclid.org/euclid.bbms/1103408547, journées Montoises (Mons, 1992)
5. Büchi, J.R.: Weak second-order arithmetic and finite automata. Z. Math. Logik Grundlagen Math. **6**, 66–92 (1960). https://doi.org/10.1002/malq.19600060105
6. Béal, M.P., Perrin, D., Restivo, A.: Recognizability of morphisms. Ergodic Theory Dyn. Syst. 1–25 (2023). https://doi.org/10.1017/etds.2022.109
7. Cassaigne, J., Nicolas, F.: Quelques propriétés des mots substitutifs. Bull. Belg. Math. Soc. Simon Stevin **10**(suppl.), 661–676 (2003). http://projecteuclid.org/euclid.bbms/1074791324
8. Cassaigne, J., Richomme, G., Saari, K., Zamboni, L.Q.: Avoiding abelian powers in binary words with bounded abelian complexity. Int. J. Found. Comput. S. **22**(4), 905–920 (2011). https://doi.org/10.1142/S0129054111008489
9. Charlier, É., Leroy, J., Rigo, M.: Asymptotic properties of free monoid morphisms. Linear Algebra Appl. **500**, 119–148 (2016). https://doi.org/10.1016/j.laa.2016.02.030
10. Charlier, É., Rampersad, N., Shallit, J.: Enumeration and decidable properties of automatic sequences. Internat. J. Found. Comput. Sci. **23**(5), 1035–1066 (2012). https://doi.org/10.1142/S0129054112400448
11. Cobham, A.: Uniform tag sequences. Math. Syst. Theory **6**, 164–192 (1972). https://doi.org/10.1007/BF01706087
12. Dekking, F.M.: The spectrum of dynamical systems arising from substitutions of constant length. Zeitschrift für Wahrscheinlichkeitstheorie und Verwandte Gebiete **41**(3), 221–239 (1978). https://doi.org/10.1007/BF00534241
13. Durand, F., Leroy, J.: The constant of recognizability is computable for primitive morphisms. Journal of Integer Sequences **20**(#17.4.5) (2017). https://cs.uwaterloo.ca/journals/JIS/VOL20/Leroy/leroy4.html
14. Erdős, P.: Some unsolved problems. Michigan Math. J. **4**, 291–300 (1958). https://doi.org/10.1307/mmj/1028997963
15. Fici, G., Puzynina, S.: Abelian combinatorics on words: a survey. Comput. Sci. Rev. **47**, 100532 (2023). https://doi.org/10.1016/j.cosrev.2022.100532
16. Goč, D., Rampersad, N., Rigo, M., Salimov, P.: On the number of abelian bordered words (with an example of automatic theorem-proving). Internat. J. Found. Comput. Sci. **25**(8), 1097–1110 (2014). https://doi.org/10.1142/S0129054114400267
17. Guo, Y.J., Lü, X.T., Wen, Z.X.: On the boundary sequence of an automatic sequence. Discrete Math. **345**(1) (2022). https://doi.org/10.1016/j.disc.2021.112632
18. Krawczyk, E., Müllner, C.: Automaticity of uniformly recurrent substitutive sequences (2023). https://doi.org/10.48550/arXiv.2111.13134

19. Mossé, B.: Reconnaissabilité des substitutions et complexité des suites automatiques. Bulletin de la Société Mathématique de France **124**(2), 329–346 (1996). https://doi.org/10.24033/bsmf.2283

20. Mossé, B.: Puissance de mots et reconnaissabilité des points fixes d'une substitution. Theor. Comput. Sci. **99**(2), 327–334 (1992). https://doi.org/10.1016/0304-3975(92)90357-L

21. Mousavi, H.: Automatic theorem proving in Walnut (2016). https://doi.org/10.48550/ARXIV.1603.06017

22. Parreau, A., Rigo, M., Rowland, E., Vandomme, É.: A new approach to the 2-regularity of the ℓ-abelian complexity of 2-automatic sequences. Electron. J. Comb. **22**(#P1.27) (2022). https://doi.org/10.37236/4478

23. Puzynina, S., Whiteland, M.A.: Abelian closures of infinite binary words. J. Comb. Theory Ser. A **185**, 105524 (2022). https://doi.org/10.1016/j.jcta.2021.105524

24. Rigo, M.: Formal languages, automata and numeration systems. Vol. 1. ISTE, London; John Wiley & Sons Inc, Hoboken, NJ (2014), introduction to combinatorics on words, With a foreword by Valérie Bethé. https://doi.org/10.1002/9781119008200

25. Rigo, M.: Formal languages, automata and numeration systems. Vol. 2. Networks and Telecommunications Series, ISTE, London; John Wiley & Sons Inc, Hoboken, NJ (2014), applications to recognizability and decidability, With a foreword by Valérie Bethé. https://doi.org/10.1002/9781119042853

26. Rigo, M., Salimov, P.: Another generalization of abelian equivalence: binomial complexity of infinite words. Theor. Comput. Sci. **601**, 47–57 (2015). https://doi.org/10.1016/j.tcs.2015.07.025

27. Rigo, M., Stipulanti, M., Whiteland, M.A.: Binomial complexities and Parikh-collinear morphisms. In: Diekert, V., Volkov, M.V. (eds.) Developments in Language Theory - 26th International Conference, DLT 2022, Tampa, FL, USA, May 9–13, 2022, Proceedings. Lecture Notes in Computer Science, vol. 13257, pp. 251–262. Springer (2022). https://doi.org/10.1007/978-3-031-05578-2_20

28. Schaeffer, L.: Deciding Properties of Automatic Sequences. Master's thesis, Univ. of Waterloo (2013). https://uwspace.uwaterloo.ca/handle/10012/7899

29. Shallit, J.: Abelian complexity and synchronization. INTEGERS: Electron. J. Comb. Number Theory (#A.36) (2021). http://math.colgate.edu/integers/v36/v36.pdf

30. Shallit, J.: The Logical Approach to Automatic Sequences: Exploring Combinatorics on Words with Walnut. London Mathematical Society Lecture Note Series, Cambridge University Press (2022). https://doi.org/10.1017/9781108775267

On the Solution Sets of Entire Systems of Word Equations

Aleksi Saarela$^{(\boxtimes)}$ (iD)

Department of Mathematics and Statistics, University of Turku, Turku, Finland
amsaar@utu.fi

Abstract. The set of all constant-free word equations satisfied by a given morphism is called an entire system of equations. We show that in the three-variable case, the set of nonperiodic solutions of any entire system can be described using parametric formulas with just one numerical parameter. We also show how the solution set of any equation can be represented as a union of solution sets of entire systems. Even though an infinite union is needed in some cases, this still points towards a stronger version of Hmelevskii's theorem about parametric solutions of three-variable word equations.

Keywords: Combinatorics on words · Word equation · Entire system · Parametric solution

1 Introduction

Word equations have been studied both from an algorithmic and algebraic point of view. Some well-known results are that the complexity of the satisfiability problem of word equations can be solved in nondeterministic linear space [7], and that every system of word equations is equivalent to a finite subsystem [1,4].

In this article, we concentrate on constant-free equations, and all equations are assumed to be constant-free from now on. For some relations between constant-free equations and equations with constants, see [16].

Equations with one or two variables are trivial. The three-variable case, on the other hand, is highly nontrivial, while simultaneously being much simpler than the four-variable case. Some examples of difficult results about three-variable equations are Hmelevskii's theorem that every three-variable equation has a parametric solution [6] (this does not hold for equations with four or more variables), and a bound 18 for the size of independent systems [12] (no finite bound is known for equations with four or more variables).

Hmelevskii's theorem, in particular, is relevant for this article. The original proof, and even the simpler modernized version of that proof in [8,14], is very long. Also, while the basic concept of parametric solutions is simple, the actual parametric formulas that arise from the proof can be very complicated. They were analyzed in [13,14], and it was proved, for example, that the number

---

Supported by the Academy of Finland under grant 339311.

© The Author(s), under exclusive license to Springer Nature Switzerland AG 2023
A. Frid and R. Mercaş (Eds.): WORDS 2023, LNCS 13899, pp. 261–273, 2023.
https://doi.org/10.1007/978-3-031-33180-0_20

of numerical parameters needed in these formulas is at most logarithmic with respect to the length of the equation, and the total length of the formulas is at most exponential.

In this article, we find out that for a large number of three-variable equations and systems of equations, namely, for so-called entire systems and unbalanced equations, the set of solutions can be described using explicitly given formulas that are quite simple. In particular, only one numerical parameter is needed to represent nonperiodic solutions. We also outline a strategy to extend our result to all three-variable equations, although one numerical parameter will no longer be sufficient in that case. This can potentially lead to a much stronger and more explicit version of Hmelevskii's theorem in the future.

2   Preliminaries

First, we go through some basic definitions and lemmas abouts words. For more, see [9,10].

Let $\mathbb{N}$ denote the set of nonnegative integers. Throughout the article, let Σ be an alphabet that contains at least two letters a and b. Let ε denote the empty word.

A word u is a *factor* (*prefix*, *suffix*) of a word w if there exist words x, y such that $w = xuy$ ($w = uy$, $w = xu$, respectively). If one of words u, v is a prefix (suffix) of the other, we use the notation $u \sim_p v$ ($u \sim_s v$, respectively).

If $w \in \Sigma^*$ and $n \in \mathbb{N}$, then w^n is called a *power* of w, or more specifically, an *n-power* of w. If u is a prefix of w, then $w^n u$ is a called a *fractional power* of w. We also use negative powers as follows: If $w = uv$, then $u^{-1}w = v$ and $wv^{-1} = u$. If x is not a prefix (suffix) of w, then $x^{-1}w$ (wx^{-1}, respectively) is not defined, so whenever we use expressions like these, we have to make sure that they represent well-defined words.

Let Ξ be another alphabet. A mapping $h : \Xi^* \to \Sigma^*$ is a *morphism* if $h(UV) = h(U)h(V)$ for all $U, V \in \Xi^*$. The morphism h is *periodic* if there exists $w \in \Sigma^*$ such that $h(U) \in w^*$ for all $U \in \Xi^*$, and *nonperiodic* otherwise.

Next, we state some well-known results that are needed later.

Lemma 1. *Let $x, y \in \Sigma^*$. The following are equivalent:*

- *x and y are powers of a common word.*
- *$xy = yx$.*
- *x and y satisfy a non-trivial relation, that is, there exist $x_1, \ldots, x_m, y_1, \ldots, y_n \in \{x, y\}$ such that $(x_1, \ldots, x_m) \neq (y_1, \ldots, y_n)$ but $x_1 \cdots x_m = y_1 \cdots y_n$.*

Lemma 2. *Let $x, y \in \Sigma^*$. If y is a fractional power of x and ends in x, then x and y are powers of a common word.*

Lemma 3. *Let $x, y, z \in \Sigma^*$ and let $xy = yz$. Then $x = z = \varepsilon$ or*

$$x = uv, \qquad y = (uv)^j u, \qquad z = vu$$

for some $u, v \in \Sigma^*$ and $j \in \mathbb{N}$.

The next result is one of the equivalent formulations of the periodicity theorem of Fine and Wilf.

Theorem 4 (Fine and Wilf [3]). *Let $x, y \in \Sigma^*$. If a power of x and a power of y have a common prefix of length $|xy| - \gcd(|x|, |y|)$, then x and y are powers of a common word.*

Let us fix an alphabet of variables Ξ and an alphabet of constants Σ. A *word equation* is a pair $E = (U, V)$, where $U, V \in \Xi^*$, and a *solution* of E is a morphism $h : \Xi^* \to \Sigma^*$ such that $h(U) = h(V)$. The equation E is *nontrivial* if $U \neq V$.

A *system of equations* is a set of equations. A *solution* of a system is a morphism that satisfies all equations in the system. A system is *nontrivial* if it contains at least one nontrivial equation.

The set of all solutions of an equation or system E is denoted by $\mathrm{Sol}(E)$, and the set of all equations satisfied by a morphism h is denoted by $\mathrm{Eq}(h)$. Then $\mathrm{Eq}(h)$ is a system of equations, and it is called an *entire system*. Equations or systems E_1, E_2 are *equivalent* if $\mathrm{Sol}(E_1) = \mathrm{Sol}(E_2)$, and morphisms h_1, h_2 are *equivalent* if $\mathrm{Eq}(h_1) = \mathrm{Eq}(h_2)$.

We are particularly interested in the three-variable case $\Xi = \{X, Y, Z\}$. Throughout the article, we let X, Y, Z be distinct variables, and we use the shorthand notation $[x, y, z]$, where $x, y, z \in \Sigma^*$, for the morphism $h : \{X, Y, Z\}^* \to \Sigma^*$ defined by $h(X) = x$, $h(Y) = y$, $h(Z) = z$.

Example 5. Consider the equation $E = (XY, YZ)$. We can easily check that the morphism $h = [uv, (uv)^j u, vu]$, where $u, v \in \Sigma^*$ and $j \in \mathbb{N}$, is a solution of E:

$$h(XY) = uv(uv)^j u = (uv)^j uvu = h(YZ).$$

It follows from Lemma 3 that all solutions of E are of this form or of the form $[\varepsilon, u, \varepsilon]$.

An equation (U, V) is *balanced* if every variable has as many occurrences in U as in V, and *unbalanced* otherwise. Results related to balanced equations can be found, for example, in [5,15]. We need the following two theorems.

Theorem 6 (Harju and Nowotka [5]). *Let $g, h : \{X, Y, Z\}^* \to \Sigma^*$ be nonperiodic morphisms such that $\mathrm{Eq}(g) \neq \mathrm{Eq}(h)$. Then $\mathrm{Eq}(g) \cap \mathrm{Eq}(h)$ does not contain any unbalanced equations.*

Theorem 7 (Harju and Nowotka [5]). *If two unbalanced equations have a common nonperiodic solution, then they have the same set of periodic solutions.*

Budkina and Markov [2] classified all three-generator subsemigroups of a free semigroup. In the next theorem, we give a reformulation of this theorem in terms of morphisms and equations. An essentially equivalent result was proved independently by Spehner [17,18]. These results have been used to study three-variable word equations in [5,11], for example. In [5], there is also a good comparison of these results.

Theorem 8 (Budkina and Markov [2]**).** *Every nonperiodic morphism from* $\{X, Y, Z\}^*$ *to* Σ^* *that satisfies a nontrivial equation is equivalent, up to a permutation of the variables, to a morphism of one of the following types:*

1. $[a, b, w]$, *where* $w \in \{a, b\}^*$.
2. $[a, b^m, b^n]$, *where* $m, n \in \mathbb{N}$ *and* $m, n \geq 1$ *and* $\gcd(m, n) = 1$.
3. $[a, a^p b a^q, a^{p'} b \prod_{i=1}^{n} (a^{k_i} b) a^{q'}]$, *where* $p, q, p', q', n, k_1, \ldots, k_n \in \mathbb{N}$ *and* $pp' = qq' = 0$ *and* $1 \leq p + q \leq k_1, \ldots, k_n$.
4. $[a, a^p b (a^k b)^m, b(a^k b)^n a^q]$, *where* $p, q, k, m, n \in \mathbb{N}$ *and* $k, m, n \geq 1$ *and* $p, q \leq k$ *and* $\gcd(m + 1, n + 1) = 1$.
5. $[a, a^p b (a^k b)^m a^q, b(a^k b)^n]$, *where* $p, q, k, m, n \in \mathbb{N}$ *and* $p, q, k, m, n \geq 1$ *and* $p, q \leq k$ *and* $\gcd(m + 1, n + 1) = 1$.
6. $[a, a^p b a^q, b \prod_{i=1}^{n} (a^{k_i} b)(a^k b \prod_{i=1}^{n} (a^{k_i} b))^m]$, *where* $p, q, k, m, n, k_1, \ldots, k_n \in \mathbb{N}$ *and* $m, p, q \geq 1$ *and* $p, q \leq k < p + q \leq k_1, \ldots, k_n$.

3   Lemmas

In this section, we prove some lemmas that are needed later.

Lemma 9. *Let* $x, y \in \Sigma^*$ *and* $m, n, p, q \in \mathbb{N}$ *and* $m, n, p + q \geq 1$. *Let* $x^p \sim_p y^m x$ *and* $x^q \sim_s xy^n$. *Then* x *and* y *are powers of a common word or*

$$x = (uv)^j u, \qquad y = x^{p-1} u v x^q = x^p v u x^{q-1}$$

for some $u, v \in \Sigma^*$ *and* $j \in \mathbb{N}$.

Proof. Throughout the proof, we assume that $p \geq q$ and, consequently, $p \geq 1$. The case $p < q$ is symmetric and can be handled in a similar way.

First, let $|y| \geq |x^{p+q-1}|$. Then $y = x^{p-1} w x^q$ for some word w. From $x^p \sim_p y^m x$ it follows that $x \sim_p w x^q (x^{p-1} w x^q)^{m-1} x$. Therefore, x is a prefix of $w^k x$ for some $k \geq 1$. It follows that $w = uv$ and $x = (uv)^j u$ for some $u, v \in \Sigma^*$ and $j \geq 0$.

Next, let $|y| < |x^{p+q-1}|$ and $q = 0$. Then $|x^p| \geq |xy|$ and $|y^m x| \geq |xy|$. From $x^p \sim_p y^m x$ it follows that x is a fractional power of y, and then from the theorem of Fine and Wilf it follows that x and y are powers of a common word.

Next, let $|y| < |x^{p+q-1}|$ and $q \geq 1$ and $|x^p| < |y|$. Then y begins and ends with powers of x that overlap by a factor of length at least $|x|$. By Lemma 2, x and y are powers of a common word.

Next, let $|y| < |x^{p+q-1}|$ and $q \geq 1$ and $|x| < |y| \leq |x^p|$. Then y is a fractional power of x and ends in x. By Lemma 2, x and y are powers of a common word.

Finally, let $|y| < |x^{p+q-1}|$ and $q \geq 1$ and $|y| \leq |x|$. Then x is a fractional power of y and ends in y. By Lemma 2, x and y are powers of a common word. □

Lemma 10. *Let* $x, y \in \Sigma^*$ *and* $m, n, \in \mathbb{N}$ *and* $\gcd(m, n) = 1$. *Let* $x^m y = y z^n$. *Then* x, y, z *are powers of a common word or*

$$x = (st)^n, \qquad y = (st)^i s, \qquad z = (ts)^m$$

for some $s, t \in \Sigma^*$ *and* $i \in \mathbb{N}$.

Proof. By Lemma 3, $x^m = uv$, $y = (uv)^j u$, $z^n = vu$ for some $u, v \in \Sigma^*$ and $j \in \mathbb{N}$. Then uv is an m-power, and because its conjugate vu is an n-power, uv is also an n-power. By $\gcd(m, n) = 1$, $uv = w^{mn}$ for some $w \in \Sigma^*$. We can write $w = st$, $u = (st)^k s$, $v = t(st)^{mn-k-1}$ for some $s, t \in \Sigma^*$ and $k \in \mathbb{N}$, $k \leq mn - 1$. Then $x = (st)^n$ and $z = (ts)^m$ and $y = (st)^{mnj+k} s$. $\qquad\square$

Lemma 11. *Let $x, y, z \in \Sigma^*$ and $i, k \in \mathbb{N}$ and $i \geq 1$ and $k \geq 2$. Let $(xz)^i x = y^k$. Then x, y, z are powers of a common word or*

$$x = (uv)^j u, \qquad y = (uv)^{j+1} u, \qquad z = vu((uv)^{j+1} u)^{k-2} uv$$

for some $u, v \in \Sigma^*$ and $j \geq 0$.

Proof. If $i \geq 2$ or $|x| \geq |y|$, then $|(xz)^i x| = |y^k| \geq |xzy|$, so xz and y are powers of a common word by the theorem of Fine and Wilf, and then x, y, z are powers of a common word.

If $i = 1$ and $|x| < |y|$, then $y = sx = xt$ and $z = ty^{k-2} s$ for some $s, t \in \Sigma^+$. By Lemma 3, $s = uv$, $t = vu$, $x = (uv)^j u$ for some $u, v \in \Sigma^*$ and $j \in \mathbb{N}$, and then $y = (uv)^{j+1} u$ and $z = vu((uv)^{j+1} u)^{k-2} uv$. $\qquad\square$

Lemma 12. *Let $h : \{X, Y, Z\}^* \to \Sigma^*$ be a nonperiodic morphism. If $E \in \mathrm{Eq}(h)$ is unbalanced, then E is equivalent to $\mathrm{Eq}(h)$.*

Proof. Every solution of $\mathrm{Eq}(h)$ is a solution of E. Every periodic solution of E is a solution of all balanced equations in $\mathrm{Eq}(h)$, because periodic morphisms satisfy all balanced equations, and also a solution of all unbalanced equations in $\mathrm{Eq}(h)$ by Theorem 7. If g is a nonperiodic solution of E, then $E \in \mathrm{Eq}(g) \cap \mathrm{Eq}(h)$, so it must be $\mathrm{Eq}(g) = \mathrm{Eq}(h)$ by Theorem 6. This means that g is a solution of $\mathrm{Eq}(h)$. We have shown that E and $\mathrm{Eq}(h)$ have the same solutions. $\qquad\square$

4 Solutions of Entire Systems

In this section, we go through all entire systems $\mathrm{Eq}(h)$, where $h : \{X, Y, Z\}^* \to \Sigma^*$ is a nonperiodic morphism that satisfies a nontrivial equation. By Theorem 8, we can concentrate on the morphisms specified in that theorem.

In each case, with the help of Lemma 12, we find that the entire system is equivalent to a relatively simple unbalanced equation. Then we proceed to find an explicit description of all nonperiodic solutions of that equation. In each such description, there are two word parameters, denoted by u and v, and possibly one numerical parameter, denoted by j.

The formulas that describe the nonperiodic solutions also give some periodic solutions when u and v are powers of a common word. In some cases, all periodic solutions are obtained this way, but in other cases, some periodic solutions are missing. If we want to represent the set of all solutions, both the nonperiodic and periodic ones, we can always do that by adding separate formulas that give all the periodic solutions, although then we need two numerical variables.

Lemma 13. *Let* $n, k_0, \ldots, k_n \in \mathbb{N}$. *Let*

$$h = [a, b, a^{k_0} \prod_{i=1}^{n} ba^{k_i}].$$

Then $\mathrm{Eq}(h)$ *is equivalent to the equation*

$$E = (X^{k_0} \prod_{i=1}^{n} YX^{k_i}, Z)$$

and $[x, y, z]$ *is a solution of* E *if and only if*

$$x = u, \qquad y = v, \qquad z = u^{k_0} \prod_{i=1}^{n} vu^{k_i} \qquad (1)$$

for some $u, v \in \Sigma^*$.

Proof. It is easy to check that h is a solution of E. Thus $E \in \mathrm{Eq}(h)$, and E is unbalanced, so E and $\mathrm{Eq}(h)$ are equivalent by Lemma 12.

Let $g = [x, y, z]$ be a solution of E. We can let x and y be arbitrary words u and v, and then g is a solution if and only if $z = u^{k_0} \prod_{i=1}^{n} vu^{k_i}$. □

Lemma 14. *Let* $m, n \in \mathbb{N}$ *and* $m, n \geq 1$ *and* $\gcd(m, n) = 1$. *Let*

$$h = [a, b^m, b^n].$$

Then $\mathrm{Eq}(h)$ *is equivalent to the equation*

$$E = (Y^n, Z^m)$$

and $[x, y, z]$ *is a solution of* E *if and only if*

$$x = u, \qquad y = v^m, \qquad z = v^n \qquad (2)$$

for some $u, v \in \Sigma^*$.

Proof. It is easy to check that h is a solution of E. Thus $E \in \mathrm{Eq}(h)$, and E is unbalanced, so E and $\mathrm{Eq}(h)$ are equivalent by Lemma 12.

Let $g = [x, y, z]$ be a solution of E. Then $y^n = z^m$ is both an n-power and an m-power, so it is also an mn-power of some word v because $\gcd(m, n) = 1$. This means that $y = v^m$ and $z = v^n$. Then x can be an arbitrary word u and this always gives a solution g. □

Lemma 15. *Let* $p, q, p', q', n, k_1, \ldots, k_n \in \mathbb{N}$ *and* $pp' = qq' = 0$ *and* $1 \leq p+q \leq k_1, \ldots, k_n$. *Let*

$$h = \left[a, a^p ba^q, a^{p'} b \prod_{i=1}^{n} (a^{k_i} b) a^{q'}\right].$$

Then Eq(h) *is equivalent to the equation*

$$E = \left(X^p Z X^q, X^{p'} Y \prod_{i=1}^{n} (X^{k_i - p - q} Y) X^{q'}\right).$$

If $[x, y, z]$ *is a nonperiodic solution of* E, *then*

$$x = (uv)^j u, \qquad y = x^{p-1} uvx^q, \qquad z = x^{p'-1} uv \prod_{i=1}^{n} (x^{k_i - 1} uv) x^{q'} \qquad (3)$$

for some $u, v \in \Sigma^*$ *and* $j \geq 0$. *Moreover, every morphism defined by these formulas is a solution of* E, *except that if* $p' = q' = 0$ *and* $k_i = 1$ *for all* i, *then we must require that* $j \leq n$.

Proof. It is easy to check that h is a solution of E. Thus $E \in$ Eq(h), and E is unbalanced, so E and Eq(h) are equivalent by Lemma 12.

Let $g = [x, y, z]$ be a nonperiodic solution of E. We have

$$x^p z x^q = x^{p'} y \prod_{i=1}^{n} (x^{k_i - p - q} y) x^{q'},$$

so $x^p \sim_{\mathrm{p}} y^m x$ for some $m \geq 1$. If $p = 0$, this is trivial, and if $p > 0$, then $p' = 0$ and m is the smallest number i such that $k_i > p - q$, or $m = n + 1$ if such i does not exist. Similarly, $x^q \sim_{\mathrm{s}} xy^{m'}$ for some $m' \geq 1$. If x and y are powers of a common word, then g is periodic, so it follows from Lemma 9 that x and y are of the claimed form. Now g is a solution of E if and only if

$$z = x^{-p+p'} y \prod_{i=1}^{n} (x^{k_i - p - q} y) x^{q' - q}$$

$$= x^{-p+p'} x^{p-1} uvx^q \prod_{i=1}^{n} (x^{k_i - p - q} x^{p-1} uvx^q) x^{q' - q}$$

$$= x^{p'-1} uv \prod_{i=1}^{n} (x^{k_i - 1} uv) x^{q'}$$

and if this is a well-defined word, despite $p' - 1$ being negative in the case $p' = 0$. If $k_i \geq 2$ for some i or if $q' \geq 1$, then $x^{p'-1}$ is followed by $(uv)^r x = x(vu)^r$ for some $r \geq 1$, making z a well-defined word. If $k_i = 1$ for all i and $p' = q' = 0$, then

$$z = x^{-1} uv \prod_{i=1}^{n} (uv) = ((uv)^j u)^{-1} (uv)^{n+1} = v(uv)^{n-j}$$

which is a well-defined word if and only if $j \leq n$ (or if $u = \varepsilon$ and $v^{n-j+1} = \varepsilon$, but that only gives periodic solutions). $\qquad\square$

Lemma 16. *Let $p, q, k, m, n \in \mathbb{N}$ and $k, m, n \geq 1$ and $p, q \leq k$ and $\gcd(m + 1, n + 1) = 1$. Let*

$$h = [a, a^p b(a^k b)^m, b(a^k b)^n a^q].$$

Then Eq(h) is equivalent to the equation

$$E = ((X^{k-p}Y)^{n+1}X^k, X^k(ZX^{k-q})^{m+1}).$$

If $[x, y, z]$ is a nonperiodic solution of E, then

$$x = (uv)^j u, \qquad y = x^{p-1}uv(x^{k-1}uv)^m, \qquad z = (vux^{k-1})^n vux^{q-1} \qquad (4)$$

for some $u, v \in \Sigma^*$ and $j \geq 0$. Moreover, every morphism defined by these formulas is a solution of E, except that if $k = 1$ and $p = 0$, then we must require $j \leq m$, and if $k = 1$ and $q = 0$, then we must require $j \leq n$.

Proof. It is easy to check that h is a solution of E. Thus $E \in$ Eq(h), and E is unbalanced, so E and Eq(h) are equivalent by Lemma 12.

Let $g = [x, y, z]$ be a nonperiodic solution of E. We have

$$(x^{k-p}y)^{n+1}x^k = x^k(zx^{k-q})^{m+1},$$

so Lemma 10 gives

$$x^{k-p}y = (st)^{m+1}, \qquad zx^{k-q} = (ts)^{n+1}, \qquad x^k = (st)^i s$$

for some $s, t \in \Sigma^*$ and $i \in \mathbb{N}$.

If $i = 0$, then $s = x^k$ and

$$y = x^{p-k}(st)^{m+1} = x^{p-k}(x^k t)^{m+1} = x^p t(x^k t)^m,$$
$$z = (ts)^{n+1}x^{q-k} = (tx^k)^{n+1}x^{q-k} = (tx^k)^n tx^q.$$

If we let $u = x$ and $v = t$, then this matches (4) with $j = 0$.

If $k \geq 2$ and $i \geq 1$, then by Lemma 11, either x, s, t are powers of a common word, making g periodic, or $s = (uv)^j u$, $x = (uv)^{j+1}u$, $t = vux^{k-2}uv$ for some $u, v \in \Sigma^*$ and $j \in \mathbb{N}$. We get

$$y = x^{p-k}(st)^{m+1} = x^{p-k}((uv)^j uvux^{k-2}uv)^{m+1} = x^{p-1}uv(x^{k-1}uv)^m$$
$$z = (ts)^{n+1}x^{q-k} = (vux^{k-2}uv(uv)^j u)^{n+1}x^{q-k} = (vux^{k-1})^n vux^{q-1}.$$

This matches (4) with $j \geq 1$. Because $uv(x^{k-1}uv)^m$ always begins with $uvx = xvu$, y is a well-defined word even if $p = 0$. Similarly, z is a well-defined word even if $q = 0$.

If $k = 1$ and $i \geq 1$, then

$$x = (st)^i s, \qquad y = x^{p-1}(st)^{m+1}, \qquad z = (ts)^{n+1}x^{q-1}.$$

If we let $u = s$ and $v = t$ and $j = i$, then this matches (4) with $j \geq 1$. Finally, we have to make sure that y and z are well-defined words even if $p = 0$ or $q = 0$. If $p = 0$, then we must require $i \leq m$, and if $q = 0$, then we must require $i \leq n$. $\square$

Lemma 17. *Let $p, q, k, m, n \in \mathbb{N}$ and $p, q, k, m, n \geq 1$ and $p, q \leq k$ and $\gcd(m + 1, n + 1) = 1$. Let*

$$h = [a, a^p b(a^k b)^m a^q, b(a^k b)^n].$$

Then Eq(h) is equivalent to the equation

$$E = ((X^{k-p} Y X^{k-q} Z)^{n+1}, (X^k Z)^{m+n+2}).$$

If $[x, y, z]$ is a nonperiodic solution of E, then

$$x = (uv)^j u, \qquad y = x^{p-1} uv(x^{k-1} uv)^m x^q, \qquad z = x^{-1} uv(x^{k-1} uv)^n \qquad (5)$$

for some $u, v \in \Sigma^*$ and $j \geq 0$. Moreover, every morphism defined by these formulas is a solution of E, except that if $k = 1$, then we must require that $j \leq n$.

Proof. It is easy to check that h is a solution of E. Thus $E \in \text{Eq}(h)$, and E is unbalanced, so E and Eq(h) are equivalent by Lemma 12.

Let $g = [x, y, z]$ be a nonperiodic solution of E. We have

$$(x^{k-p} y x^{k-q} z)^{n+1} = (x^k z)^{m+n+2},$$

and $\gcd(n + 1, m + n + 2) = 1$, so

$$x^{k-p} y x^{k-q} z = w^{m+n+2}, \qquad x^k z = w^{n+1}$$

for some $w \in \Sigma^*$. We get

$$w^{m+1} = w^{m+n+2} w^{-n-1} = x^{k-p} y x^{k-q} z(x^k z)^{-1} = x^{k-p} y x^{-q}.$$

If $k = 1$, then from $xz = w^{n+1}$ it follows that $w = uv$ and $x = (uv)^j u$ for some $u, v \in \Sigma^*$ and $j \in \mathbb{N}$, $j \leq n$. Then

$$y = x^{p-1} w^{m+1} x^q = x^{p-1} uv(uv)^m x^q, \qquad z = x^{-1} w^{n+1} = x^{-1} uv(uv)^n.$$

This matches (5).

If $k \geq 2$ and $|x^{k-1}| \leq |w|$, then from $x^k z = w^{n+1}$ it follows that $w = x^{k-1} t$ for some $t \in \Sigma^*$, and that x^k is a prefix of $x^{k-1} tx$, so x is a prefix of tx. This means that $t = uv$ and $x = (uv)^j u$ for some $u, v \in \Sigma^*$ and $j \in \mathbb{N}$. Then

$$y = x^{p-k} w^{m+1} x^q = x^{p-1} uv(x^{k-1} uv)^m x^q, \qquad z = x^{-k} w^{n+1} = x^{-1} uv(x^{k-1} uv)^n.$$

This matches (5). Because $uv(x^{k-1} uv)^n$ always begins with $uvx = xvu$, z is a well-defined word.

If $k \geq 2$ and $|w| < |x^{k-1}|$, then from $x^k z = w^{n+1}$ it follows that x^k and w^{n+1} have a common prefix of length $|x^k| > |xw|$. By the theorem of Fine and Wilf, x and w are powers of a common word, and this leads to g being periodic. $\square$

Lemma 18. *Let $p, q, k, m, n, k_1, \ldots, k_n \in \mathbb{N}$ and $m, p, q \geq 1$ and $p, q \leq k < p + q \leq k_1, \ldots, k_n$. Let*

$$h = \left[a, a^p b a^q, b \prod_{i=1}^{n} (a^{k_i} b) \left(a^k b \prod_{i=1}^{n} (a^{k_i} b)\right)^m\right].$$

Then Eq(h) *is equivalent to the equation*

$$E = \left((X^k Z)^{m+2}, \left(X^{k-p}Y \prod_{i=1}^{n}(X^{k_i-p-q}Y)X^{k-q}Z\right)^{m+1}\right).$$

If $[x, y, z]$ *is a nonperiodic solution of* E, *then*

$$x = (uv)^j u, \qquad y = x^{p-1}uvx^q, \qquad z = x^{-k}\left(x^{k-1}uv \prod_{i=1}^{n}(x^{k_i-1}uv)\right)^{m+1} \qquad (6)$$

for some $u, v \in \Sigma^*$ *and* $j \geq 0$. *Moreover, every morphism defined by these formulas is a solution of* E, *except that if* $k = 1$ *and* $n = 0$, *then we must require that* $j \leq m$.

Proof. It is easy to check that h is a solution of E. Thus $E \in$ Eq(h), and E is unbalanced, so E and Eq(h) are equivalent by Lemma 12.

Let $g = [x, y, z]$ be a nonperiodic solution of E. We have

$$(x^k z)^{m+2} = \left(x^{k-p}y \prod_{i=1}^{n}(x^{k_i-p-q}y)x^{k-q}z\right)^{m+1}, \qquad (7)$$

and therefore,

$$|x^k z| \leq |x^{k-p}y \prod_{i=1}^{n}(x^{k_i-p-q}y)x^{k-q}z|.$$

Thus x^k is a prefix of $x^{k-p}y \prod_{i=1}^{n}(x^{k_i-p-q}y)x^{k-q}$, and consequently, x^p is a prefix of $y \prod_{i=1}^{n}(x^{k_i-p-q}y)x^{k-q}$. It follows that $x^p \sim_p y^{m'}x$ for some $m' \geq 1$. Similarly, we see that $x^q \sim_s xy^{m''}$ for some $m'' \geq 1$. If x and y are powers of a common word, then g is periodic, so it follows from Lemma 9 that x and y are of the claimed form.

The left-hand side and right-hand side of (7) is both an $(m + 2)$-power and an $(m + 1)$-power, so it is an $(m + 2)(m + 1)$-power of some word w. Now g is a solution of E if and only if

$$x^k z = w^{m+1}, \qquad x^{k-p}y \prod_{i=1}^{n}(x^{k_i-p-q}y)x^{k-q}z = w^{m+2},$$

so

$$w = w^{m+2}w^{-m-1} = x^{k-p}y \prod_{i=1}^{n}(x^{k_i-p-q}y)x^{-q}$$

and

$$z = x^{-k}w^{m+1} = x^{-k}\left(x^{k-p}y\prod_{i=1}^{n}(x^{k_i-p-q}y)x^{-q}\right)^{m+1}$$

$$= x^{-k}\left(x^{k-p}x^{p-1}uvx^q\prod_{i=1}^{n}(x^{k_i-p-q}x^{p-1}uvx^q)x^{-q}\right)^{m+1}$$

$$= x^{-k}\left(x^{k-1}uv\prod_{i=1}^{n}(x^{k_i-1}uv)\right)^{m+1}.$$

This is always a well-defined word, except that in the case $k = 1$ and $n = 0$, we get $z = x^{-1}(uv)^{m+1}$, and we must additionally require that $j \leq m$.   □

Let us take a closer look at the lemmas proved in this section. We see that the formulas (1)–(6) that describe the nonperiodic solutions contain at most one free numerical variable j. The other numbers in these formulas, denoted by symbols such as k_i, p, q and so on, are actually constants defined by the morphism h. The next example illustrates this in the case of the last lemma.

Example 19. Consider Lemma 18.
  If $p = q = m = 1$ and $k = 2$ and $n = 0$, then a nonperiodic solution $[x, y, z]$ of E is of the form

$$x = (uv)^j u, \qquad y = uv(uv)^j u, \qquad z = vuuv$$

for some $u, v \in \Sigma^*$ and $j \geq 0$.
  If $p = q = k = m = 1$ and $n = 0$, then a nonperiodic solution $[x, y, z]$ of E is of the form

$$x = (uv)^j u, \qquad y = uv(uv)^j u, \qquad z = v(uv)^{1-j}$$

for some $u, v \in \Sigma^*$ and $j \in \{0, 1\}$.

5 Connections to Hmelevskii's Theorem and Future Work

In Sect. 4, we found an explicit representation for the nonperiodic solutions of every nontrivial entire system of three-variable equations. By the next theorem, this also gives a representation for the nonperiodic solutions of every unbalanced three-variable equation.

Theorem 20. *The family of the sets* $\mathrm{Sol}(\mathrm{Eq}(h))$, *where* $h : \{X, Y, Z\}^* \to \Sigma^*$ *is a nonperiodic morphism that satisfies a nontrivial equation, is the same as the family of the sets* $\mathrm{Sol}(E)$, *where* E *is an unbalanced three-variable equation with a nonperiodic solution.*

Proof. In Sect. 4, we proved that every such entire system Eq(h) is equivalent to an unbalanced equation. On the other hand, every unbalanced equation E with a nonperiodic solution h is equivalent to Eq(h) by Lemma 12.   □

As was mentioned in the introduction, Hmelevskii proved that every three-variable equation has a parametric solution (for a precise definition of parametric words and parametric solutions, see, for example, [6] or [14]). The representations we found for entire systems are much simpler than the ones guaranteed by Hmelevskii's theorem. In particular, the best known upper bound for the number of numerical parameters in parametric solutions of three-variable equations is logarithmic with respect to the length of the equation. The formulas we found in the previous section, on the other hand, use at most one numerical parameter (although if we want to represent also periodic solutions, we need two numerical parameters).

We would like to prove a similar result also for balanced three-variable equations. The next theorem points towards such a result.

Theorem 21. *Let E be a three-variable equation. Let H be a set of representatives of all equivalence classes of morphisms $\{X, Y, Z\}^* \to \Sigma^*$. Then*

$$\mathrm{Sol}(E) = \bigcup_{h \in \mathrm{Sol}(E) \cap H} \mathrm{Sol}(\mathrm{Eq}(h)).$$

Proof. Every $g \in \mathrm{Sol}(E)$ is equivalent to some $h \in H$, and then $h \in \mathrm{Sol}(E) \cap H$ and $g \in \mathrm{Sol}(\mathrm{Eq}(g)) = \mathrm{Sol}(\mathrm{Eq}(h))$.

On the other hand, if $f \in \mathrm{Sol}(\mathrm{Eq}(h))$ for some $h \in \mathrm{Sol}(E) \cap H$, then $E \in \mathrm{Eq}(h)$ and thus $f \in \mathrm{Sol}(\mathrm{Eq}(h)) \subseteq \mathrm{Sol}(E)$.   □

If the set $\mathrm{Sol}(E) \cap H$ is finite, then this theorem, together with the results in Sect. 4, gives a simple parametric solution for E. Unfortunately, the set $\mathrm{Sol}(E) \cap H$ can be infinite, but the results in [11] give some restrictions on how complicated this set can be. This could allow us to prove a stronger version of Hmelevskii's theorem.

In particular, we expect that every three-variable equation has a parametric solution that uses only three numerical parameters, instead of a logarithmic number. However, proving this kind of result requires additional work in the future.

References

1. Albert, M.H., Lawrence, J.: A proof of Ehrenfeucht's conjecture. Theoret. Comput. Sci. **41**(1), 121–123 (1985). https://doi.org/10.1016/0304-3975(85)90066-0
2. Budkina, L.G., Markov, A.A.: *F*-semigroups with three generators. Akademiya Nauk SSSR. Matematicheskie Zametki **14**, 267–277 (1973)
3. Fine, N.J., Wilf, H.S.: Uniqueness theorems for periodic functions. Proc. Am. Math. Society **16**, 109–114 (1965). https://doi.org/10.1090/S0002-9939-1965-0174934-9

4. Guba, V.S.: Equivalence of infinite systems of equations in free groups and semi-groups to finite subsystems. Matematicheskie Zametki **40**(3), 321–324 (1986). https://doi.org/10.1007/BF01142470

5. Harju, T., Nowotka, D.: On the independence of equations in three variables. Theoret. Comput. Sci. **307**(1), 139–172 (2003). https://doi.org/10.1016/S0304-3975(03)00098-7

6. Hmelevskiĭ, J.I.: Equations in free semigroups. American Mathematical Society (1976), translated by G. A. Kandall from the Russian original: Trudy Mat. Inst. Steklov. **107** (1971)

7. Jeż, A.: Word equations in non-deterministic linear space. J. Comput. Syst. Sci. **123**, 122–142 (2022). https://doi.org/10.1016/j.jcss.2021.08.001

8. Karhumäki, J., Saarela, A.: An analysis and a reproof of Hmelevskii's Theorem. In: Ito, M., Toyama, M. (eds.) DLT 2008. LNCS, vol. 5257, pp. 467–478. Springer, Heidelberg (2008). https://doi.org/10.1007/978-3-540-85780-8_37

9. Lothaire, M.: Combinatorics on Words. Addison-Wesley (1983)

10. Lothaire, M.: Algebraic Combinatorics on Words. Cambridge University Press (2002) http://www-igm.univ-mlv.fr/berstel/Lothaire/AlgCWContents.html

11. Nowotka, D., Saarela, A.: One-variable word equations and three-variable constant-free word equations. Int. J. Found. Comput. Sci. **29**(5), 935–950 (2018). https://doi.org/10.1142/S0129054118420121

12. Nowotka, D., Saarela, A.: An optimal bound on the solution sets of one-variable word equations and its consequences. SIAM J. Comput. **51**(1), 1–18 (2022). https://doi.org/10.1137/20M1310448

13. Saarela, A.: On the complexity of Hmelevskii's theorem and satisfiability of three unknown equations. In: Diekert, V., Nowotka, D. (eds.) DLT 2009. LNCS, vol. 5583, pp. 443–453. Springer, Heidelberg (2009). https://doi.org/10.1007/978-3-642-02737-6_36

14. Saarela, A.: Word equations and related topics: independence, decidability and characterizations (2012) http://urn.fi/URN:ISBN:978-952-12-2737-0 doctoral dissertation, University of Turku

15. Saarela, A.: Systems of word equations, polynomials and linear algebra: a new approach. Eur. J. Comb. **47**, 1–14 (2015). https://doi.org/10.1016/j.ejc.2015.01.005

16. Saarela, A.: Hardness results for constant-free pattern languages and word equations. In: Proceedings of the 47th ICALP. LIPIcs, vol. 168, pp. 140:1–140:15. Schloss Dagstuhl-Leibniz-Zentrum fuer Informatik (2020). https://doi.org/10.4230/LIPIcs.ICALP.2020.140

17. Spehner, J.C.: Quelques problémes d'extension, de conjugaison et de présentation des sous-monoïdes d'un monoïde libre. Ph.D. thesis, Univ. Paris (1976)

18. Spehner, J.-C.: Les systemes entiers d'equations sur un alphabet de 3 variables. In: Jürgensen, H., Lallement, G., Weinert, H.J. (eds.) Semigroups Theory and Applications. LNM, vol. 1320, pp. 342–357. Springer, Heidelberg (1988). https://doi.org/10.1007/BFb0083443

On Arch Factorization and Subword Universality for Words and Compressed Words

Philippe Schnoebelen[(⊠)] and Julien Veron

LMF, CNRS & ENS, Paris-Saclay, France
phs@lsv.fr

Abstract. Using arch-jumping functions and properties of the arch factorization of words, we propose a new algorithm for computing the subword circular universality index of words. We also introduce the subword universality signature for words, that leads to simple algorithms for the universality indexes of SLP-compressed words.

1 Introduction

A *subword* of a given word is obtained by removing some letters at arbitrary places. For example, abba is a subword of ab̲r̲ac̲ad̲ab̲r̲a, as witnessed by the underlined letters. Subwords are a fundamental notion in formal language theory and in algorithmics but they are not as well-behaved as *factors*, a special case of subwords where the kept letters correspond to an interval inside the original word.[1]

Words and languages can be characterised or compared via their subwords. For example, we can distinguish u_1 = nationalists from u_2 = antinationalists by the subword x = ino. Indeed, only u_2 has x as a subword. We say that x is a *distinguisher* (also, a *separator*) between u_1 and u_2. Observe that ino is a *shortest* distinguisher between the two words.[2] In applications one may want to distinguish between two similar DNA strings, or two traces of some program execution: in these situations where inputs can be huge, finding a short distinguishing subword requires efficient algorithms [Sim03]. When considering the usual first-order logic of words (i.e., labelled linear orders), a distinguisher x can be seen as a Σ_1 formula separating the two words.

Definability by Subwords. These considerations led Imre Simon to the introduction of *piecewise-testable* languages in his 1972 Phd thesis [Sim72,Sim75]: these

---

[1] Some papers use the terminology "subwords" for factors, and "scattered subwords" or "scattered factors" for subwords. We follow [SS83].

[2] This is a very rare situation with the English lexicon, where different words almost always admit a length-2 distinguisher. To begin with, two words can already admit a length-1 distinguisher unless they use exactly the same set of letters.

Work partially supported by Labex DigiCosme (project ANR-11-LABEX-0045-DIGICOSME) operated by ANR as part of the program «Investissement d'Avenir» Idex Paris-Saclay (ANR-11-IDEX-0003-02).

© The Author(s), under exclusive license to Springer Nature Switzerland AG 2023
A. Frid and R. Mercaş (Eds.): WORDS 2023, LNCS 13899, pp. 274–287, 2023.
https://doi.org/10.1007/978-3-031-33180-0_21

languages can be defined entirely in terms of forbidden and required subwords. In logical terms, this corresponds to $\mathcal{B}\Sigma_1$-definability, see [DGK08]. Piecewise testability is an important and fundamental concept, and it has been extended to, among others, trees [BSS12, GS16], picture languages [Mat98], or words over arbitrary scattered linear orderings [CP18].

From a descriptive complexity point of view, a relevant measure is the *length of subwords* used in defining piecewise-testable languages, or in distinguishing between two individual words. Equivalently, the required length for these subwords is the required number of variables for the $\mathcal{B}\Sigma_1$ formula. This measure was investigated in [KS19] where it is an important new tool for bounding the complexity of decidable logic fragments.

Subword Universality. Barker, Day *et al.* introduced the notion of subword universality: a word u is k-*universal* if all words of length at most k are subwords of u [BFH+20, DFK+21]. They further define the *subword universality index* $\iota(u)$ as the largest k such that u is k-universal. Their motivations come, among others, from works in reconstructing words from subwords [DPFD19] or computing edit distance [DFK+21], see also the survey in [KKMS22]. In [BFH+20], the authors prove several properties of $\iota(u)$, e.g., when u is a palindrome, and further introduce the *circular* subword universality index $\zeta(u)$, which is defined as the largest $\iota(u')$ for u' a conjugate of u. Alternatively, $\zeta(u)$ can be seen as the subword universality index $\iota([u]_\sim)$ for a *circular word* (also called necklace, or cyclic word), i.e., an equivalence of words modulo conjugacy.

While it is easy to compute $\iota(u)$, computing $\zeta(u)$ is trickier but [BFH+20] proves several bounds relating $\zeta(u)$ to the values of $\iota(u^n)$ for $n \in \mathbb{N}$. This is leveraged in [FGN21] where an $O(|u| \cdot |A|)$ algorithm computing $\zeta(u)$ is given. That algorithm is quite indirect, with a delicate and nontrivial correctness proof. Further related works are [KKMS21] where, given that $\iota(u) = k$, one is interested in all the words of length $k+1$ that do not occur as subwords of u, [FHH+22] where one considers words that are just a few subwords away from k-universality, and [KKMP22] where the question whether u has a k-universal factor of given length is shown to be NP-complete.

Our Contribution. In this paper we introduce new tools for studying subword (circular) universality. First we focus on the arch factorizations (introduced by Hébrard [Héb91]) and show how *arch jumping* functions lead to simple proofs of combinatorial results on subword universality indexes, allowing a new and elegant algorithm for computing $\zeta(u)$. These arch-jumping functions are implicit in some published constructions and proofs (e.g., in [FK18, FGN21, KKMS21]) but studying them explicitly brings simplifications and improved clarity.

In a second part we give bilinear-time algorithms that compute the universality indexes ι and ζ for compressed words. This is done by introducing a compact *subword universality signature* that can be computed compositionally. These algorithms and the underlying ideas can be useful in the situations we mentioned earlier since long DNA strings or program execution traces are usually very repetitive, so that handling them in compressed form can entail huge savings in both memory and communication time.

More generally this is part of a research program on algorithms and logics for computing and reasoning about subwords [KS15, HSZ17, KS19, GLHK+20]. In that area, handling words in compressed form raises additional difficulties. For example it is not known whether one can compute efficiently the length of a shortest distinguisher between two compressed words. Let us recall here that reasoning on subwords is usually harder than reasoning on factors, and this is indeed true for compressed words: While deciding whether a compressed X is a *factor* of a compressed Y is polynomial-time, deciding whether X is a *subword* of Y is intractable (in PSPACE and PP-hard, see [Loh12, Sect. 8]). However, in the special case where one among X or Y is a *power word*, i.e., a compressed word with restricted nesting of concatenation and exponentiation, the subword relation is polynomial-time, a result crucial for the algorithms in [Sch21] where one handles exponentially long program executions in compressed forms.

Outline of the Paper. Section 2 recalls all the necessary definitions for subwords and universality indexes. Section 3 introduces the arch-jumping functions, relates them to universality indexes and proves some basic combinatorial results. Then Sect. 4 provides a simple algorithm for the circular universality index. In Sect. 5 we introduce the subword universality signature of words and show how they can be computed compositionally. Finally Sect. 6 considers SLP-compressed words and their subword universality indexes.

2   Basic Notions

Words and Subwords. Let $A = \{a, b, \ldots\}$ be a finite alphabet. We write $u, v, w, s, t, x, y \ldots$ for words in A^*. Concatenation is denoted multiplicatively while ε denotes the empty word. When $u = u_1 u_2 u_3$ we say that u_1 is a *prefix*, u_2 is a *factor*, and u_3 is a *suffix*, of u. When $u = vw$ we may write $v^{-1}u$ to denote w, the suffix of u one obtains after removing its v prefix. When $u = v_0 w_1 v_1 w_2 \cdots w_n v_n$, the concatenation $w_1 w_2 \cdots w_n$ is a *subword* of u, i.e., a subsequence obtained from u by removing some of its letters (possibly none, possibly all). We write $u \preccurlyeq v$ when u is a subword of v.

A word $u = a_1 \cdots a_\ell$ has length ℓ, written $|u| = \ell$, and we let $A(u) \overset{\text{def}}{=} \{a_1, \ldots, a_\ell\}$ denote its alphabet, a subset of A. We let $Cuts(u) = \{0, 1, \ldots, \ell\} \subseteq \mathbb{N}$ denote the set of *cutting positions inside* u, i.e., positions between u's letters, where u can be split: for $0 \leq i \leq j \leq \ell$, we let $u(i, j)$ denote the factor $a_{i+1} a_{i+2} \cdots a_j$. With this notation, $u(0, j)$ is u's prefix of length j, and $u(i, \ell)$ is the suffix $(u(0, i))^{-1}u$. Note also that $u(i, i) = \varepsilon$ and $u(i, j) = u(i, k)\, u(k, j)$ whenever the factors are defined. If $u = u_1 u_2$, we say that $u_2 u_1$ is a *conjugate* of u. For $i \in Cuts(u)$, the i-th conjugate of u is $u(i, \ell)\, u(0, i)$ and is denoted by $u^{\sim i}$. Finally $u^{\mathrm{R}} \overset{\text{def}}{=} a_\ell \cdots a_1$ denotes the *mirror* of u.

Rich Words and Arch Factorizations. A word $u \in A^*$ is *rich* if it contains at least one occurrence of each letter $a \in A$, otherwise we say that it is *incomplete*. A rich word having no rich strict prefix is an *arch*. The mirror of an arch is called a *co-arch* (it is generally not an arch). Observe that an arch (or a co-arch) necessarily ends (respectively, starts) with a letter that occurs only once in it.

The *arch factorization* of u, introduced by Hebrard [Héb91], is a decomposition $u = s_1 \cdots s_m \cdot r$ of u into $m + 1$ factors given by the following:

- if u is not rich then $m = 0$ and $r = u$,
- otherwise let s_1 be the shortest prefix of u that is rich (it is an arch) and let $s_2, \ldots, s_m, r$ be the arch factorization of the suffix $(s_1)^{-1}u$.

We write $r(u)$ for the last factor in u's factorization, called the *rest* of u. For example, with $A = \{a, b, c\}$, the arch factorization of $u_{ex} = \mathtt{baccabbcbaabacba}$ is $\mathtt{bac} \cdot \mathtt{cab} \cdot \mathtt{bcba} \cdot \mathtt{abac} \cdot \mathtt{ba}$, with $m = 4$ and $r(u_{ex}) = \mathtt{ba}$. Thus the arch factorization is a leftmost decomposition of u into arches, with a final rest $r(u)$.

There is a symmetric notion of co-arch factorization where one factors u as $u = r' \cdot s'_1 \cdots s'_m$ such that r' is incomplete and every s'_i is a co-arch, i.e., a rich factor whose first letter occurs only once.

All the above notions assume a given underlying alphabet A, and we should speak more precisely of "A-rich" words, "A-arches", or "rest $r_A(u)$". When A is understood, we retain the simpler terminology and notation.

Subword Universality. In [BFH+20], Barker et al. define the *subword universality index* of a word u, denoted $\iota_A(u)$, or just $\iota(u)$, as the largest $m \in \mathbb{N}$ such that any word of length m in A^* is a subword of u.

It is clear that $\iota(u) = m$ iff the arch factorization of u has m arches. Hence one can compute $\iota(u)$ in linear time simply by scanning u from left to right, keeping track of letter appearances in consecutive arches, and counting the arches [BFH+20, Prop. 10]. Using that scanning algorithm for ι, one sees that the following equalities hold for all words u, v:

$$\iota(u\,v) = \iota(u) + \iota\big(r(u)v\big)\,, \qquad r(u\,v) = r\big(r(u)v\big)\,. \qquad (1)$$

Barker et al. further define the *circular subword universality index* of u, denoted $\zeta(u)$, as the largest $\iota(u')$ for u' a conjugate of u. Obviously, one always has $\zeta(u) \geq \iota(u)$. Note that $\zeta(u)$ can be strictly larger that $\iota(u)$, e.g., with $A = \{a, b\}$ and $u = \mathtt{aabb}$ one has $\iota(u) = 1$ and $\zeta(u) = 2$. These descriptive complexity measures are invariant under mirroring of words, i.e., $\iota(u^R) = \iota(u)$ and $\zeta(u^R) = \zeta(u)$, and monotonic w.r.t. the subword ordering:

$$u \preccurlyeq v \implies \iota(u) \leq \iota(v) \wedge \zeta(u) \leq \zeta(v)\,. \qquad (2)$$

The behaviour of ζ can be deceptive. For example, while ι is superadditive, i.e., $\iota(uv) \geq \iota(u) + \iota(v)$ —just combine eqs. (1) and (2)— we observe that $\zeta(uv) < \zeta(u) + \zeta(v)$ can happen, e.g., with $u = \mathtt{ab}$ and $v = \mathtt{bbaa}$.

3   Arch-Jumping Functions and Universality Indexes

Let us fix a word $w = a_1 a_2 \cdots a_L$ of length L. We now introduce the α and β *arch-jumping functions* that describe the reading of an arch starting from some position inside w. For $i \in Cuts(w)$, we let

$$\alpha(i) = \min\{j \mid A(w(i,j)) = A\}, \qquad \beta(j) = \max\{i \mid A(w(i,j)) = A\}.$$

These are partial functions: $\alpha(i)$ and $\beta(j)$ are undefined when $w(i, L)$ or, respectively, $w(0, j)$, does not contain all the letters from A. See Fig. 1 for an illustration.

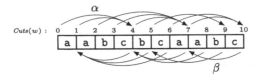

Fig. 1. Arch-jumping functions α, β for $A = \{a, b, c\}$ and $w = \mathtt{aabcbcaabc}$.

The following properties are easily seen to hold for all $i, j \in dom(\alpha)$:

$$\alpha(i) \geq i + |A|, \qquad i \leq j \implies \alpha(i) \leq \alpha(j), \tag{3}$$

$$\beta(\alpha(i)) \geq i, \qquad \alpha(\beta(\alpha(i))) = \alpha(i). \tag{4}$$

Since β is a mirror version of α, it enjoys similar properties that we won't spell out here.

Remark 3.1. As will be seen in the rest of this section, the arch jumping functions are a natural and convenient tool for reasoning about arch factorizations. Similar concepts can certainly be found in the literature. Already in [Héb91], Hébrard writes $p(n)$ for what we write $\alpha^n(0)$, i.e., the n-times iteration $\alpha(\alpha(\cdots(\alpha(0))\cdots))$ of α on 0: the starting point for the $p(n)$'s is fixed, not variable. In [FK18], Fleischer and Kufleitner use rankers like X_a and Y_b to jump from a current position in a word to the next (or previous) occurrence of a given letter, here a and b: this can specialise to our α and β if one knows what is the last letter of the upcoming arch. In [KKMS21] $\mathtt{minArch}$ corresponds exactly to our α, but there $\mathtt{minArch}$ is a data structure used to store information, not a notational tool for reasoning algebraically about arches.

3.1   Subword Universality Index via Jumping Functions

The connection between the jumping function α and the subword universality index $\iota(w)$ is clear:

$$\iota(w) = \max\{n \mid \alpha^n(0) \text{ is defined}\}. \tag{5}$$

For example, w in Fig. 1 has $\alpha^3(0) = 10 = |w|$ so $\iota(w) = 3$.

We can generalise Eq. 5: $\iota(w) = n$ implies $\alpha^p(0) \leq \beta^{n-p}(|w|)$ for all $p = 0, \ldots, n$, and the reciprocal holds. We can use this to prove the following:

Proposition 3.2. $\iota(uv) \leq \iota(u) + \iota(v) + 1$.

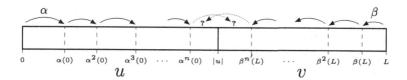

Fig. 2. Comparing $\iota(u\,v)$ with $\iota(u) + \iota(v)$.

Proof. Write n and n' for $\iota(u)$ and $\iota(v)$. Thus, on $w = u\,v$ with $L = |u| + |v|$, one has $\alpha^{n+1}(0) > |u|$ and $\beta^{n'+1}(L) < |u|$. See Fig. 2. Hence $\iota(w) < n + n' + 2$.

We can also reprove a result from [BFH+20]:

Proposition 3.3. $\iota(u\,u^{\mathrm{R}}) = 2\iota(u)$.

Proof. Write n for $\iota(u)$. When $w = u\,u^{\mathrm{R}}$ and $L = |w|$, the factor $w\big(\alpha^n(0), \beta^n(L)\big)$ is $r(u) \cdot r(u)^{\mathrm{R}}$ hence is not rich. Thus $\alpha^{n+1}(0) > |u| + |r(u)^{\mathrm{R}}| = \beta^n(L)$, entailing $\iota(u\,u^{\mathrm{R}}) < 2n + 1$.

3.2   Subword Circular Universality Index via Jumping Functions

The jumping functions can be used to study the circular universality index $\zeta(u)$. For this we consider the word $w = u\,u$ obtained by concatenating two copies of u, so that $L = 2\ell$. Now, instead of considering the conjugates of u, we can consider the factors $w(i, i + \ell)$ of w: see Fig. 3.

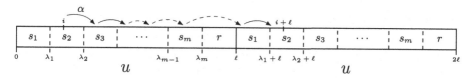

Fig. 3. Computing $\iota(u^{\sim i})$ on $w = u^2$.

This leads to a characterisation of $\zeta(u)$ in terms of α on $w = u\,u$:

$$\zeta(u) = \max_{0 \le i < \ell} \max\{n \mid \alpha^n(i) \le i + \ell\} \tag{6}$$

or, using $u^{\sim \ell} = u^{\sim 0}$,

$$= \max_{0 < i \le \ell} \max\{n \mid \alpha^n(i) \le i + \ell\} . \tag{7}$$

Bounding $\zeta(u)$. For $k = 0, \ldots, m$, we write λ_k for the cumulative length $|s_1 \cdots s_k|$ of the k first arches of u, i.e., we let $\lambda_k \stackrel{\text{def}}{=} \alpha^k(0)$.

The following Lemma and its corollary are a version of Lemma 20 from [BFH+20] but we give a different proof.

Lemma 3.4. *Let u and u' be two conjugate words.*

(a) $\iota(u) - 1 \le \iota(u') \le \iota(u) + 1$.
(b) *If furthermore $r(u) = \varepsilon$ then $\iota(u') \le \iota(u)$.*

Proof. Let $s_1 \cdots s_m \cdot r$ be the arch factorization of u and assume that $u' = u^{\sim i}$ as depicted in Fig. 3.

(a) If the position i falls inside some arch s_p of u (or inside the rest r) we see that $s_{p+1} \cdots s_m \cdot s_1 \cdots s_{p-1}$ is a subword of u' hence $\iota(u') \ge m - 1$. This gives $\iota(u) - 1 \le \iota(u')$, and the other inequality is obtained by exchanging the roles of u and u'.

(b) If furthermore $r = \varepsilon$, then $\lambda_{p-1} \le i < \lambda_p$ for some p. Looking at u' as a factor of $w = u^2$ (and assuming that $\alpha^{m+1}(i)$ is defined) we deduce $\alpha^{m+1}(i) \ge \alpha^{m+1}(\lambda_{p-1}) = \lambda_p + \ell > i + \ell$. This proves $\iota(u^{\sim i}) < m + 1$. ∎

Corollary 3.5. *(a)* $\iota(u) \le \zeta(u) \le \iota(u) + 1$.
(b) *Furthermore, if $r(u) = \varepsilon$, then $\zeta(u) = \iota(u)$.*

4   An $O(|u| \cdot |A|)$ Algorithm for $\zeta(u)$

The following crucial lemma shows that computing $\zeta(u)$ does not require checking all the conjugates $u^{\sim i}$ for $0 \le i < \ell$.

Lemma 4.1. *Let $u = a_1 \cdots a_\ell$ be a rich word with arch factorization $s_1 \cdots s_m \cdot r$.*
(a) There exists some $0 < d \le \lambda_1 \stackrel{\text{def}}{=} |s_1|$ such that $\zeta(u) = \iota(u^{\sim d})$.
(b) Furthermore, there exists $a \in A$ such that $d = \min\{i \mid a_i = a\}$, i.e., d can be chosen as a position right after a first occurrence of a letter in u.

Proof. Let $n = \zeta(u)$. For (a) it is enough to show that $\iota(u^{\sim d}) \ge n$ for some $d \in (0, \lambda_1]$.

By Eq. 7 there exists some $0 < i_0 \le \ell$ such that $\alpha^n(i_0) \le i_0 + \ell$. We consider the sequence $i_0 < i_1 < \cdots < i_n$ given by $i_{k+1} = \alpha(i_k)$. If $i_n \le \ell$ then taking $d = 1$ works: monotonicity of α entails $\alpha^n(d) \le \alpha^n(i_0) \le \ell$ and we deduce $\iota(u^{\sim d}) \ge n$. Clearly $d = 1$ fulfils (b).

So assume $i_n > \ell$ and let k be the largest index such that $i_k \le \ell$ (hence $k < n$). Since $\alpha(\ell) = \ell + \lambda_1$ (recall $\lambda_1 \stackrel{\text{def}}{=} |s_1|$), monotonicity of α entails $i_{k+1} = \alpha(i_k) \le \ell + \lambda_1$, i.e., i_{k+1} lands inside the first arch of the second copy of u in w.

Let now $d \stackrel{\text{def}}{=} i_{k+1} - \ell$ so that $u^{\sim d} = w(d, d + \ell) = w(d, i_{k+1})$. Since $\alpha^{n-k-1}(i_{k+1}) = i_n \le i_0 + \ell$, one has $\alpha^{n-k-1}(d) \le i_0$ hence $\iota(w(d, i_0)) \ge n - k - 1$. We also have $\iota(w(i_0, i_{k+1})) = k + 1$ since $i_{k+1} = \alpha^{k+1}(i_0)$. This yields

$$\iota(u^{\sim d}) = \iota(w(d, d + \ell)) \ge (n - k - 1) + (k + 1) = n,$$

entailing (a). For (b) observe that $w(i_{k+1} - 1, i_{k+1})$ is the last letter of an arch across the end of the first u in w to the beginning of the second u in w. Since it is the first occurrence of this letter in this arch, it is also in u. Since d is i_{k+1} shifted to the first copy of u, (b) is fulfilled.

Algorithm 4.2 (Computing $\zeta(u)$). For each position d such that $u(d-1, d)$ is the first occurrence of a letter in u, one computes $\iota(u^{\sim d})$ (in time $O(|u|)$ for each d), and returns the maximum value found.   □

The correctness of this algorithm is given by Lemma 4.1 (if u is not rich, $\zeta(u) = 0$ and this will be found out during the computation of $\iota(u^{\sim 1})$). It runs in time $O(|A| \cdot |u|)$ since there are at most $|A|$ values for d, starting with $d = 1$.

There are two heuristic improvements that can speed up the algorithm[3]:

- As soon as we have encountered two different values $\iota(u^{\sim d}) \neq \iota(u^{\sim d'})$, we can stop the search for a maximum in view of corollary 3.5.(a).
  For example, for $u = $ aabaccb, the first occurrences of a, b, and c, are with $d = 1, 3$ and 5. So one starts with computing $\iota(u^{\sim 1}) = \iota($abaccb a$) = 2$. Then one computes $\iota(u^{\sim 3}) = \iota($accb aab$) = 1$. Now, and since we have encountered two different values, we may conclude immediately that $\zeta(u) = 2$ without the need to compute $\iota(u^{\sim 5})$.
- When computing some $\iota(u^{\sim d})$ leads us to notice $r(u^{\sim d}) = \varepsilon$, we can stop the search in view of corollary 3.5.(b).
  For example, and again with $u = $ aabaccb, the computation of $\iota(u^{\sim 1})$ led us to the arch-factorization $u^{\sim 1} = $ abac $\cdot$ cba $\cdot$ ε, with 2 arches and with $r(u^{\sim 1}) = \varepsilon$. We may conclude immediately that $\zeta(u) = \iota(u^{\sim 1}) = 2$ without trying the remaining conjugates.

Observe that the above algorithm does not have to explicitly build $u^{\sim d}$. It is easy to adapt any naive algorithm for $\iota(u)$ so that it starts at some position d and wraps around when reaching the end of u.

5   Subword Universality Signatures

In this section, we write $\iota_*(u)$, $r_*(u)$, etc., to denote the values of $\iota(u)$, $r(u)$, etc., *when one assumes that $A(u)$ is the underlying alphabet*. This notation is less heavy than writing, e.g., $\iota_{A(u)}(u)$, but it is needed since we shall consider simultaneously $\iota_*(u)$ and $\iota_*(v)$ when $A(u) \neq A(v)$, i.e., when the two universality indexes have been obtained in different contexts.

When u is a word, we define a function S_u on words via:

$$S_u(x) = \langle \iota_*(x\,u), A(r_*(x\,u)) \rangle \quad \text{for all } x \text{ such that } A(u) \not\subseteq A(x). \tag{8}$$

In other words, $S_u(x)$ is a summary of the arch factorization of $x\,u$: it records the number of arches in $x\,u$ and the letters of the rest $r_*(x\,u)$, assuming that the alphabet is $A(x\,u)$.

---

[3] They do not improve the *worst-case* complexity.

Note that $S_u(x)$ is only defined when $A(u) \nsubseteq A(x)$, i.e., when at least one letter from u does not appear in x. With this restriction, $S_u(x)$ and $S_u(x')$ coincide (or are both undefined) whenever $A(x) = A(x')$. For this reason, we sometimes write $S_u(B)$, where B is a set of letters, to denote any $S_u(x)$ with $A(x) = B$.

We are now almost ready to introduce the main new object: a compact data structure with enough information for computing S_u on arbitrary arguments.

With a word u we associate $e(u)$, a word listing the letters of u in the order of their first appearance in u. For example, by underlining the first occurrence of each letter in $u = \underline{cc}\underline{a}\underline{cab}\underline{c}bba$ we show $e(u) = \text{cab}$. We also write $f(u)$ for the word listing the letters of u in order of their last occurrence: in the previous example $f(u) = \text{cba}$.

Definition 5.1. *The* subword universality signature *of a word u is the pair $\Sigma(u) = \langle e(u), s_u \rangle$ where s_u is S_u restricted to the strict suffixes of $e(u)$.*

Example 5.2. With $u = \underline{aab}\underline{ac}$ we have:

$$\Sigma(u) = \begin{cases} e(u) = \text{abc} \\ s_u = \begin{cases} \varepsilon \mapsto \langle 1, \varnothing \rangle \\ \text{c} \mapsto \langle 1, \{a, c\} \rangle \\ \text{bc} \mapsto \langle 2, \varnothing \rangle \end{cases} \text{in view of:} \quad \begin{array}{l} \varepsilon \cdot u = \text{aabac} \cdot \varepsilon \\ \text{c} \cdot u = \text{caab} \cdot \text{ac} \\ \text{bc} \cdot u = \text{bca} \cdot \text{abac} \cdot \varepsilon \end{array} \end{cases}$$

NB: the strict suffixes of $e(u)$ are ε, c and bc.

While finite (and quite small) $\Sigma(u)$ contains enough information for computing S_u on any argument x on any alphabet. One can use the following algorithm:

Algorithm 5.3 (Computing $S_u(x)$ from $\Sigma(u)$).
Given inputs x and $\Sigma(u) = \langle e(u), s_u \rangle$ we proceed as follows:
(a) Retrieve $A(u)$ from $e(u)$. Check that $A(u) \nsubseteq A(x)$, since otherwise $S_u(x)$ is undefined.
(b) Now with $x \in dom(S_u)$, let y be the longest suffix of $e(u)$ with $A(y) \subseteq A(x)$ —necessarily y is a *strict* suffix of $e(u)$— and extract $\langle n_y, B_y \rangle$ from $s_u(y)$.
(c.1) If $A(x) \subseteq A(u)$, return $S_u(x) = \langle n_y, B_y \rangle$.
(c.2) Similarly, if $n_y = 1$ return $S_u(x) = \langle n_y, B_y \rangle$.
(c.3) Otherwise return $S_u(x) = \langle 1, A(u) \rangle$.

Proof (of correctness). Assume $x \in dom(S_u)$. Since u contains a letter not appearing in x, the first arch of $x\,u$ ends inside u, so let us consider the factorization $u = u_1 u_2$ such that $x\,u_1$ is the first arch of $x\,u$ (see picture below, where $e(u)$ is underlined).

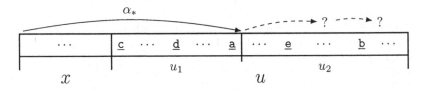

Now u_1 has a last letter, say a, that appears only once in u_1 and not at all in x. Observe that a letter b appears after a in $e(u)$ iff it does not appear in u_1, and thus must appear in x. Hence the y computed in step (b) is the suffix of $e(u)$ after a (in the above picture y would be eb).

If $A(x) \subseteq A(u)$ then $y\, u_1$ is rich, and is in fact an arch since its last letter, a, appears only once. So $S_u(x)$ and $S_u(y)$ coincide and step (c.1) is correct.

In case $A(x) \not\subseteq A(u)$, both x and u contain some letters that are absent from the other word, so necessarily $\iota_*(x\, u) = 1$ and $r_*(x\, u) = u_2$. There only remains to compute $A(u_2)$ from $\Sigma(u)$. We know that $\mathbf{s}_u(y) = \langle n_y, B_y \rangle$. If $n_y > 1$ this means that u_2 contains at least another $A(u)$-arch, so $A(u_2) = A(u)$ and step (c.3) is correct. If $n_y = 1$ this means that $y\, u$ only has one arch, namely $y\, u_1$, and B_y provides $A(u_2)$: step (c.2) is correct in this case.

Remark 5.4 (Space and time complexity for Algorithm 5.3). For simplifying our complexity evaluation, we assume that there is a fixed maximum size for alphabets so that storing a letter $a \in A$ uses space $O(1)$, e.g., 64 bits. When storing $\Sigma(u)$, the $e(u)$ part uses space $O(|A|)$. Now $\mathbf{s}_u$ can be represented in space $O(|A| \log |u|)$ when $e(u)$ and $f(u)$ are known: it contains at most $|A|$ pairs $\langle n_x, B_x \rangle$ where x is a suffix of $e(u)$ and B_x is always the alphabet of a strict suffix of $f(u)$: x and B_x can thus be represented by a position (or a letter) in $e(u)$ and $f(u)$. The n_x values each need at most $\log |u|$ bits.

Regarding time, the algorithm runs in time $O\big(|x| + |\Sigma(u)| + |A(u)|\big)$. $\square$

5.1   Universality Indexes from Signatures

Obviously the signature $\Sigma(u)$ contains enough information for retrieving $\iota_*(u)$: this is found in $\mathbf{s}_u(\varepsilon)$. More interestingly, one can also retrieve $\zeta_*(u)$:

Proposition 5.5. *Let u be a word with $\iota_*(u) = m$. Then $\zeta_*(u) = m + 1$ iff there exists a strict suffix x of $e(u)$ with $\mathbf{s}_u(x) = \langle n_x, B_x \rangle$ such that $n_x = m + 1$ and $A(x) \subseteq B_x$. Otherwise $\zeta_*(u) = m$.*

Proof. ($\Leftarrow$): assume $\mathbf{s}_u(x) = \langle m + 1, B_x \rangle$ with $A(x) \subseteq B_x$. Thus $\iota_*(x\, u) = m + 1$. Factor u as $u = u_1 u_2 r$ such that $x\, u_1$ is the first arch of $x\, u$ and such that $r = r_*(x\, u)$ is its rest. Then u_2 contains m arches and $B_x = A(r)$. Let now $u' \overset{\text{def}}{=} r\, u_1 u_2$. We claim that $\iota_*(u') = m + 1$. Indeed $r\, u_1$ is rich since $x\, u_1$ is rich and $A(x) \subseteq A(r)$, so $\iota_*(r\, u_1 u_2) \geq m + 1$. Since u' and u are conjugates, we deduce $\zeta_*(u) = \iota_*(u') = m + 1$ from Corollary 3.5.(a).

($\Rightarrow$): assume $\zeta_*(u) = m + 1$. By Lemma 4.1 we know that $\iota_*\big(u^{\sim i}\big) = m + 1$ for some position $0 < i \leq \lambda_1$ falling just after a first occurrence of a letter in u. Looking at factors of $w = u\, u$ as we did before, we have $\alpha^{m+1}(i) \leq i + \ell$, leading to $j \overset{\text{def}}{=} \alpha^m(i) \leq \ell$ (see picture below).

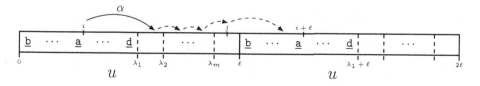

Define now x as the suffix of $e(u)$ that contains all letters in $u(i, \lambda_1)$, that is, all underlined letters to the right of i. This is a strict suffix since $i > 0$. Now $x\,u(0, i)$ is rich, and $u(i, j)$ is made of exactly m arches, so $\iota_*(x\,u) = n_x = m + 1$ and $r_*(x\,u) = u(j, \ell)$.

Then $B_x = A\big(u(j, \ell)\big)$ and $w(j, i + \ell)$ is rich, so $w(j, \ell)$ contains all letters missing from $w(i, i + \ell) = u(0, i)$. In other words $B_x \supseteq A(x)$, concluding the proof.

Corollary 5.6 (Computing universality indexes from signatures). *One can compute $\iota_*(u)$ and $\zeta_*(u)$ from $\Sigma(u)$ in time $(|A| + \log |u|)^{O(1)}$.*

Actual implementations can use heuristics based on Lemma 3.4.(b): if $\mathbf{s}_u(\varepsilon) = \langle m, \varnothing \rangle$ then $\zeta_*(u) = m$.

5.2 Combining Signatures

Subword universality signatures can be computed compositionally.

Algorithm 5.7 (Combining signatures). The following algorithm takes as input the signatures $\Sigma(u)$ and $\Sigma(v)$ of any two words and computes $\Sigma(u\,v)$:

(a) Retrieve $A(u)$ and $A(v)$ from $e(u)$ and $e(v)$, then compute $e(u\,v)$ as $e(u)\,e'$ where e' is the subword of $e(v)$ that only retains the letters from $A(v) \smallsetminus A(u)$.

(b) Consider now any strict suffix x of $e(u\,v)$ and compute $\mathbf{s}_{u\,v}(x)$ as follows:

(b.1) If $A(v) \not\subseteq A(x) \cup A(u)$ then let $\mathbf{s}_{u\,v}(x) \overset{\text{def}}{=} S_v\big(x\,e(u)\big)$, using Algorithm 5.3.

(b.2) If $A(v) \subseteq A(x) \cup A(u)$, then $A(u) \not\subseteq A(x)$. Write $\langle n, B \rangle$ for $\mathbf{s}_u(x)$:

(b.2.1) If now $A(v) \cup B \neq A(x) \cup A(u)$ then let $\mathbf{s}_{u\,v}(x) \overset{\text{def}}{=} \langle n, A(v) \cup B \rangle$.

(b.2.2) Otherwise retrieve $\mathbf{s}_v(B) = \langle n', B' \rangle$ and let $\mathbf{s}_{u\,v}(x) \overset{\text{def}}{=} \langle n + n', B' \rangle$.

Proof (of correctness). Step (a) for $e(u\,v)$ is correct.

In step (b) we want to compute $S_{u\,v}(x)$. Now $x\,(u\,v) = (x\,u)\,v$ so $S_{u\,v}(x)$ coincides with $S_v(x\,u)$ *when the latter is defined* . This is the case in step (b.1) where one computes $S_v(x\,u)$ by replacing $x\,u$ with $x\,e(u)$, an argument with same alphabet (recall that the algorithm does not have access to u itself).

In step (b.2) where $S_v(x\,u)$ is not defined, computing $S_u(x)$ provides n and $B = A(r)$ for the arch factorization $x\,u = s_1 \cdots s_n \cdot r$ of $x\,u$.

We can continue with the arch factorization of $r\,v$ and combine the two sets of arches if these factorizations rely on the same alphabet: this is step (b.2.2).

Otherwise, $r\,v$ only uses a subset of the letters of $x\,u$. There won't be a new arch, only a longer rest: $r_*(x\,u\,v) = r\,v$. Step (b.2.1) is correct.

Note that Algorithm 5.7 runs in time $O\big(|A(u\,v)| + |\Sigma(u)| + |\Sigma(v)|\big)$ and that the result has linear size $|\Sigma(u\,v)| = O(|\Sigma(u)| + |\Sigma(v)|)$.

6   Universality Indexes for SLP-Compressed Words

We are now ready to compute the universality indexes of SLP-compressed words. Recall that an SLP X is an acyclic context-free grammar in Chomsky normal form where furthermore each non-terminal has only one production rule, i.e., the grammar is deterministic (see survey [Loh12]). SLPs are the standard mathematical model for compression of texts and files and, modulo polynomial-time encodings, it encompasses most compression schemes used in practice.

Formally, an SLP X with m rules is a list $\langle N_1 \to \rho_1; \cdots ; N_m \to \rho_m \rangle$ of production rules where each right-hand side ρ_i is either a letter a from A or a concatenation $N_j N_{j'}$ of two nonterminals with $j, j' < i$. It has size $|X| = O(m \log m)$ when A is fixed.

Each nonterminal N_i encodes a word, its *expansion*, given inductively via:

$$exp(N_i) \stackrel{\text{def}}{=} \begin{cases} a & \text{if } \rho_i = a, \\ exp(N_j)exp(N_{j'}) & \text{if } \rho_i = N_j\, N_{j'}. \end{cases}$$

Finally, the expansion $exp(X)$ of the SLP itself is the expansion $exp(N_m)$ of its last nonterminal. This is a word (or file) of length $2^{O(|X|)}$ and one of the main goals in the area of compressed data science is to develop efficient methods for computing relevant information about $exp(X)$ directly from X, i.e., without actually decompressing the word or file.

In this spirit we can state:

Theorem 6.1. *The universality indexes $\iota\big(exp(X)\big)$ and $\zeta\big(exp(X)\big)$ can be computed from an SLP X in bilinear time $O\big(|A| \cdot |X|\big)$.*

Proof. One just computes $\Sigma\big(exp(N_1)\big), \ldots, \Sigma\big(exp(N_k)\big)$ for the non-terminals $N_1, \ldots, N_k$ of X. If N_i is associated with a production rule $N_i \to N_{i_1} N_{i_2}$, we compute $\Sigma\big(exp(N_i)\big)$ by combining $\Sigma\big(exp(N_{i_1})\big)$ and $\Sigma\big(exp(N_{i_2})\big)$ via Algorithm 5.7 (recall that $i_1, i_2 < i$ since the grammar is acyclic). If N_i is associated with a production $N_i \to a$ for some $a \in A$, then $\Sigma\big(exp(N_i)\big) = \Sigma(a)$ is trivial. In the end we can extract the universality indexes of $exp(X)$, defined as $exp(N_k)$, from $\Sigma\big(exp(N_k)\big)$ using Corollary 5.6. Note that all signatures have size $O(|A| \cdot |X|)$ since for any $u = exp(N_i)$, $\log |u|$ is in $O(|X|)$. With the analysis of Algorithm 5.7 and Corollary 5.6, this justifies the claim about complexity.

7   Conclusion

We introduced arch-jumping functions and used them to describe and analyse the subword universality and circular universality indexes $\iota(u)$ and $\zeta(u)$. In particular, this leads to a simple and elegant algorithm for computing $\zeta(u)$.

In a second part we defined the subword universality signatures of words, a compact data structure with enough information for extracting $\iota(u)$ and $\zeta(u)$. Since one can efficiently compute the signature of $u\,v$ by composing the signatures of u and v, we obtain a polynomial-time algorithm for computing $\iota(X)$ and $\zeta(X)$ when X is a SLP-compressed word. This raises our hopes that one can compute some subword-based descriptive complexity measures on compressed words, despite the known difficulties encountered when reasoning about subwords.

References

[BFH+20] Barker, L., Fleischmann, P., Harwardt, K., Manea, F., Nowotka, D.: Scattered factor-universality of words. In: Jonoska, N., Savchuk, D. (eds.) DLT 2020. LNCS, vol. 12086, pp. 14–28. Springer, Cham (2020). https://doi.org/10.1007/978-3-030-48516-0_2

[BSS12] Bojańczyk, M., Segoufin, L., Straubing, H.: Piecewise testable tree languages. Logical Meth. Comp. Science 8(3) (2012)

[CP18] Carton, O., Pouzet, M.: Simon's theorem for scattered words. In: Hoshi, M., Seki, S. (eds.) DLT 2018. LNCS, vol. 11088, pp. 182–193. Springer, Cham (2018). https://doi.org/10.1007/978-3-319-98654-8_15

[DFK+21] Day, J.D., Fleischmann, P., Kosche, M., Koß, T., Manea, F., Siemer, S.: The edit distance to k-subsequence universality. In: Proceedings of 38th International Symposium Theoretical Aspects of Computer Science (STACS 2021), vol. 187. LIPiCS, pp. 25:1–25:19. Leibniz-Zentrum für Informatik (2021)

[DGK08] Diekert, V., Gastin, P., Kufleitner, M.: A survey on small fragments of first-order logic over finite words. Int. J. Foundat. Comput. Sci. 19(3), 513–548 (2008)

[DPFD19] Day, J.D., Fleischmann, P., Manea, F., Nowotka, D.: k-spectra of weakly-c-balanced words. In: Hofman, P., Skrzypczak, M. (eds.) DLT 2019. LNCS, vol. 11647, pp. 265–277. Springer, Cham (2019). https://doi.org/10.1007/978-3-030-24886-4_20

[FGN21] Fleischmann, P., Germann, S.B., Nowotka, D.: Scattered factor universality - the power of the remainder. arXiv:2104.09063 [cs.CL], (April 2021)

[FHH+22] Fleischmann, P., Haschke, L., Huch, A., Mayrock, A., Nowotka, D.: Nearly k-universal words - investigating a part of Simon's congruence. In: Proceedings 24th International Conference Descriptional Complexity of Formal Systems (DCFS 2022), vol. 13439. LNCS, pp. 57–71. Springer (2022). https://doi.org/10.1007/978-3-031-13257-5_5

[FK18] Fleischer, L., Kufleitner, M.: Testing Simon's congruence. In: Proceedings of 43rd International Symposium on Mathematical Foundations of Computer Science (MFCS 2018), vol. 117. LIPiCS, pp. 62:1–62:13. Leibniz-Zentrum für Informatik (2018)

[GLHK+20] Goubault-Larrecq, J., Halfon, S., Karandikar, P., Narayan Kumar, K., Schnoebelen, P.: The ideal approach to computing closed subsets in well-quasi-orderings. In: Schuster, P.M., Seisenberger, M., Weiermann, A. (eds.) Well-Quasi Orders in Computation, Logic, Language and Reasoning. TL, vol. 53, pp. 55–105. Springer, Cham (2020). https://doi.org/10.1007/978-3-030-30229-0_3

[GS16] Goubault-Larrecq, J., Schmitz, S.: Deciding piecewise testable separability for regular tree languages. In: Proc. 43rd International Colloquium on Automata, Languages and Programming (ICALP 2016), vol. 55. LIPiCS, pp. 97:1–97:15. Leibniz-Zentrum für Informatik (2016)

[Héb91] Hébrard, J.-J.: An algorithm for distinguishing efficiently bit-strings by their subsequences. Theoret. Comput. Sci. 82(1), 35–49 (1991)

[HSZ17] Halfon, S., Schnoebelen, Ph., Zetzsche, G.: Decidability, complexity, and expressiveness of first-order logic over the subword ordering. In: Proceedings of 32nd ACM/IEEE Symposium on Logic in Computer Science (LICS 2017), pp. 1–12. IEEE Comp. Soc. Press (2017)

[KKMP22] Kosche, M., Koß, T., Manea, F., Pak, V.: Subsequences in bounded ranges: Matching and analysis problems. In: Proceedings of 16th International Conference on Reachability Problems (RP 2022), vol. 13608. LNCS, pp. 140–159. Springer (2022). https://doi.org/10.1007/978-3-031-19135-0_10

[KKMS21] Kosche, M., Koß, T., Manea, F., Siemer, S.: Absent subsequences in words. In: Bell, P.C., Totzke, P., Potapov, I. (eds.) RP 2021. LNCS, vol. 13035, pp. 115–131. Springer, Cham (2021). https://doi.org/10.1007/978-3-030-89716-1_8

[KKMS22] Kosche, M., Koß, T., Manea, F., Siemer, S.: Combinatorial algorithms for subsequence matching: A survey. In: Proceedings of 12th International Workshop Non-Classical Models of Automata and Applications (NCMA 2022), vol. 367. EPTCS, pp. 11–27 (2022)

[KS15] Karandikar, P., Schnoebelen, P.: Generalized Post embedding problems. Theory Comput. Syst. 56(4), 697–716 (2015)

[KS19] Karandikar, P., Schnoebelen, Ph.: The height of piecewise-testable languages and the complexity of the logic of subwords. Logical Meth. Comp. Sci. 15(2) (2019)

[Loh12] Lohrey, M.: Algorithmics on SLP-compressed strings: A survey. Groups Complexity Cryptol. 4(2), 241–299 (2012)

[Mat98] Matz, O.: On piecewise testable, starfree, and recognizable picture languages. In: Nivat, M. (ed.) FoSSaCS 1998. LNCS, vol. 1378, pp. 203–210. Springer, Heidelberg (1998). https://doi.org/10.1007/BFb0053551

[Sch21] Schnoebelen, Ph.: On flat lossy channel machines. In: Proceedings of 29th EACSL Conference on Computer Science Logic (CSL 2021), vol. 183. LIPiCS, pp. 37:1–37:22. Leibniz-Zentrum für Informatik (2021)

[Sim72] Simon, I.: Hierarchies of Event with Dot-Depth One. PhD thesis, University of Waterloo, Dept. Applied Analysis and Computer Science, Waterloo, ON, Canada (1972)

[Sim75] Simon, I.: Piecewise testable events. In: Brakhage, H. (ed.) GI-Fachtagung 1975. LNCS, vol. 33, pp. 214–222. Springer, Heidelberg (1975). https://doi.org/10.1007/3-540-07407-4_23

[Sim03] Simon, I.: Words distinguished by their subwords. In: Proceedings of 4th International Conference on Words (WORDS 2003) (2003)

[SS83] Sakarovitch, J., Simon, I.: Subwords. In: Lothaire, M., (ed.) Combinatorics on Words, vol. 17. Encyclopedia of Mathematics and Its Applications, chapter 6, pp. 105–142. Cambridge Univ. Press (1983)

Characteristic Sequences of the Sets of Sums of Squares as Columns of Cellular Automata

Pierre-Adrien Tahay[(⊠)]

FNSPE, Czech Technical University in Prague, Prague, Czech Republic
pierre.adrien.tahay@cvut.cz

Abstract. A classical result due to Lagrange states that any natural number can be written as a sum of four squares. Characterizations of integers that are a sum of two and three squares were established by Fermat, Euler, Legendre and Gauss. In this paper we denote by s_1, s_2 and s_3 the characteristic functions of the integers which are respectively sums of one, two and three squares. We recall the already known results about the nonautomaticity of s_1 and about the 2-automaticity of s_3 and we prove the nonautomaticity of s_2. In the second part, we recall a construction of s_1 as a column of a cellular automaton and we give a construction for s_3 as an immediate application of a result of Rowland and Yassawi about the construction of p-automatic sequences when p is a prime number [17]. Finally we show that s_2 is also constructible as a column of a cellular automaton and we provide an explicit construction.

Keywords: sum of squares · cellular automata · automatic sequences · nonautomatic sequences

1 Introduction

For all integers $k \geq 1$ we define:

$$s_k(n) = \begin{cases} 1 & \text{if } n \text{ is a sum of } k \text{ squares,} \\ 0 & \text{otherwise.} \end{cases}$$

The function s_1 is simply the characteristic sequence of the squares and by definition $s_1(n) = 1$ if and only if there exists $m \in \mathbb{Z}$ such that $n = m^2$.

We refer to [9] to have a complete survey of the representations of integers as sums of squares.

In 1632, Girard conjectured that an odd prime number is the sum of two squares if and only if it is of the form $4k + 1$. Fermat proved this result in 1654 and the complete characterisation for all integers was obtained by Euler in the following century and gives this expression for s_2:

$$s_2(n) = \begin{cases} 1 & \text{if all prime divisors } q \equiv 3 \ (\text{mod } 4) \text{ of } n \text{ occur in } n \text{ to an even power} \\ 0 & \text{otherwise.} \end{cases}$$

The research received funding from the Ministry of Education, Youth and Sports of the Czech Republic through the CAAS Project.

© The Author(s), under exclusive license to Springer Nature Switzerland AG 2023
A. Frid and R. Merca\c{s} (Eds.): WORDS 2023, LNCS 13899, pp. 288–300, 2023.
https://doi.org/10.1007/978-3-031-33180-0_22

The binary sequence $(s_2(n))_{n \geq 0} = 1, 1, 1, 0, 1, 1, 0, 0, 1, 1, 1, 0, 0, 1, 0, 0, 1, \ldots$ is equal to 1 at integers of the sequence A001481 in the OEIS [18] and 0 otherwise.

The fact that every natural number can be written as a sum of four squares was already conjectured by Bachet. The first proof of this result is due to Lagrange in 1770 and so, we have trivially for all $k \geq 4$ $s_k(n) = 1$ for all $n \geq 0$.

The most difficult case is the three squares theorem. It was established by Legendre in 1798 and Gauss in 1801. Their results give this expression for s_3:

$$s_3(n) = \begin{cases} 0 \text{ if } n = 4^a(8m + 7) \text{ with nonnegative integers } a \text{ and } m \\ 1 \text{ otherwise.} \end{cases}$$

The binary sequence $(s_3(n))_{n \geq 0} = 1, 1, 1, 1, 1, 1, 1, 1, 0, 1, 1, 1, 1, 1, 1, 1, 0, 1, \ldots$ is equal to 0 at integers of the sequence A004215 in the OEIS [18] and 1 otherwise.

By Langrange's four-square theorem, only the functions s_1, s_2 and s_3 are of interest to be studied. The results about the nonautomaticity of s_1 are already well-known [15,16] and a first construction in a column of a cellular automaton has been obtained by Mazoyer and Terrier in 1999 [14] and a second by Delacourt, Poupet, Sablik and Theyssier in 2011 [6] which the author has generalized with Marcovici and Stoll in 2018 to any polynomial $P \in \mathbb{Q}(X)$ with $P(\mathbb{N}) \subset \mathbb{N}$ [13]. The automaticity of s_3 is due to Cobham [4] and we use this fact to build a cellular automaton with a method developed by Rowland and Yassawi [17]. Our main results concern the study of s_2.

2 Preliminaries

In this section we fix some notations and we recall some basic results of the theory of finite automata and automatic sequences on the one hand and cellular automata on the other hand.

2.1 Words and Morphisms

An *alphabet* $\mathcal{A}$ is a finite set of symbols called *letters*. The set $\mathcal{A}^*$ refers to the set of finite words over $\mathcal{A}$ which is the free monoid having neutral element the empty word ε. The *length* of a word $w = a_0 a_1 \cdots a_{n-1}$, with $a_i \in \mathcal{A}$ is the integer $|w| = n$. For an integer $k \geq 2$, we denote by Σ_k the alphabet $\{0, 1, \ldots, k-1\}$. For all $n \in \mathbb{N}$ we denote by $(n)_k$ the standard base-k representation of n. For two alphabets $\mathcal{A}$ and $\mathcal{B}$, a morphism is a map $h : \mathcal{A}^* \longrightarrow \mathcal{B}^*$ such that for all words $x, y \in \Sigma^*$ we have $h(xy) = h(x)h(y)$. If $\mathcal{A} = \mathcal{B}$ we can iterate the morphism h. For all $a \in \mathcal{A}$ we define $h^0(a) = a$ and $h^i(a) = h(h^{i-1}(a))$. For a morphism $h : \mathcal{A} \longrightarrow \mathcal{A}$, if there is an integer k such that $|h(a)| = k$ for all $a \in \mathcal{A}$, we said that h is k-*uniform*. A 1-uniform morphism is called a *coding*. We can naturally extend the notion of morphism to infinite words. We said that a k-uniform morphism $h : \mathcal{A} \longrightarrow \mathcal{A}$ is *prolongable* if there exists a letter $a \in \mathcal{A}$ and a word $w \in \mathcal{A}^* \backslash \{\varepsilon\}$ such that $h(a) = aw$. In this case, the sequence $(h^n(a))_{n \geq 0}$ converges to the infinite word $h^\omega(a) = awh(w)h^2(w) \cdots$. Moreover, $h^\omega(a)$ is the unique fixed point of h which starts with a.

2.2   Finite Automata and Automatic Sequences

We refer to the book of Allouche and Shallit [1, Sections 5, 6] for a complete survey of the theory of finite automata and automatic sequences.

Definition 1. *A deterministic finite automaton with output (DFAO) is a 6-tuple* $(Q, \Sigma_k, \delta, q_0, \mathcal{A}, \omega)$ *where Q is a finite set of states, $q_0 \in Q$ is the initial state, $\omega : Q \longrightarrow \mathcal{A}$ is the output function, and $\delta : Q \times \Sigma_k \longrightarrow Q$ is the transition function. We say that a sequence $(u_n)_{n \geq 0}$ of elements in $\mathcal{A}$ is k-automatic if there exists a DFAO $(Q, \Sigma_k, \delta, q_0, \mathcal{A}, \omega)$ such that $u_n = \omega(\delta(q_0, (n)_k))$ for all $n \geq 0$.*

Example 1. One of the most famous automatic sequences is the Prouhet–Thue–Morse or Thue–Morse sequence. There are several equivalent ways to define it. For example, if we denote by $(t_n)_{n \geq 0}$ this sequence, it can be defined by:

$$t_n = \begin{cases} 0, & \text{if the number of 1's in } (n)_2 \text{ is even;} \\ 1, & \text{otherwise.} \end{cases}$$

With this definition it is clear that $(t_n)_{n \geq 0}$ is a 2-automatic sequence generated by the following finite automaton:

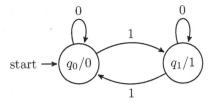

To prove that a sequence is automatic, the most natural way is to explicitly construct a finite automaton that generates it, as for the Thue–Morse sequence in the previous example. However, in practice there are many criteria for proving the automaticity of a sequence without building the automaton. One of the interests is that they make it possible to show, on the contrary, that a sequence is nonautomatic using the contraposition of one of these criteria. In a recent paper, Allouche, Shallit and Yassawi give a large number of methods to prove that a sequence is nonautomatic [2]. See also the paper of Coons [5] where the author establishes the nonautomaticity of several number theoretic functions with various criteria.

2.3   Cellular Automata

Cellular automata are another model for calculation. They were first introduced in the 1940s by von Neumann to study a self-reproduction phenomenon. In 1966, Burks took up and completed von Neumann's work posthumously [20]. Cellular automata were definitely popularised by the famous Conway's Game of Life in 1970.

These objects have a very different nature from deterministic finite automata, but rather surprisingly, Rowland and Yassawi established in 2015 a very strong link between both [17]. Before recalling their result, let us begin with some definitions.

Definition 2. *Let $\mathcal{A}$ be a finite set (typically an alphabet) endowed with the discrete topology and let $\sigma : (\mathcal{A}^d)^{\mathbb{Z}} \longrightarrow \mathcal{A}^{\mathbb{Z}}$ be the shift map. A cellular automaton with memory d is a continuous map $\Phi : (\mathcal{A}^d)^{\mathbb{Z}} \longrightarrow \mathcal{A}^{\mathbb{Z}}$ with respect to the product topology such that $\sigma \circ \Phi = \Phi \circ \sigma$. The case $d = 1$ is the classical definition of a cellular automaton.*

By the Curtis–Hedlund–Lyndon theorem [10] the previous definition is equivalent to the following

Definition 3. *A map $\Phi : (\mathcal{A}^d)^{\mathbb{Z}} \longrightarrow \mathcal{A}^{\mathbb{Z}}$ is a cellular automaton if and only if there is a local rule $\phi : (\mathcal{A}^d)^{l+r+1} \longrightarrow \mathcal{A}$ for some $l \geq 0$ (left radius of ϕ) and some $r \geq 0$ (right radius of ϕ), such that for all $R \in (\mathcal{A}^{\mathbb{Z}})^d$ and for all $m \in \mathbb{Z}$,*

$$(\Phi(R))(m) = \phi(R(m - l), R(m - l + 1), \ldots, R(m + r)).$$

Now, we define the *spacetime diagram* of a cellular automaton.

Definition 4. *If $\Phi : (\mathcal{A}^d)^{\mathbb{Z}} \longrightarrow \mathcal{A}^{\mathbb{Z}}$ is a cellular automaton with memory d, a spacetime diagram for Φ with initial conditions $R_0, \ldots, R_{d-1}$ is the sequence $(R_n)_{n \geq 0}$ defined recursively by $R_n = \Phi(R_{n-d}, \ldots, R_{n-1})$ for $n \geq d$.*

Consequently, for a cellular automaton with memory d, each row is determined by the previous d rows.

Definition 5. *Let q be a power of a prime number. We denote by $\mathbb{F}_q$ the finite field with q elements. In the special case where $\mathcal{A} = \mathbb{F}_q$ we say that a cellular automaton $\Phi : (\mathbb{F}_q^d)^{\mathbb{Z}} \longrightarrow \mathbb{F}_q^{\mathbb{Z}}$ with memory d is linear if Φ is a $\mathbb{F}_q$-linear map.*

By the Curtis–Hedlund–Lyndon theorem, a cellular automaton Φ with memory d is linear if and only if there exist coefficients $f_{j,i} \in \mathbb{F}_q$ for $-l \leq j \leq r$ and $0 \leq i \leq d - 1$ such that $(\Phi(R_0, \ldots, R_{d-1}))(m) = \sum_{i=0}^{d-1} \sum_{j=-l}^{r} f_{j,i} R_i(m + j)$ for all $R_0, \ldots, R_{d-1} \in \mathbb{F}_q^{\mathbb{Z}}$ and $m \in \mathbb{Z}$.

We can now recall the result of Rowland and Yassawi.

Theorem 1 (Rowland and Yassawi [17]). *Let p a prime number and q a power of p. A sequence of elements in $\mathbb{F}_q$ is p-automatic if and only if it is a column of a spacetime diagram of a linear cellular automaton with memory over $\mathbb{F}_q$ whose initial conditions are eventually periodic in both directions.*

Remark 1. The fact that every column of a linear cellular automaton on $\mathbb{F}_q$ is p-automatic was already known since 1993 by Dumas and Litow [12]. Rowland and Yassawi established the converse by giving a complete characterization of p-automatic sequences. Moreover their proof is constructive and they give a method to have an explicit cellular automaton which generates a given p-automatic sequence. The main ingredient is Christol's theorem.

Theorem 2 (Christol [3]). *Let q a power of a prime number. A sequence $(u_n)_{n\geq 0}$ of elements in $\mathbb{F}_q$ is q-automatic if and only if its generating series $F(t) = \sum_{n\geq 0} u_n t^n$ is algebraic over $\mathbb{F}_q(t)$.*

The principle of the method of Rowland and Yassawi is to find a polynomial $P \in \mathbb{F}_q(t, x)$ such that $P(t, F(t)) = 0$ whose existence is guaranteed by Christol's theorem and which defines the rule of the linear cellular automaton that computes the given p-automatic sequence from the coefficients of P. Several examples of explicit constructions of classical automatic sequences in columns of cellular automata are given directly in the article of Rowland and Yassawi or in the thesis manuscript of the author [17, 19].

One of the first results about the construction of sequences as a column of cellular automata was obtained by Fischer in 1965, who built the characteristic sequence of prime numbers, using a cellular automaton with more than 30,000 states [8]. In 1997, Korec improved this result with another cellular automaton with only 11 states [11]. The geometric construction of increasing functions and closure properties has been established by Mazoyer and Terrier in 1999, which they call Fischer constructible [14]. Their constructions use signals which are a way to transmit information by connecting two cells in a cellular automaton. Marcovici, Stoll and Tahay use these kinds of contructions with signals to provide other constructions for characteristic sequences of polynomials and a construction for the Fibonacci word which is an emblematic nonautomatic sequence and which is also a Sturmian word. Recently, Dolce and Tahay extended the construction for the Fibonacci word to all Sturmian words with a quadratic slope also using signals [7]. We refer to [7, 14] for a formal definition of signals in cellular automata and examples of contructions.

3   Study of the Automaticity of s_1, s_2 and s_3

3.1   Nonautomaticity of the Characteristic Sequence of the Squares

We recall the following well-known result

Proposition 1 ([15, 16]). *The sequence $(s_1(n))_{n\geq 0}$ is nonautomatic.*

Ritchie proved in 1963 the fact that $(s_1(n))_{n\geq 0}$ is not a 2-automatic sequence [16]. Minsky and Papert gave an elegant criterion that can be applied to prove that $(s_1(n))_{n\geq 0}$ is not k-automatic for any $k \geq 2$.

Proposition 2 (Minsky and Papert [15]). *Let $f : \mathbb{N} \to \mathbb{N}$ be an increasing function and we define the set $\pi_f(x) = \#\{n : f(n) \leq x\}$. If the two conditions:*

1. $\lim_{x \to \infty} \dfrac{\pi_f(x)}{x} = 0$ *2.* $\lim_{n \to \infty} \dfrac{f(n+1)}{f(n)} = 1$

are satisfied, then the sequence $u = \mathbf{1}_{f(\mathbb{N})}$ is nonautomatic.

Proposition 1 can be directly deduced from Minsky and Papert's criterion with $f : \mathbb{N} \to \mathbb{N}$, $x \to x^2$.

Remark 2. We can find an alternative proof of Proposition 1 using language theory, see Example 5.5.1 and Example 8.6.2 in [1].

Remark 3. Let $\mathcal{A} = \{a, b, c\}$. Let $f : \mathcal{A}^* \longrightarrow \mathcal{A}^*$ and $\pi : \mathcal{A} \longrightarrow \{0, 1\}$ be the morphism and the coding defined respectively by:

$$f : \begin{cases} a \mapsto abcc \\ b \mapsto bcc \\ c \mapsto c \end{cases} \quad \text{and} \quad \pi : \begin{cases} a, b \mapsto 1 \\ c \mapsto 0 \end{cases}$$

It is clear that $\pi(f^\omega(a))$ generates $(s_1(n))_{n \geq 0}$. So, $(s_1(n))_{n \geq 0}$ is a morphic sequence which is nonautomatic.

3.2   Automaticity of $(s_3(n))_{n \geq 0}$

We give the result on the automaticity of $(s_3(n))_{n \geq 0}$ already known by Cobham [4] before the one on $(s_2(n))_{n \geq 0}$ which we establish in the next section.

Proposition 3 ([4]). *The sequence $(s_3(n))_{n \geq 0}$ is 2-automatic.*

Proof. Using the observation that a number is not representable in the form $4^a(8m + 7)$ if and only if its binary representation does not terminate with three successive 1's followed by an even number of 0's, Cobham gives an explicit construction of a finite automaton with 6 states, that generates the sequence $(s_3)_{n \geq 0}$ (see Fig. 1 where we use the convention to start with the most significant digit of the binary expansion to read the automaton).

Cobham deduces from the finite automaton generating $(s_3(n))_{n \geq 0}$ in Fig. 1 the following 2-uniform morphism g and the coding σ that generates the characteristic sequence of the set of sums of three squares.

Let $\mathcal{A} = \{a, b, c, d, e, f\}$. Let $g : \mathcal{A}^* \longrightarrow \mathcal{A}^*$ and $\sigma : \mathcal{A} \longrightarrow \{0, 1\}$ be the morphism and the coding defined by:

$$g : \begin{cases} a \mapsto ab \\ b \mapsto ac \\ c \mapsto ad \\ d \mapsto ed \\ e \mapsto fb \\ f \mapsto eb \end{cases} \quad \text{and} \quad \sigma : \begin{cases} a, b, c, e \mapsto 1 \\ d, f \mapsto 0 \end{cases}$$

Then $\sigma(g^\omega(a))$ generates $(s_3(n))_{n \geq 0}$.

We give now an alternative proof using Christol's theorem which will be used to build $(s_3(n))_{n \geq 0}$ in a column of a cellular automaton in Sect. 4.3.

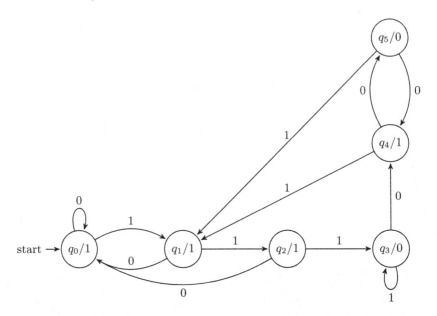

Fig. 1. A finite automaton generating the sequence $(s_3(n))_{n\geq 0}$

Proposition 4. *Let $F(t) = \sum_{n\geq 0} s_3(n)t^n$ be the generating series of $(s_3(n))_{n\geq 0}$ defined on $\mathbb{F}_2(t)$. Let $P \in \mathbb{F}_2(t,x)$ be the polynomial $P(t,x) = t + t^2 + t^3 + t^5 + t^6 + (1+t^8)x + (1+t^8)x^4$. Then $P(t,F(t)) = 0$.*

Proof. With Legendre and Gauss characterization for $(s_3(n))_{n\geq 0}$ we have clearly for all $n \geq 0$, $s_3(4n) = s_3(n)$, $s_3(8n+7) = 0$ and $s_3(8n+1) = s_3(8n+2) = s_3(8n+3) = s_3(8n+5) = s_3(8n+6) = 1$. Then, we have

$$F(t) = \sum_{n\geq 0} s_3(n)t^n$$

$$= \sum_{n\geq 0} s_3(4n)t^{4n}$$

$$+ \sum_{n\geq 0} s_3(8n+1)t^{8n+1} + \sum_{n\geq 0} s_3(8n+2)t^{8n+2} + \sum_{n\geq 0} s_3(8n+3)t^{8n+3}$$

$$+ \sum_{n\geq 0} s_3(8n+5)t^{8n+5} + \sum_{n\geq 0} s_3(8n+6)t^{8n+6} + \sum_{n\geq 0} s_3(8n+7)t^{8n+7}$$

$$= \sum_{n\geq 0} s_3(n)t^{4n} + \sum_{n\geq 0} t^{8n+1} + \sum_{n\geq 0} t^{8n+2} + \sum_{n\geq 0} t^{8n+3} + \sum_{n\geq 0} t^{8n+5} + \sum_{n\geq 0} t^{8n+6}$$

$$= (F(t))^4 + (t + t^2 + t^3 + t^5 + t^6)\frac{1}{1+t^8}$$

which shows the result.

Then, the generating series of $(s_3(n))_{n\geq 0}$ is algebraic over $\mathbb{F}_2(t)$, which proves its 2-automaticity by Christol's theorem.

3.3   Nonautomaticity of the Set of Sums of Two Squares

The nonautomaticity of $(s_1(n))_{n\geq 0}$ and the 2-automaticity of $(s_3(n))_{n\geq 0}$ are well known results with different ways to prove them. A natural question now is to study the automaticity or the nonautomaticity of $(s_2(n))_{n\geq 0}$. Let us begin by recalling this definition.

Definition 6. *A sequence $(a(n))_{n\geq 1}$ is called* multiplicative *if, for all integers $m, n \geq 1$ coprime we have $a(mn) = a(m)a(n)$.*

Proposition 5. *The sequence $(s_2(n))_{n\geq 1}$ is multiplicative.*

Proof. With the identity $(a^2 + b^2)(c^2 + d^2) = (ac + bd)^2 + (ad - bc)^2$ we have clearly for all $m, n \geq 0, s_2(mn) \geq s_2(m)s_2(n)$.

Now, let $m, n \geq 1$ coprime.
If $s_2(mn) = 0$ we have trivially $s_2(mn) \leq s_2(m)s_2(n)$. If $s_2(mn) = 1$, by the two squares theorem this means that every prime divisor $q \equiv 3 \pmod 4$ of mn occurs in mn to an even power. Because m and n are coprime, it is necessarily also the case for all prime divisors $q \equiv 3 \pmod 4$ respectively in m and in n and then $s_2(m) = s_2(n) = 1$, which completes the proof.

We give now a criterion of nonautomaticity for multiplicative sequences.

Theorem 3 ([2,21]). *Let $v > 1$ be an integer and f a multiplicative function. Assume that for some integer $h \geq 1$ there exist infinitely many primes q_1 such that $f(q_1^h) \equiv 0 \pmod v$. Furthermore assume that there exist relatively prime integers b and c such that for all primes $q_2 \equiv c \pmod b$ we have $f(q_2) \not\equiv 0 \pmod v$. Then the sequence $(f(n))_{n\geq 1} \pmod v$ is not k-automatic for any $k \geq 2$.*

Proposition 6. *The sequence $(s_2(n))_{n\geq 1}$ is not k-automatic for any $k \geq 2$.*

Proof. Because $(s_2(n))_{n\geq 1}$ is a binary sequence it equals $(s_2(n))_{n\geq 1} \pmod 2$. Moreover $(s_2(n))_{n\geq 1}$ is multiplicative by Proposition 5, and by the Girard-Fermat's theorem, an odd prime number p is a sum of two squares if and only if $p \equiv 1 \pmod 4$. So, we can apply the previous theorem to $(s_2(n))_{n\geq 1}$ with $v = 2, h = 1, c = 1, b = 4$, the infinity of prime numbers $q_1 \equiv 3 \pmod 4$ and for q_2 we can take the prime numbers such that $q_2 \equiv 1 \pmod 4$.

4   Construction by Cellular Automata of s_1, s_2 and s_3

In all our constructions, time axis is oriented upward.

4.1 Cellular Automaton of the Characteristic Sequence of the Square

Here, we recall the construction of Delacourt, Poupet, Sablik and Theyssier [6] to obtain the characteristic sequence of squares by the hitting of one single signal in column 0. Vertical signals are called *walls* and represented by a straight line. Signals of other slopes are represented by arrows (see Fig. 2). The two first lines are initial conditions. We start by sending a signal of slope 1 to the northeast-direction (blue arrow). Every time it meets the wall (represented by a vertical green straight line), the wall gets shifted by one cell to the right and the blue signal of slope 1 changes its direction to a blue signal of slope -1. When the blue signal of slope -1 meets the 0-column, then a 1 is marked, and a new signal of slope 1 is sent.

4.2 Cellular Automaton of the Set of Sums of Two Squares

Theorem 4. *The sequence* $(s_2(n))_{n \geq 0}$ *can be obtained as a column of a cellular automaton.*

Proof. We will use a geometric construction similar to the one for $(s_1(n))_{n \geq 0}$. First, we use the fact that for all $n \geq 0$ "$s_1(n) = 1 \Rightarrow s_2(n) = 1$" which corresponds to the fact that if an integer is a square $n = m^2$, then it is a sum of two squares, $n = m^2 + 0^2$. So, we start by taking up the cellular automaton which builds the characteristic of squares.

Using exactly the same construction but with each cell shifted one line above, we obtain all the integers that are sums of two squares of the form $n^2 + 1^2$. We represent the signals of slope 1 and slope -1 by red arrows in Fig. 3. In comparison to Fig. 2, the green wall propagates vertically one more cell before being shifted to the right when it meets a red signal.

To finish the construction, we need to mark the columns whose horizontal coordinate is a perfect square. We use the general method developed by Marcovici, Stoll and Tahay to build the polynomial sequences (see [13, Proposition 2]). We define a new signal of slope 1 in the diagonal of the spacetime diagram (grey arrows on Fig. 3). If F denotes the cellular automaton of Fig. 2, we build the cellular automaton $\sigma \circ F$. The blue arrows of slope -1 in Fig. 2 become blue walls in the columns which correspond to the perfect squares in Fig. 3. When these blue walls meet the grey diagonal signals, they continue to spread through the columns.

Now, it just remains to define signals of slope $\dfrac{1}{2}$ from each perfect square m^2 in the colums 0 at the same time as the blue signals of slope 1 (black arrows to the northeast direction in Fig. 3). When one of these signals meets a blue wall in a column that is a perfect square n^2, it changes its direction and we define a signal of slope $-\dfrac{1}{2}$ which meets column 0 at the line $m^2 + n^2$.

4.3   Linear Cellular Automaton of the 2-automatic Sequence $(s_3)_{n \geq 0}$

In this last case, we can use the result of Rowland and Yassawi [17] by using the polynomial obtained in Proposition 4.

Let $F(t) = \displaystyle\sum_{n \geq 0} s_3(n)t^n$ and $P(t,x) = t+t^2+t^3+t^5+t^6+(1+t^8)x+(1+t^8)x^4$ the polynomial such that $P(t, F(t)) = 0$. We encode the spacetime diagram of a cellular automaton by the series $\displaystyle\sum_{n \geq 0} \sum_{m \in \mathbb{Z}} a_{n,m} t^n x^m$, where $a_{n,m}$ represents the cell on row $n \in \mathbb{N}$ and column $m \in \mathbb{Z}$.

In order to apply the method of Rowland and Yassawi, we use the transformation $x \longrightarrow 1 + t + tx$ to the polynomial $P(t,x)$. We have

$$
\begin{aligned}
P(t, 1+t+tx) &= t + t^2 + t^3 + t^5 + t^6 + 1 + t + tx + t^8 + t^9 + t^9 x \\
&\quad + 1 + t^4 + t^4 x^4 + t^8 + t^{12} + t^{12} x^4 \\
&= tx + t^2 + t^3 + t^4(1+x^4) + t^5 + t^6 + t^9(1+x) + t^{12}(1+x^4)
\end{aligned}
$$

We can divide all the coefficients in the last row by t and we define $Q(t,x) = x + t + t^2 + (1+x^4)t^3 + t^4 + t^5 + (1+x)t^8 + (1+x^4)t^{11}$ and we have for $G(t) = \displaystyle\sum_{n \geq 1} s_3(n+1)t^n$, $F(t) = 1 + t + tG(t)$ and $Q(t, G(t)) = 0$.

We denote by $Q^{(0,1)}$ the derivative function of Q with respect to its second argument. Now, we have $G(0) = 0$ and $Q^{(0,1)}(t,x) = 1 + t^8$ which satisfies $Q^{(0,1)}(0,0) \neq 0$.

By Rowland and Yassawi theorem, the power series $\dfrac{Q^{(0,1)}(t,x)}{Q(t,x)} =$ $\displaystyle\sum_{n \geq 0} R_n(x)t^n$ encodes a spacetime diagram of a cellular automaton where for all $n \in \mathbb{N}$ and all $m \in \mathbb{Z}$ the coefficient x^m in $R_n(x)$ represents the cell at row n and column m and where the sequence $(s_3(n+1))_{n \geq 1}$ occurs in the column -2.

By using the relation $Q^{(0,1)}(t,x) = Q(t,x) \displaystyle\sum_{n \geq 0} R_n(x)t^n$ and by collecting the terms by common powers of t, we deduce that $R_n(x)$ satisfies the recurrence :

$$
\begin{aligned}
R_n(x) =& \frac{1}{x}R_{n-1}(x) + \frac{1}{x}R_{n-2}(x) + \left(\frac{1}{x} + x^3\right)R_{n-3}(x) + \frac{1}{x}R_{n-4}(x) + \frac{1}{x}R_{n-5}(x) \\
&+ \left(\frac{1}{x} + 1\right)R_{n-8}(x) + \left(\frac{1}{x} + x^3\right)R_{n-11}(x),
\end{aligned}
$$

for all $n \geq 12$. Let $R_{-1}(x) = s_3(0)x^{-2} = x^{-2}$ and $R_0(x) = s_3(1)x^{-2} = x^{-2}$. Then, we define a cellular automaton with memory 14 $(12 + 1 + 1)$, where the sequence $(s_3(n))_{n \geq 0}$ occurs in the column -2 which is highlighted in red in the spacetime diagram in Fig. 4.

5   Open Questions

1. Let $s_{k,p}$ be the characteristic sequence of sums of k pth powers. For every p, does there exist K such that $s_{k,p}$ is automatic for $k \geq K$ and nonautomatic for $k < K$? Are all these sequences representable by CA? If not, minimal p and a value of k such that $s_{k,p}$ is not representable?
2. Is it possible to generate s_3 by using a geometric construction with signals similarly to that for s_1 and s_2?
3. Sequences s_1 and s_3 are both morphic sequences. Is it also the case for s_2? Because of the nonautomaticity of s_2, of course in this case the morphism would be nonuniform like for s_1.

Acknowledgment. The author thanks the anonymous referees for the many corrections and suggestions for open questions.

A   Appendices

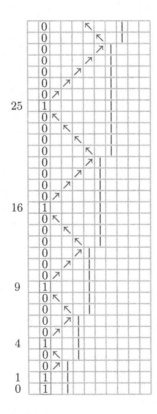

Fig. 2. Construction of $(s_1(n))_{n \geq 0}$ by a cellular automaton

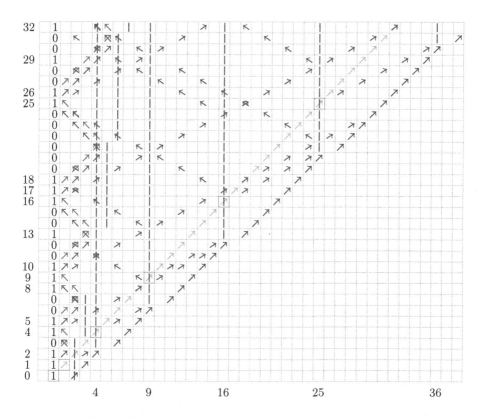

Fig. 3. Construction of $(s_2(n))_{n\geq 0}$ by a cellular automaton

Fig. 4. Construction of $(s_3(n))_{n\geq 0}$ by a cellular automaton

References

1. Allouche, J.-P., Shallit, J.: Automatic Sequences: Theory, Applications, Generalizations. Cambridge University Press, Cambridge (2003)
2. Allouche, J.-P., Shallit, J., Yassawi, R.: How to prove that a sequence is not automatic. Expo. Math. **40**, 1–22 (2022)
3. Christol, G., Kamae, T., Mendès France, M., Rauzy, G.: Suites algébriques, automates et substitutions. Bul. Soc. Math. France **79**, 401–419 (1980)
4. Cobham, A.: Uniform tag sequences. Math. Syst. Theory **6**, 164–192 (1972)
5. Coons, M.: (Non)automaticity of number theoretic functions. J. Théor. Nombres Bordeaux **22**, 339–352 (2010)
6. Delacourt, M., Poupet, V., Sablik, M., Theyssier, G.: Directional dynamics along arbitrary curves in cellular automata. Theor. Comput. Sci. **412**, 3800–3821 (2011)
7. Dolce, F., Tahay, P.-A.: Column representation of sturmian words in cellular automata. Lect. Notes Comput. Sci. **13257**, 127–138 (2022)
8. Fischer, P.C.: Generation of primes by a one-dimensional real-time iterative array. J. ACM **12**, 388–394 (1965)
9. Grosswald, E.: Representations of Integers as Sums of Squares. Springer, New York (1985). https://doi.org/10.1007/978-1-4613-8566-0
10. Hedlund, G.A.: Endomorphisms and automorphisms of the shift dynamical system. Math. Syst. Theory **3**, 320–375 (1969)
11. Korec, I.: Real-time generation of primes by a one-dimensional cellular automaton with 11 states. In: Prívara, I., Ružička, P. (eds.) MFCS 1997. LNCS, vol. 1295, pp. 358–367. Springer, Heidelberg (1997). https://doi.org/10.1007/BFb0029979
12. Litow, B., Dumas, P.: Additive cellular automata and algebraic series. Theor. Comput. Sci. **119**, 345–354 (1993)
13. Marcovici, I., Stoll, T., Tahay, P.-A.: Construction of some nonautomatic sequences by cellular automata. Lect. Notes Comput. Sci. **10875**, 113–126 (2018)
14. Mazoyer, J., Terrier, V.: Signals in one-dimensional cellular automata. Theor. Comput. Sci. **217**, 53–80 (1999)
15. Minsky, M., Papert, S.: Unrecognizable sets of numbers. J. ACM **13**, 281–286 (1966)
16. Ritchie, R.W.: Finite automata and the set of squares. J. ACM **10**, 528–531 (1963)
17. Rowland, E., Yassawi, R.: A characterization of p-automatic sequences as columns of linear cellular automata. Adv. Appl. Math. **63**, 68–89 (2015)
18. Sloane, N.J.A.: On-line Encyclopedia of Integer Sequences (2003). https://oeis.org/
19. Tahay, P.-A.: Colonnes dans les automates cellulaires et suites généralisées de Rudin-Shapiro. PhD thesis, Université de Lorraine (2020). https://hal.univ-lorraine.fr/tel-03184510/document
20. von Neumann, J.: The Theory of Self Reproducing Automata. University of Illinois Press, Cambridge (1966)
21. Yazdani, S.: Multiplicative functions and k-automatic sequences. J. Théor. Nombres Bordeaux **13**, 651–658 (2001)

Author Index

© The Editor(s) (if applicable) and The Author(s), under exclusive license
to Springer Nature Switzerland AG 2023
A. Frid and R. Mercaş (Eds.): WORDS 2023, LNCS 13899, p. 301, 2023.
https://doi.org/10.1007/978-3-031-33180-0

Printed in the United States
by Baker & Taylor Publisher Services